STO
ALLEN COUNTY PUBLIC LIBRARY
ACPL ITEM
DISCARDED
3 1833 00088 6140
Y0-BYU-151
G

DEC 12 '72

*PROGRESS IN SURFACE SCIENCE*

**VOLUME 1**

# Progress in Surface Science

## VOLUME 1

Edited by

SYDNEY G. DAVISON

*Department of Physics, Institute of Colloid and Surface Science*
*Clarkson College of Technology, Potsdam, New York*

PERGAMON PRESS

OXFORD · NEW YORK · TORONTO

SYDNEY · BRAUNSCHWEIG

Pergamon Press Ltd., Headington Hill Hall, Oxford
Pergamon Press Inc., Maxwell House, Fairview Park, Elmsford, New York 10523
Pergamon of Canada Ltd., 207 Queen's Quay West, Toronto 1
Pergamon Press (Aust.) Pty. Ltd., 19a Boundary Street, Rushcutters Bay, N.S.W. 2011, Australia
Vieweg & Sohn GmbH, Burgplatz 1, Braunschweig

---

Copyright © 1972 Pergamon Press Ltd.

*All Rights Reserved. No part of this publication may be reproduced, stored in a retrieval system, or transmitted, in any form or by any means, electronic, mechanical, photocopying, recording or otherwise, without the prior permission of Pergamon Press Ltd.*

First edition 1972

Library of Congress Catalog Card No. 77–141188

*Printed in Great Britain by A. Wheaton & Co., Exeter*

08 016878 7

# CONTENTS



1715470

# PREFACE

By separating matter into its three phases, surfaces (or interfaces) play a leading role in nature. It is not surprising, therefore, that phenomena associated with surfaces are of fundamental interest to a wide variety of scientists. As a result of this, surface science has evolved as a fragmented interdisciplinary subject, which today is the concern of most scientific disciplines. With this in mind, the *Progress in Surface Science* series will endeavour to publish review articles on as broad a spectrum of surface phenomena as possible. By adopting such an all-embracing policy, it is hoped that the series will help to promote the exchange of ideas between scientists interested in studying the different aspects of surfaces, and thus lead to a greater unification of the area. To facilitate the flow of ideas, the articles will be written at a level which can be read by graduate students and research workers who are newcomers to the field.

In editing the present volume, I have received a great deal of assistance from the members of the international Advisory Board. I am particularly indebted to Drs. Peter Mark and Leonard Weiss, who not only contributed articles, but have on many occasions given me the benefit of consultations with them. I also wish to express my sincere thanks to all the other authors for their contributions and patience with the editing and production processes. Regretfully, I have to report the death of the author Dr. Thomas A. Goodwin, who died of Hodgkins disease on June 13, 1971.

Finally, I would like to thank my wife, Prudence, for her constant encouragement and assistance in proof reading.

*Potsdam, New York*
*October*, 1971

Sydney G. Davison

# PREFACE

# ADVISORY BOARD

D. G. DERVICHIAN,
Dept. of Biophysics,
Pasteur Institute,
25 Rue Du Docteur Roux,
XV Arrond,
Paris, France

D. D. ELEY,
Department of Chemistry,
University of Nottingham,
Nottingham, NG7 2RD,
England

H. EYRING,
Department of Chemistry,
University of Utah,
Salt Lake City,
Utah 84112, U.S.A.

H. E. FARNSWORTH,
Department of Physics,
Brown University,
Providence,
Rhode Island 02912, U.S.A.

A. N. FRUMKIN,
Institute of Electrochemistry,
Academy of Sciences,
Leninsky Prospekt 31,
Moscow V-71, U.S.S.R.

H. C. GATOS,
Department of Metallurgy and Materials Science,
Massachusetts Institute of Technology,
Cambridge,Massachusetts 02139,U.S.A.

K. H. HAUFFE,
Institute of Physical Chemistry,
University of Göttingen,
Burgerstrasse 50, 34 Göttingen,
Germany

J. KOUTECKÝ,
Belfer Graduate School of Science,
Yeshiva University,
Amsterdam Ave. & 186th Street,
New York, N.Y. 10033, U.S.A.

J. D. LEVINE,
David Sarnoff Research Center,
R.C.A. Laboratories,
Princeton, New Jersey 08540,
U.S.A.

A. A. MARADUDIN,
Department of Physics,
University of California,
Irvine, California 92664,
U.S.A.

P. MARK,
Electrical Engineering Dept.,
Princeton University,
Princeton, New Jersey 08540,
U.S.A.

R. PARSONS,
School of Chemistry,
University of Bristol,
Bristol BS8 1TS, England

B. A. PETHICA,
Unilever Research,
Port Sunlight, Wirral,
Cheshire L62 4XN, England

O. SINANOĞLU,
Sterling Chemistry Laboratory,
Yale University, New Haven,
Connecticut 06520, U.S.A.

R. W. STAEHLE,
Corrosion Center,
Department of Metallurgical Engineering,
Ohio State University,
Columbus, Ohio 43210, U.S.A.

E. A. STERN,
Physics Department,
University of Washington,
Seattle, Washington 98105, U.S.A.

D. TABOR,
Cavendish Laboratory,
University of Cambridge,
Cambridge CB2 3RQ, England

F. C. TOMPKINS,
Department of Chemistry,
Imperial College of Science and Technology,
London, S.W.7, England

H. H. UHLIG,
Corrosion Laboratory,
Massachusetts Institute of Technology,
Cambridge,
Massachusetts 02139, U.S.A.

L. L. M. VAN DEENEN,
Biochemisch Laboratorium,
Der Rijksuniversiteit,
Vondellaan 26, Utrecht,
The Netherlands

F. A. VANDENHEUVEL,
Animal Research Institute,
Central Experimental Farm,
Dept. of Agriculture, Ottawa,
Ontario, Canada

L. WEISS,
Department of Experimental Pathology,
Roswell Park Memorial Institute,
666 Elm Street, Buffalo,
New York 14203, U.S.A.

TH. WOLKENSTEIN,
Institute of Physical Chemistry,
Academy of Sciences,
Leninsky Prospekt 31,
Moscow V-71, U.S.S.R.

# LIST OF CONTRIBUTORS

G. J. DOOLEY, III, Aerospace Research Laboratories, Wright Patterson Airforce Base, Ohio

T. A. GOODWIN, Department of Electrical Engineering, Princeton University

J. T. GRANT, Aerospace Research Laboratories, Wright Patterson Airforce Base, Ohio

T. W. HAAS, Aerospace Research Laboratories, Wright Patterson Airforce Base, Ohio

J. P. HARLOS, Roswell Park Memorial Institute, Buffalo, New York

M. P. HOOKER, Systems Research Laboratories Inc., Dayton, Ohio

N. I. IONOV, A. F. Ioffe Physico-Technical Institute, Academy of Sciences of the U.S.S.R., Leningrad K-21

A. G. JACKSON, Systems Research Laboratories Inc., Dayton, Ohio

P. MARK, Department of Electrical Engineering, Princeton University

S. R. MORRISON, Stanford Research Institute, Menlon Park, California

L. WEISS, Roswell Park Memorial Institute, Buffalo, New York

K. F. WOJCIECHOWSKI, Institute of Experimental Physics, University of Wrocław, Poland

# FUTURE ARTICLES

VOLUME 2

| | | |
|---|---|---|
| Part 1: | A. M. BRODSKY and Y. V. PLESKOV. | Electron Photoemission at Metal /Electrolyte Solution Interface. |
| Part 2: | A. A. LUCAS and M. SUNJIC. | Fast-electron Spectroscopy of Collective Excitations in Solids. |
| Part 3: | G. POSTE and C. MOSS. | The Study of Surface Reactions in Biological Systems by Ellipsometry. |
| Part 4: | T. WOLFRAM and R. E. DEWAMES. | Surface Dynamics of Magnetic Materials. |

# FUTURE ARTICLES

Y. C. CHENG and H. D. HARLAND. Electronic States at the Silicon-Silicon Dioxide Interface.

E. DRAUGLIS. Thermal Accommodation.

W. MEHL and F. LOHMANN. The Organic Crystal/Electrolyte Interface.

E. W. MUELLER and T. T. TSONG. Field-Ion Microscopy of Metal Surfaces.

D. C. MEARS. Surface Corrosion of Metal Implants.

P. W. PALMBERG. Application of the Auger Effect to the Clean Surface Problem.

P. D. PAPAHADJOPOULOS. Liposomes as Models for Membranes.

L. A. PETERMANN. Thermal Desorption Kinetics of Chemisorbed Gases.

A. E. B. PRESLAND, D. L. TRIMM and G. L. PRICE. Thermal Reorganization of Evaporated Thin Films.

P. REHBINDER and E. SHCHUKIN. Surface and Interface Phenomena in Solids during Deformation and Fracture.

D. O. SHAH. Monolayers of Lipids in Relation to Membranes.

S. TOXVAERD. Surface Structure of Simple Fluids.

R. F. WALLIS. Lattice Dynamics of Crystals Surfaces.

T. WOLKENSTEIN. Photoadsorptive Effect of Semiconductors.

# THE INFLUENCE OF CHEMISORPTION ON THE ELECTRICAL CONDUCTIVITY OF THIN SEMICONDUCTORS*

THOMAS A. GOODWIN† and PETER MARK

*Department of Electrical Engineering,*
*Princeton University, Princeton, New Jersey, 08540*

## CONTENTS



*Supported by the Office of Naval Research under Contract No. N00014-67-A-0151-0014.

†Submitted in partial fulfillment of the requirement of the Ph.D. degree, Princeton University. Present address: Bell Telephone Laboratories, Murray Hill, N.J.

# INTRODUCTION

In recent years many semiconductor physicists have become interested in chemisorption phenomena, because of the surface barrier produced in the adsorbent by electron exchange with the adsorbate. Particular emphasis has been placed on the effect of the adsorbate on the electrical properties of the adsorbent (conductivity, photovoltage, contact potential) and on the kinetics of the adsorbate–adsorbent interactions. From a practical viewpoint, perhaps the most important consequence of chemisorption is the ability of many semiconductors, particularly compound semiconductors, to catalyse gas phase chemical reactions on their surfaces[1]. Chemisorption is also the mechanism by which conventional photographic and electrophotographic materials can be dye sensitized to tailor the spectral response[2]. Chemisorption of ambient constituents can also have a dominant effect on the electrical properties of a variety of thin film substrates used in microelectronics, including single crystals, evaporated films, chemically sprayed layers, and sintered and pressed powders. For example, the adsorption of oxygen on the surfaces of a variety of *n*-type materials (ZnO[2–4], BaO[5], CdSe[6,7], CdS[8], Ge[9], PbSe[10], PbS[11], PbTe[12], Zn doped $SnO_2$[13] and ZnS[14]) causes a large reduction (frequently by several decades) in the electrical conductivity of the adsorbent. The most common model used to account qualitatively for this effect considers the chemisorbed oxygen as an acceptor-like surface state (trap), whose energy lies in the gap between the conduction and valence bands. Oxygen chemisorption can also influence the characteristics of field-effect[15,16] and photo-voltaic[17,18] devices. Finally, in an attempt to determine the effects of very small partial pressures of oxygen ($10^{-11}$ torr and lower) on the electrical properties of thin CdS crystals, it was found that negatively charged oxygen ions (charged by a mass spectrometer ion source) raised the conductivity of the CdS[19]. However, at the beginning of the present work, no literature known to the authors described the effect of a pure oxygen ambient on the conductivity of CdS as a function of both temperature and ambient oxygen pressure.

Part I of this paper examines some of the models that might account for the effect of a reversible chemisorption state on the equilibrium conductivity of a thin, low conductivity, semiconductor. Since the experiments discussed in Part II deal with oxygen adsorption on CdS, an *n*-type material, the calculations will be in terms of an *n*-type bulk. The computational discussion, however, is general and the results are not restricted to CdS. With appropriate changes, the results could be applied to *p*-type materials. The influence of both acceptor-like and donor-like chemisorption states are examined.

Part II discusses the effects of chemisorbed oxygen upon the equilibrium conductivity of thin insulating platelets of CdS, as a function of both ambient temperature and ambient oxygen pressure. The results will be related to the calculations of Part I. We will show that the concentration of chemisorption surface states $N_s$ is a function of both temperature and adsorbate pressure, and that the experimental results are consistent with the present calculations, provided one takes into account the variation of $N_s$ with temperature.

# PART I MODELS

## 1. Scope of the Analysis and Assumptions

The aim of this discussion is to examine in what ways the electrical conductivity of a thin, resistive, semiconductor can be influenced by the reversible chemisorption of an ambient constituent from the vapor phase, and under what circumstances measurable conductivity changes can be expected. The adsorbent bulk is characterized by the defect states (of structural or impurity origin) in the principle band gap and the adsorbent–adsorbate interaction by the surface state produced by the adsorbate. Several models are examined that differ from each other by the choice of defect and surface states. For each choice of defect and surface state configuration, the simultaneous satisfaction of overall charge neutrality, occupation statistics and Poisson equation is used to compute the equilibrium conductivity of the adsorbent (actually the average adsorbent mobile carrier concentration) as a function of temperature, with the surface state concentration as parameter. Further, our major goal is a survey of the effects derived from various physically reasonable defect and surface state models and not an attempt to achieve extremely accurate numerical solution over a wide temperature range, based on precise, but perhaps as yet unwarranted, physical models. Hence, certain simplifying assumptions are made. First, the effective concentration of conduction band states $N_c$ is assumed constant with temperature, rather than varying as the three-halves power of the temperature. Also, in computing conductivity, the electron mobility is assumed constant with temperature. In reality, the mobility is temperature-dependent; for example, the Hall mobility $\mu$ of electrons in CdS has been shown to vary from 400 cm$^2$/volt-sec at 273°K to 4000 cm$^2$/volt-sec at 100°K[20]. These variations can be neglected, because the exponential variation of conductivity with reciprocal temperature is far greater than the temperature dependence of $\mu$ and of $N_c$. Also, these two temperature dependencies are opposing and tend to cancel. Further, in these calculations, the concentration of chemisorption states $N_s$ is temperature-independent. In reality, there is no way of guaranteeing this. One might expect that $N_s$ is related to the concentration of physically adsorbed species and would, therefore, predict that $N_s$ would increase as the temperature decreases, due to a condensation of gas on the surface. After deriving the effects with a temperature-independent $N_s$, we will show how the results are affected by the introduction of a temperature dependence. It is possible, in this way, to explain the basic behavior without being unduly hampered with details that are not crucial to the problem.

## 2. Partially Compensated Shallow Bulk Donor Level and a Band of Acceptor-like Surface States

### (i) *Formulation of the problem*

It is quite unlikely that any one model can explain completely the relation between a chemisorbed gas and the resultant equilibrium conductivity of the adsorbent. In order to account for measurements which are history dependent, associated with irreversible effects, and made on a sample not necessarily in equilibrium, it is desirable to begin with extremely simple models and to add complexity after the fundamentals are established. Consider first the model of Fig. 1. It shows the simplified energy band diagram of a semiconductor in which a single, discrete, bulk donor level is partially compensated by bulk acceptors in the lower-half of the band gap. The chemisorption state is assumed to be a band of acceptor-like states with constant density per unit energy $N_s/\Delta E$. For a discrete surface state, $\Delta E \to 0$. The charge on the surface causes a Schottky barrier which may, or may not, extend through to the center of the

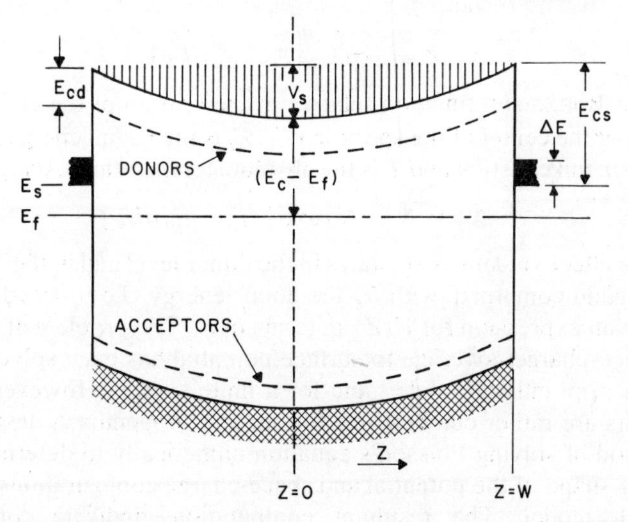


FIG. 1. Band diagram of an $n$-type crystal of finite thickness, which is partially compensated by acceptor states, and has some charge located in acceptor-like surface states.

sample. The exact shape of the space-charge barrier as a function of distance from the center of the sample is specified by $V(Z)$. The values of the potential at the center and the surface are, respectively, $V(0) = 0$ and $V(W) \equiv V_s$.

Overall charge neutrality in the extrinsic range yields:

$$\int_0^W n(Z)dZ + \int_0^W n_d(Z)dZ = W N_{d'} - n_s;\ N_{d'} = N_d - N_a. \quad (1)$$

In this equation, $n(Z)$ and $n_d(Z)$ are, respectively, the volume concentrations of conduction band electrons and electron occupied donors, $n_s$ is the surface concentration of electron occupied surface states, and $N_d$ and $N_a$ are, respectively, the total bulk donor and acceptor volume concentrations. Because the semiconductor is compensated, the donor state is never entirely filled. Hence, the donor level may be assumed always above the Fermi level and Boltzmann statistics may be used on the left side of eqn. (1) to obtain:

$$X N_{c'} \int_0^W \exp\left[-V(Z)/kT\right]dZ = W N_{d'} - n_s. \quad (2)$$

Here

$$X \equiv \exp\left[-(E_c - E_f)/kT\right] \quad (3)$$

is the Boltzmann function relative to the bottom of the conduction band at the center of the specimen $E_c$, $E_f$ is the Fermi energy, $k$ is the Boltzmann constant and $T$ is the absolute temperature. Also,

$$N_{c'} = N_c + N_d \exp\left[(E_c - E_d)/kT\right] \quad (4)$$

is the effective density of states in the donor level and in the conduction band combined, with $E_d$ the donor energy. To proceed, we require an expression for $V(Z)$ in terms of $n_s$. The problem of relating surface charge coverage to surface potential has been solved, both for a semi-infinite solid[21] and for a finite slab.[22] However, these results are rather cumbersome for our use. Appendix A describes a method of solving Poisson's equation numerically to determine the exact shape of the potential and space-charge configurations for the above model. The resultant computations indicate conditions under which the potential barrier may be approximated by a parabolic potential.

Finally, consider the occupation $n_s$ of the surface states $N_s$. We assume that, if the states are banded, they will have a constant density of states per unit energy:[23]

$$N_s/\Delta E = 4\pi m_s^*/h^2 \tag{5}$$

where $m_s^*$ is the effective mass of the two-dimensional surface band and $h$ is Planck's constant. The constant density of states arises from the assumption that the surface lattice potential is a two-dimensional periodic function and that the surface state bands have a parabolic energy–momentum relation. Although a sparsely covered surface is unlikely to have the adsorbate atoms arranged in a periodic manner, it is also unlikely that the surface states would then be banded. On the other hand, if the density of chemisorbed states is high enough to cause banding, the assumption that the chemisorbed atoms are ordered periodically becomes more plausible, since the lowest energy for the states will position the ions at virtual adsorbent lattice sites.[24]

The electron occupation of the surface states is given by

$$n_s = \int_{E_s}^{E_s+\Delta E} (N_s/\Delta E) f(E-E_f)\, dE \tag{6}$$

where $E_s$ is the bottom of the surface state band and $f(E-E_f) = [1+\exp(E-E_f)/kT]^{-1}$, the Fermi function. Direct integration yields:[23]

$$n_s = (N_s/\Delta E)\{\ln[1+X\exp(\epsilon_{cs}-\Psi_s)]-\ln[1+X\exp(\epsilon_{cs}-\Psi_s -\Delta\epsilon)]\} \tag{7}$$

where $\Psi_s \equiv V_s/kT$, $\Delta\epsilon \equiv \Delta E/kT$ and $\epsilon_{cs} \equiv E_{cs}/kT$. In the case of a discrete surface state, $\Delta\epsilon \to 0$, and eqn. (7) becomes

$$n_s = [N_s X\exp(\epsilon_{cs}-\Psi_s)]/[1+X\exp(\epsilon_{cs}-\Psi_s)]. \tag{8}$$

To obtain the specimen conductivity parallel to the adsorbing surface (perpendicular to $Z$), we must compute the average conduction band electron concentration

$$\langle n_c\rangle \equiv (1/W)\int_0^W n_c(Z)\,dZ = (N_c X/W)\int_0^W \exp[-\Psi(Z)]\,dZ. \tag{9}$$

This is achieved by numerically solving eqn. (2) and either eqns. (7) or (8) to find $X(E_f)$ and $\Psi(Z)$ and subsequently inserting these into eqn. (9) to obtain $\langle n_c \rangle$. The results of several typical calculations appear in Figs. 2 through 5. The discussion of these numerical results is deferred briefly, in order to present some limiting cases which can be formulated analytically.

*(ii) Analytical formulation of limiting cases*

Although it is desirable to have complete computer solutions to the problem, it is revealing, at this point, to derive approximate solutions for certain regions of the problem in closed form. In this way, insight is gained into the way in which the solutions vary with changes in the parameters of the model. It should be noted at first that for the adsorbate free case $N_s = 0$ the bands are flat, and the integrands of eqn. (1) are constant. If the surface state concentration is very small, and the distribution of these states in energy is also small, the surface states will be totally filled and will lie below the Fermi energy. Then we can define a new effective donor density $N_{d''} \equiv N_{d'} - N_s/W$, and we may obtain the average concentration of conduction band electrons $\langle n_c \rangle$ directly from eqns. (2) and (9):

$$\begin{aligned} \langle n_c \rangle &= (N_c/W)(WN_{d'} - N_s)/(XN_{c'}) = N_c N_{d''}/N_{c'} \\ &= N_c N_{d''}/(N_c + N_d \exp \epsilon_{cd}) \end{aligned} \qquad (10)$$

where $\epsilon_{cd} \equiv E_{cd}/kT$. Thus, when $N_s \ll WN_{d'}$, the *activation energy* of the ln $\langle n_c \rangle$ vs. $1/T$ curve is determined by the bulk doping and we may label this a bulk-controlled regime. However, the *magnitude* of the current is affected by the surface states through $N_{d''}$. Hence, so long as $N_s \ll WN_{d'}$, progressively increasing $N_s$ generates a family of parallel curves on the semi-logarithmic field, all with the same bulk-determined activation, but lying lower than the larger $N_s$.

For the opposite extreme case $N_s \gg WN_{d'}$, the surface states will be well above the Fermi energy and the bulk of the sample will be essentially depleted of electrons. Since $X$ exp $(\epsilon_{cs} - \Psi_s)$ will be very small, eqns. (7) and (8) reduce, respectively, to

$$n_s = (N_s/\Delta\epsilon)\, X \exp(\epsilon_{cs} - \Psi_s)[1 - \exp(-\Delta\epsilon)]; \Delta\epsilon \gtrsim 1 \qquad (11)$$

and

$$n_s = N_s X \exp(\epsilon_{cs} - \Psi_s);\ \Delta\epsilon \ll 1. \tag{12}$$

For the case of a discrete surface state $[\Delta\epsilon \to 0]$, eqn. (12) may be substituted into eqn. (2) and the $X$ factored out:

$$X = WN_{d'}/\left\{N_{c'} \int_0^W \exp[-\Psi(Z)]\,dZ + N_s \exp(\epsilon_{cs} - \Psi_s)\right\} \tag{13}$$

and this may be inserted into eqn. (9) to obtain $\langle n_c \rangle$:

$$\langle n_c \rangle = N_c N_{d'} \int_0^W \exp[-\Psi(Z)]\,dZ/\left\{N_{c'} \int_0^W \exp[-\Psi(Z)]\,dZ + \right.$$
$$\left. + N_s \exp(\epsilon_{cs} - \Psi_s)\right\}. \tag{14}$$

Since the bulk is essentially (and thus uniformly) depleted, it follows that the space-charge concentration is constant. Thus, from eqn. (A.8) of Appendix A the integrals of eqn. (14) are

$$\int_0^W \exp[-\Psi(Z)]dZ = \int_0^W \exp[-\Psi_s Z^2/W^2]dZ = W\pi^{1/2}$$
$$\operatorname{erf} \Psi_s^{1/2}/2\Psi_s^{1/2} \tag{15}$$

and

$$\Psi_s = en_s W/2\epsilon kT = eN_{d'}W^2/2\epsilon kT \tag{16}$$

where $e$ is the magnitude of the electronic charge and $\epsilon$ is the static electric permittivity of the semiconductor (in farads/cm). With eqns. (15) and (16), eqn. (14) reads

$$\langle n_c \rangle = \frac{N_c N_{d'} W \pi^{1/2} \operatorname{erf} \Psi_s^{1/2}/(2\Psi_s^{1/2})}{(\pi^{1/2} W N_{c'} \operatorname{erf} \Psi_s^{1/2}/(2\Psi_s^{1/2}) + N_s \exp(\epsilon_{cs} - \Psi_s)} \tag{17}$$

As $1/kT$ becomes large, the second term in the denominator strongly dominates and (17) becomes

$$\langle n_c \rangle \simeq [WN_c N_{d'} \exp(-(\epsilon_{cs} - \Psi_s))/N_s][\pi^{1/2} \operatorname{erf} \Psi_s^{1/2}/(2\Psi_s^{1/2})]. \tag{18}$$

Although $\Psi_s(kT)$ appears three times in this equation, the dominant temperature dependence is contributed by the exponential term. Hence, in the limit of very large surface state concentrations, there exists a surface-controlled regime, where $\ln \langle n_c \rangle$ vs. $1/T$ is very nearly a straight line, but now with the activation energy given by the surface-determined quantity $E_{cs} - V_s$.

Consider the case where $N_{d'}$ and $W$ are large enough, so that the potential barrier, produced by complete depletion of the conduction band electrons, is greater than $E_{cs}$. Then eqn. (18) indicates that, if complete depletion were to occur, the activation energy of the surface-controlled regime would be negative. Actually, there will not be complete depletion and the surface-controlled regime will not be observed, since the surface states will not become further populated when they become energetically higher than the lowest portion of the conduction band. This may be stated mathematically as follows. In order to observe the surface state-controlled regime, we must have:

$$E_{cs} > V_s = eN_{d'}W^2/2\epsilon \tag{19a}$$

so that

$$N_{d'}W^2 < 2\epsilon\, E_{cs}/e. \tag{19b}$$

*(iii) Computer Solutions*

The results of several typical computer solutions are shown in Figs. 2 through 5. Figure 2a shows the logarithm of the average density of conduction band electrons in the sample vs. $1/T$ (°K). A family of curves is displayed for which the surface state concentration $N_s$ is the parameter. The scale on the right-hand side gives the logarithm of the sample current when one volt is applied to the sample, assuming a geometry 1 mm long, 1 mm wide, and 10 microns half-thickness; an electron mobility of 200 cm²/volt-sec; and a dielectric constant of 10. These values pertain to the CdS samples discussed in Part II. Figure 2b shows the position of the Fermi level relative to the bottom of the conduction band at the center of the crystal as a function of $1/T$. Figure 2c illustrates the variation of potential barrier height with $1/T$. The family of curves in Fig. 2b and Fig. 2c are for the same values of $N_s$, respectively, as in Fig. 2a and the values of $N_s$ for each curve are tabulated in Table 1. The position of the Fermi level and the barrier height is plotted for one example only, because the results are similar in the other examples. For instance, in the model used for Fig. 3, the plot of potential barrier height would be that of Fig. 2c scaled up by a

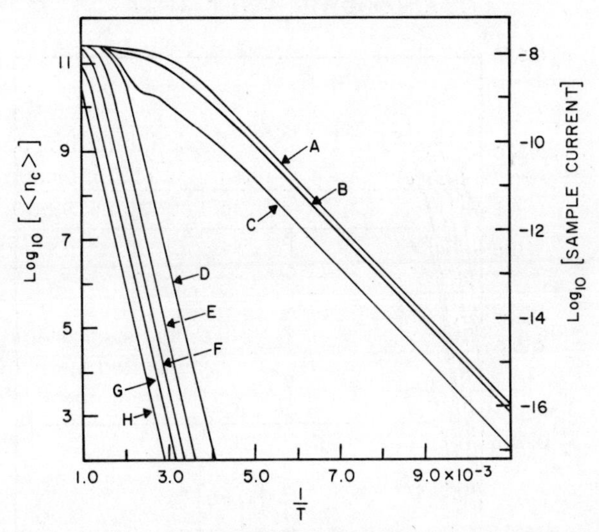


FIG. 2a. Logarithm of conductivity vs. $1/T$ for a crystal with a shallow donor level and a discrete acceptor-like surface state. [$N_{d'} = 2.0 \times 10^{11}$ per cm$^3$, $N_d = 1.0 \times 10^{16}$ per cm$^3$, $E_{cd} = 0.2$ eV, $E_{cs} = 0.9$ eV, $W = 1.0 \times 10^{-3}$ cm.]

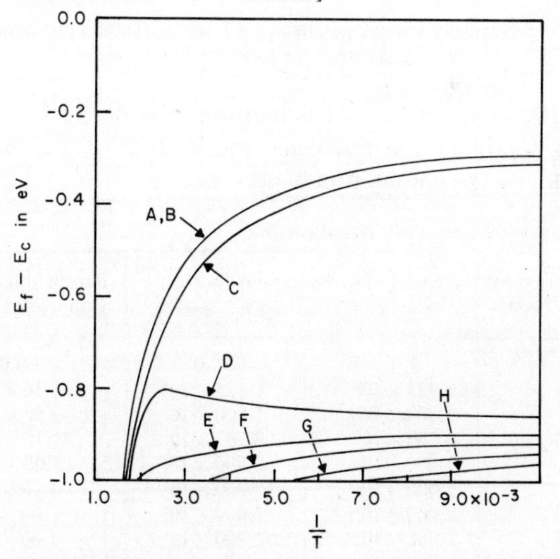


FIG. 2b. Position of Fermi level vs. $1/T$ for a crystal with a shallow donor level and a discrete acceptor-like surface state.

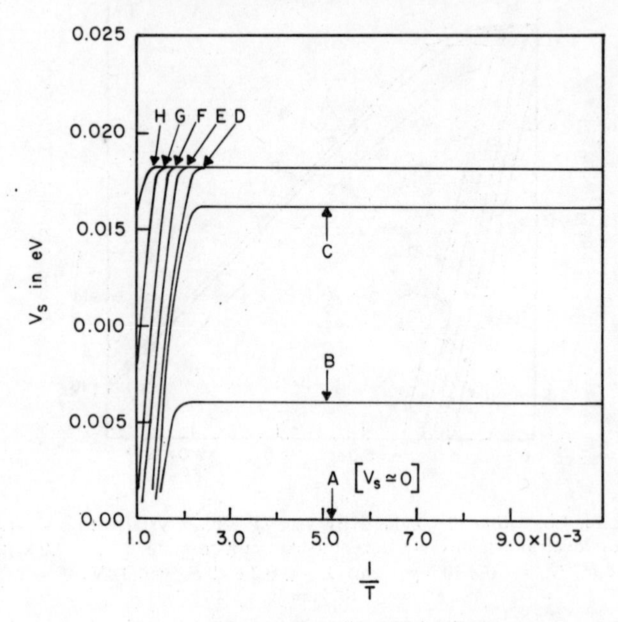


FIG. 2c. Schottky barrier height vs. $1/T$ for a crystal with a shallow donor level and a discrete acceptor-like surface state.

factor of 10. In all of these calculations, the donor level is 0.2 eV below the conduction band, while the bottom of the surface state is 0.9 eV below the conduction band.

TABLE 1. KEY TO FIGS. 2 THROUGH 5

| Symbol used to define curve | Density of surface states (per cm²) Fig. 2 | Figs. 3 to 5 | Bandwidth of surface states (eV) for Fig. 5 |
|---|---|---|---|
| A | 0.0 | 0.0 | 0.0 |
| B | $6.325 \times 10^{7}$ | $6.325 \times 10^{8}$ | $3.16 \times 10^{-3}$ |
| C | $1.800 \times 10^{8}$ | $1.800 \times 10^{9}$ | $9.00 \times 10^{-3}$ |
| D | $2.200 \times 10^{8}$ | $2.200 \times 10^{9}$ | $1.10 \times 10^{-2}$ |
| E | $2.000 \times 10^{9}$ | $2.000 \times 10^{10}$ | $1.00 \times 10^{-1}$ |
| F | $2.000 \times 10^{10}$ | $2.000 \times 10^{11}$ | 1.00[a] |
| G | $2.000 \times 10^{11}$ | $2.000 \times 10^{12}$ | $1.00 \times 10^{1}$[a] |
| H | $2.000 \times 10^{12}$ | $2.000 \times 10^{13}$ | $1.00 \times 10^{2}$[a] |

[a]Overlaps conduction band.

In Fig. 2, the donor concentration $N_d$ is $10^{16}$ per cm$^3$, while the effective doping concentration $N_{d'}$ is $2 \times 10^{11}$ per cm$^3$. The surface state is discrete. In curve "A" of Fig. 2a, the surface state concentration is zero (adsorbate free case). In the higher temperature portion, which we shall call the "saturation regime", there is complete ionization of the donors. Note that curve "B", for a surface state concentration of $6.325 \times 10^7$ per cm$^3$ [$N_s = W N_{d'}(10)^{-1/2}$], is parallel to curve "A". Curve "B" lies below curve "A" by the quantity $\log_{10}[(1-10^{-1/2})^{-1}]$. Curve "C", for a surface state concentration $N_s = 0.9\,W N_{d'} = 1.8 \times 10^8$ per cm$^2$ (this is 90% of the density needed to completely deplete the conduction band at absolute zero), also lies parallel to "A" in the lower temperature region. In this region, the curve lies below curve "A" by the quantity $\log_{10}[(1-0.9)^{-1}] = \log_{10}(10) = 1$ (or exactly one decade). Increasing the concentration of surface states to $N_s = 1.1\,W N_{d'} = 2.2 \times 10^8$ per cm$^2$ (curve "D") causes the slope of the lower temperature region to become 0.9 eV, which is the surface state energy. In the higher temperature region, both curves "C" and "D" follow nearly the same path and have a slope of 0.55 eV, the average of the donor and surface state energy. If one were to plot the special case of $N_s = W N_{d'}$, the resultant curve would have a constant slope of 0.55 eV. Since it is highly improbable that a crystal will have completely uniform doping and thickness, it is equally improbable that this precise relation will be valid over the entire crystal, hence, such a curve is not likely to be observed experimentally. Also, the basic assumption that $N_s$ is constant with temperature may break down at this point, for at higher temperatures we could have $N_s < W N_{d'}$, while at lower temperatures we could have $N_s > W N_{d'}$. The fact that such a large change takes place, when the concentration of surface states crosses this equality, indicates that there may be regions where the experimental data will exhibit a large scattering of points. Curves "E" through "H" all have constant slopes of approximately 0.9 eV. Since each successive curve corresponds to ten times the surface state concentration of the preceding one, the curves are spaced exactly one decade apart. Referring to Fig. 2c, the maximum surface potential (corresponding to complete depletion of conduction electrons) is 0.018

eV. Exact computation confirms that the slope of curves "E" through "H" is 0.9–0.882 eV, as indicated by eqn. (18).

Curves "A", "B", and "C" of Fig. 2b show the position of the Fermi level for the cases where $N_s < W N_{d'}$. For these three cases, the donor level is always partially occupied, but, because the crystal is partially compensated, the donors are never filled. Therefore, as the temperature approaches 0°K, the Fermi energy approaches the energy of the donor level and curves "A", "B", and "C" (curve "A" lies nearly on top of curve "B") all rise to $-0{\cdot}2$ eV from the bottom of the conduction band as $1/T$ increases. A similar argument shows that the Fermi energy will approach $-0{\cdot}882$ eV as $1/T$ increases when $N_s > W N_{d'}$. There is one case (curve "D") where the curve is not monotonic. This unusual behavior occurs when $W N_{d'} < N_s < 2W N_{d'}$. Referring to Fig. 2c, the potential barrier (and therefore the occupation of $N_s$) rises rapidly to the maximum value at the same temperature that shows a peak in the Fermi energy of curve "D". Since $N_s < 2W N_{d'}$, the surface state must be more than half filled at this temperature, which means that the Fermi energy must decrease as $1/T$ increases, so that it may approach its asymptotic limit of $E_s$.

Figure 3 is for a doping concentration $N_{d'} = 2 \times 10^{12}$, or ten times that of the previous example. All of the curves appear very similar to those of Fig. 2. The differences caused by increasing the doping concentration are as follows: (1) the adsorbate free curve of Fig. 3 is raised by a factor of 10; (2) the maximum potential barrier height is increased by a factor of 10 to 0.18 eV; (3) as a result of the increased barrier height, the slopes of curves "E" through "H" in Fig. 3 are reduced to 0.9 eV − 0.18 eV = 0.72 eV.

Increasing the doping concentration again by a factor of 10 would cause all of the curves to have slopes of about 0.2 eV, since the potential barrier necessary for total depletion (1.8 eV) would be greater than the surface state depth (0.9 eV). This is a case where eqns. (19) dictate that the surface-controlled regime will not be observed.

Keeping the doping concentration at $N_{d'} = 2 \times 10^{12}$, but reducing the donor concentration $N_d$ to $10^{14}$ per cm$^3$, produces the curves shown in Fig. 4. Comparing the curves of Fig. 4 with those of Fig. 3,

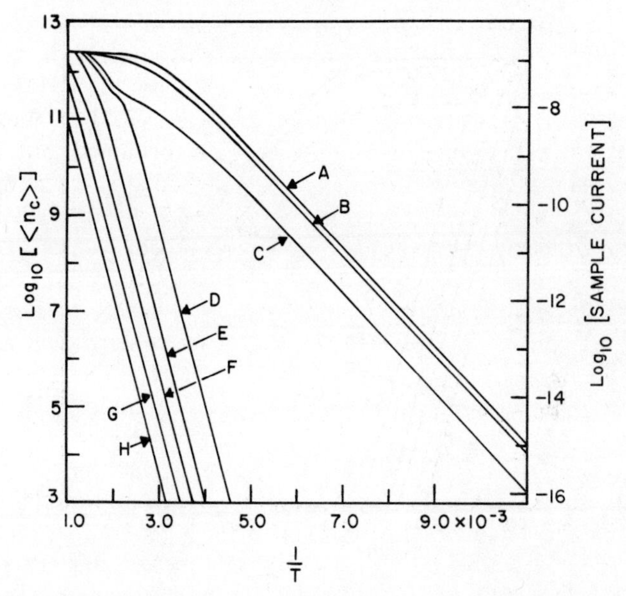


FIG. 3. Logarithm of conductivity vs. $1/T$ for a crystal with a shallow donor level and a discrete acceptor-like surface state. [$N_{d'} = 2.0 \times 10^{12}$ per cm$^3$, $N_d = 1.0 \times 10^{16}$ per cm$^3$, $E_{cd} = 0.2$ eV, $E_{cs} = 0.9$ eV, $W = 1.0 \times 10^{-3}$ cm.]

we find that curves "D" through "H" are identical, respectively, in the two figures. The major difference lies in curves "A" through "C", where we see that reducing the donor concentration causes the saturation regime to move further toward the low-temperature region.

The effect of banding on the surface states is shown in Fig. 5, for which calculation we have assumed a surface state concentration of $2 \times 10^{11}$ per eV-cm$^2$. Increasing the total number of surface states by a factor of ten increases the width of the band by a factor of 10. So long as the bandwidth is small, the results will be the same as for the discrete surface state case of Fig. 3. The bandwidth becomes 0.1 eV in curve "E" and a further increase of the surface state concentration will not effect the conductivity of the crystal significantly, since only the lowermost portion of the surface state band will be occupied. This "saturation" effect causes curves "F"

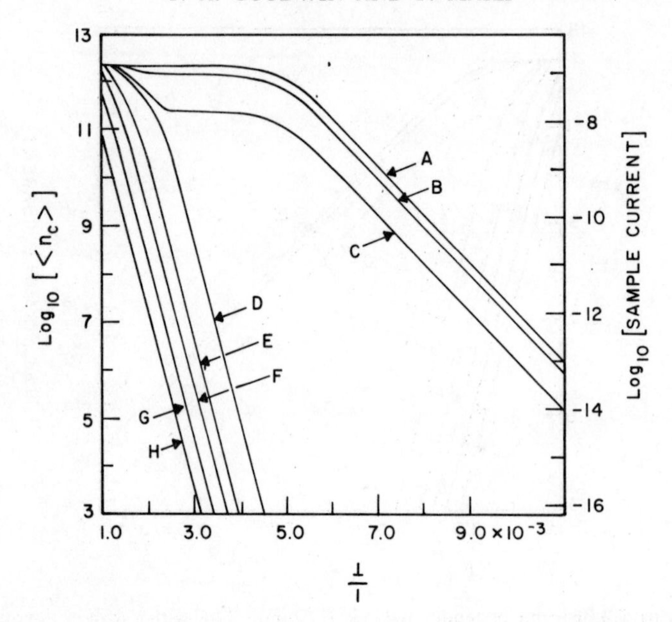


FIG. 4. Logarithm of conductivity vs. $1/T$ for a crystal with a shallow donor level and a discrete acceptor-like surface state. [$N_{d'} = 2.0 \times 10^{12}$ per cm$^3$, $N_d = 1.0 \times 10^{14}$ per cm$^3$, $E_{cd} = 0.2$ eV, $E_{cs} = 0.9$ eV, $W = 1.0 \times 10^{-3}$ cm.]

through "H" to be identical. This points to a fairly simple test for banding. If banding occurs, there should be a point beyond which adding more chemisorption states will not produce a noticeable effect on the conductivity unless, of course, there is transport in the two-dimensional surface bands themselves.

If some of the assumptions were relaxed in deriving the density of states for the banded case, one may find that the density of states is no longer constant in energy. For example, if increasing the total number of states not only increases the bandwidth, but also increases the density of states per unit energy, one would find some change in conductivity as the concentration of surface states is increased. The change, however, would be sublinear and would progress much more slowly after reaching the concentration at which the surface state bandwidth overlaps the conduction band.

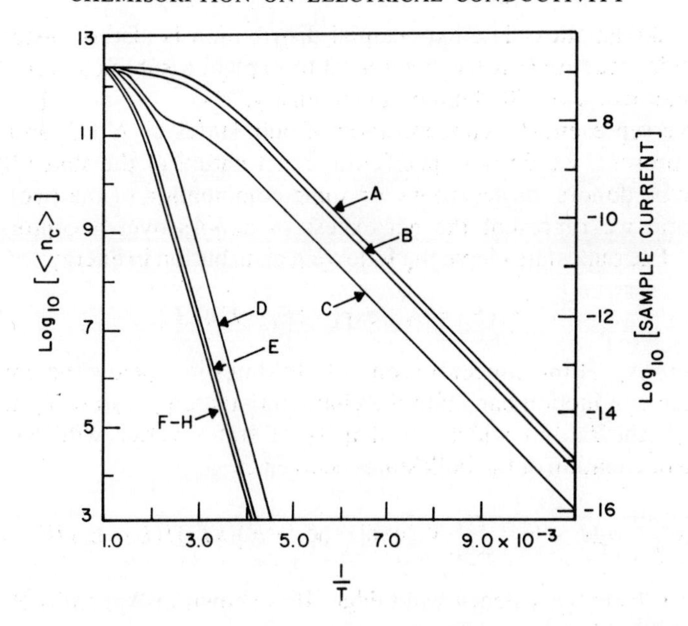


FIG. 5. Logarithm of conductivity vs. $1/T$ for a crystal with a shallow donor level and a banded acceptor-like surface state. [$N_{d'} = 2.0 \times 10^{12}$ per cm$^3$, $N_d = 1.0 \times 10^{16}$ per cm$^3$, $E_{cd} = 0.2$ eV, $E_{cs} = 0.9$ eV, $W = 1.0 \times 10^{-3}$ cm, $N_s/\Delta E = 2.0 \times 10^{11}$ per cm$^2$-eV.]

Therefore, it should be possible to detect banding, even if it does not arise from a periodic two-dimensional surface potential.

### 3. Exponential Distribution of Bulk States and a Discrete Acceptor-like Surface State

The major limitation of the preceding model is that it can be used to explain only two activation energies, one determined by the donor level and one determined by the surface state. A distribution of bulk states, however, might produce a spectrum of activation energies as the surface state concentration is varied. Also, some curvature in the conductivity vs. $1/kT$ curves might be expected. In the present section, an exponential distribution has been chosen, to facilitate the mathematics. The exponential distribution is also of historical significance, since it has been used to explain a sublinear variation of photocurrent with illumination intensity.[25]

We represent the concentration of bulk states by $N_b(E)$ and, for the present, we do not specify the exact nature of the states (they may be donors, or acceptors, or some combination of the two). As before, we represent the net excess of donors over acceptors as $N_{d'}$. The bulk states have the following distribution in energy:

$$N_b(E) = N_{bo}\exp[-(E_c - E)/kT] \tag{20}$$

where $N_{bo}$ is the concentration of bulk states just below the bottom of the conduction band and the characteristic temperature $T_c$ determines the rate at which the density of states varies with energy. The occupation of the bulk states is given by

$$n_b = \int_{E_v}^{E_c} n_b(E)dE = \int_{E_v}^{E_c} N_{bo}\exp[-(E_c - E)/kT_c]f(E - E_f)dE \tag{21}$$

where $E_v$ is the valence-band edge. It is shown in Appendix B that this integral is approximately

$$n_b \simeq N_{ef}(T)\exp[-(E_c - E_f)/kT_c] \tag{22}$$

where

$$N_{ef}(T) = N_{bo}kT_c\{1 + (T/T_c)^2(\pi^2/6)[1 + (T/T_c)^2(7\pi^2/60)]\}. \tag{23}$$

The charge neutrality equation [eqn. (1)] now becomes

$$\begin{aligned} XN_c \int_0^W \exp[-V(Z)/kT]dZ + N_{ef}(T)\exp[-(E_c - E_f)/kT_c] \\ \times \int_0^W \exp[-V(Z)/kT_c]dZ \\ = WN_{d'} - n_s. \end{aligned} \tag{24}$$

As before, $E_c$ applies to the center of the crystal and $V(Z)$ is the potential due to charge on the surface.

If the crystal is sufficiently insulating, the Debye length will be greater than the sample thickness and we may assume that the space charge is constant across the sample and that the potential is parabolic. Hence, we obtain:

$$\frac{XN_c\pi^{1/2}\text{erf}\,\Psi_s^{1/2}}{2\Psi_s^{1/2}}+\frac{X^{(T/T_c)}N_{ef}(T)\pi^{1/2}\text{erf}(\Psi_s^{1/2}T/T_c)}{2\Psi_s^{1/2}T/T_c}=N_{d'}-n_s/W \tag{25}$$

where the surface potential $\Psi_s$ is given by:

$$\Psi_s = eWn_s/2\epsilon kT \tag{26}$$

If the conductivity is large enough, so that uniform space charge cannot be assumed, the exact shape of the potential can be obtained as in Appendix C.

A computer was used to solve eqns. (25) and (26). For the first sample calculation, we let the bulk states have a characteristic temperature $T_c = 4000°K$ (the significance of this temperature is discussed later) and we set the density of states near the bottom of

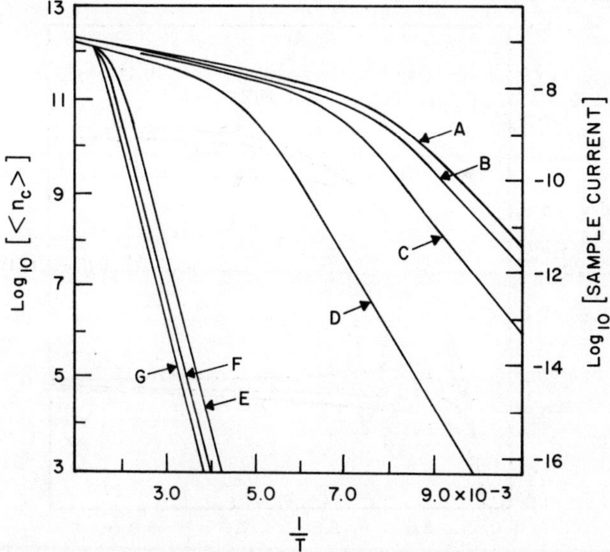


FIG. 6a. Logarithm of conductivity vs. $1/T$ for a crystal with an exponential distribution of bulk states and a discrete acceptor-like surface state. [$W = 1.0\times10^{-3}$ cm, $E_{cs} = 0.9$ eV, $N_{d'} = 1.93\times10^{12}$ per cm³, $N_{bo} = 1.0\times10^{13}$ per eV-cm³, $T_c = 4000°K$.]

the conduction band $N_{bo} = 1.0 \times 10^{13}$ per eV-cm$^3$. The effective doping concentration $N_{d'}$ is given a value such that the bulk states will be filled to within 0.2 eV of the conduction band at 0°K. This corresponds to $N_{d'} = 1.93 \times 10^{12}$ per cm$^3$. Figure 6a illustrates the average conduction-band electron concentration for the values of $N_s$ listed in Table 2. The surface state is assumed discrete and 0.9 eV deep. The current is computed with the transport parameter and dimensions used in the previous section. Curve "A" is the adsorbate free case [$N_s = 0.0$], while curve "B" is for $N_s = WN_{d'}(10)^{-3/2}$. Each subsequent curve is for a value of $N_s$ which is $(10)^{1/2}$ times the previous $N_s$. Thus, for curve "E", $N_s = WN_{d'}$. Note that curves "A" through "D" exhibit a great deal of curvature, and that the slopes for the linear (lower temperature) regions gradually increase, from 0.2 eV for the absorbate free case, to about 0.31 eV for curve "D". Figure 6b shows the position of the Fermi level with respect to the bottom of the conduction band at $Z = 0$.

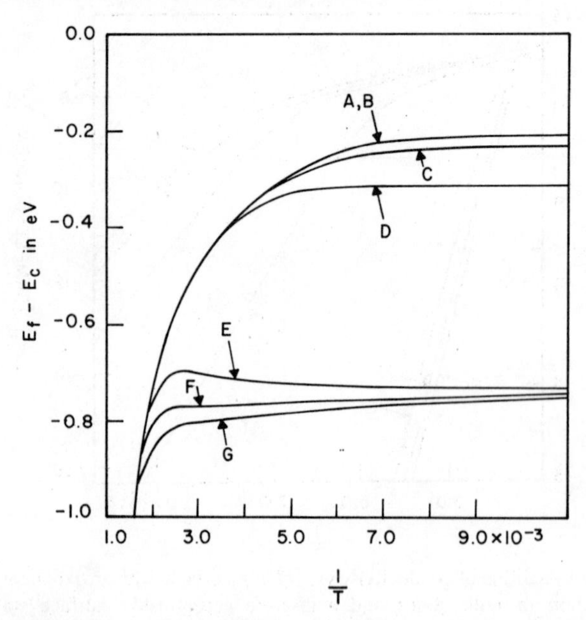


FIG. 6b. Position of Fermi level vs. $1/T$ for a crystal with an exponential distribution of bulk states and a discrete acceptor-like surface state.

TABLE 2. KEY TO FIGS. 6 THROUGH 12

| Symbol used to define curves | $N_s/WN_{d'}$ | Concentration of surface states (per $cm^3$) | | | | |
|---|---|---|---|---|---|---|
| | | Figs. 6, 11, and 12 | Fig. 7 | Fig. 8 | Fig. 9 | Fig. 10 |
| A | 0 | 0.0 | 0.0 | 0.0 | 0.0 | 0.0 |
| B | $10^{-1.5}$ | $6.1 \times 10^7$ | $1.7 \times 10^7$ | $1.7 \times 10^8$ | $3.1 \times 10^8$ | $2.2 \times 10^7$ |
| C | $10^{-1.0}$ | $1.9 \times 10^8$ | $5.4 \times 10^7$ | $5.4 \times 10^8$ | $9.6 \times 10^8$ | $6.8 \times 10^7$ |
| D | $10^{-0.5}$ | $6.1 \times 10^8$ | $1.7 \times 10^8$ | $1.7 \times 10^9$ | $3.1 \times 10^9$ | $2.2 \times 10^8$ |
| E | 1 | $1.9 \times 10^9$ | $5.4 \times 10^8$ | $5.4 \times 10^9$ | $9.6 \times 10^9$ | $6.8 \times 10^8$ |
| F | $10^{0.5}$ | $6.1 \times 10^9$ | $1.7 \times 10^9$ | $1.7 \times 10^{10}$ | $3.1 \times 10^{10}$ | $2.2 \times 10^9$ |
| G | 10 | $1.9 \times 10^{10}$ | $5.4 \times 10^9$ | $5.4 \times 10^{10}$ | $9.6 \times 10^{10}$ | $6.8 \times 10^9$ |

These two features are the main differences between the exponential bulk model and the discrete bulk donor model. The present calculation shows that it is possible, for certain distributions of bulk states to produce a gradual increase in activation energy as the concentration of acceptor-like surface states increases. Notice, however, that as $N_s$ approaches $WN_{d'}$, the activation energy increases very rapidly to the activation energy of the surface-controlled regime and, once $N_s > WN_{d'}$, the activation energy remains constant. This indicates that, if an increase in activation energy with increasing $N_s$ is observed, the surface-controlled regime has not yet been reached. Applying this to a physical experiment, suppose we were to measure the sample current vs. $1/T$, keeping the ambient pressure of an acceptor-like gas constant with temperature. If increasing the ambient pressure produced curves of increasing activation energy, then we could infer the following conclusion:

1. We have not yet reached the surface state-controlled regime; OR
2. The density of surface states $N_s$ is a function of temperature; OR
3. One or more of the crystal parameters are pressure dependent.

In curves "E" through "G", the sample becomes completely depleted at the lower temperatures. The height of the surface potential becomes 0.15 eV, and the slopes of these curves are 0.75 eV = 0.9 eV − 0.15 eV. Figure 6b shows that the behavior of the Fermi level is very similar to that of the previous section. As one might expect, curve "A" goes asymptotically to −0.2 eV, while curves "E" through "G" go to −0.75 eV. The major difference is that the asymptotic limits of curves "C" and "D" lie slightly lower than 0.2 eV. This is consistent with the increased activation energy as $N_s$ increases.

In the next example, we have kept $N_{bo}$ at $1.0 \times 10^{13}$ per eV-cm$^3$, but have reduced $T_c$ to 2000°K. In order to have the bulk states filled to within 0.2 eV of the bottom of the conduction band at 0°K, we must reduce the effective doping concentration $N_{d'}$ to $5.4 \times 10^{11}$ per cm$^3$. The major changes produced by reducing $T_c$ and, consequently, the effective doping concentration, are as follows:

1. Curves "A" through "D" of Fig. 7 have slightly more curvature than the respective curves of Fig. 6.
2. Curves "A" and "D" lie much closer to each other in Fig. 7 than in Fig. 6.
3. The high temperature conductivity is reduced.
4. The maximum surface potential is reduced to 0.048 eV.
5. The slope of the surface state controlled regime is 0.852 eV.

In the next example, $T_c$ is kept at 2000°K, while $N_{bo}$ and $N_{d'}$ are both increased by a factor of 10. As shown in Fig. 8, the saturation conductivity is one decade higher than in Fig. 7, while the low temperature portions of the adsorbate free curves approach the same values. The reason for the low temperature coincidence is that, in both cases, $N_{d'}$ has been adjusted, so that the Fermi level goes to 0.2 eV below the bottom of the conduction band as the

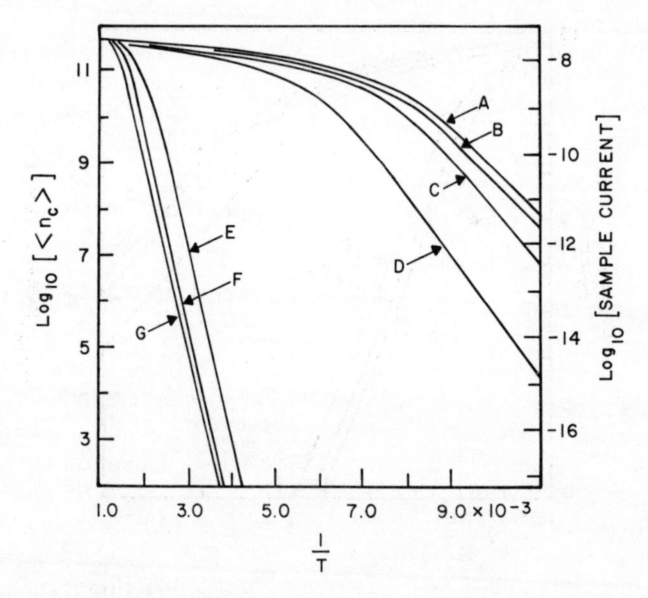


FIG. 7. Logarithm of conductivity vs. $1/T$ for a crystal with an exponential distribution of bulk states and a discrete acceptor-like surface state. [$W = 1.0 \times 10^{-3}$ cm, $E_{cs} = 0.9$ eV, $N_{d'} = 5.4 \times 10^{11}$ per cm$^3$, $N_{bo} = 1.0 \times 10^{13}$ per eV-cm$^3$, $T_c = 2000$°K.]

temperature approaches 0°K. In Fig. 8 the curves "A" through "D" are even more closely spaced than the respective curves in Fig. 7. The most drastic change appears in curves "E" through "G". Since the maximum surface potential is now about 0.44 eV, the slope of these curves is reduced to $0.9-0.44=0.46$ eV.

The next comparison we shall make is to maintain $N_{bo}$ at $1.0 \times 10^{14}$ per eV-cm³ and $T_c$ at 2000°K, while raising $N_{d'}$ to $9.6 \times 10^{12}$ (the bulk states should now fill up to within 0.1 eV of the conduction band at 0°K for the adsorbate free case). In Fig. 9, curves "A" through "C" are nearly coincident. The gradual increase in slope, which was typical of the previous figures, has now been lost and there is a sudden jump in slope between curves "D" and "E". The slope of the surface-controlled regime is now only 0.2 eV (the potential barrier is nearly 0.7 eV).

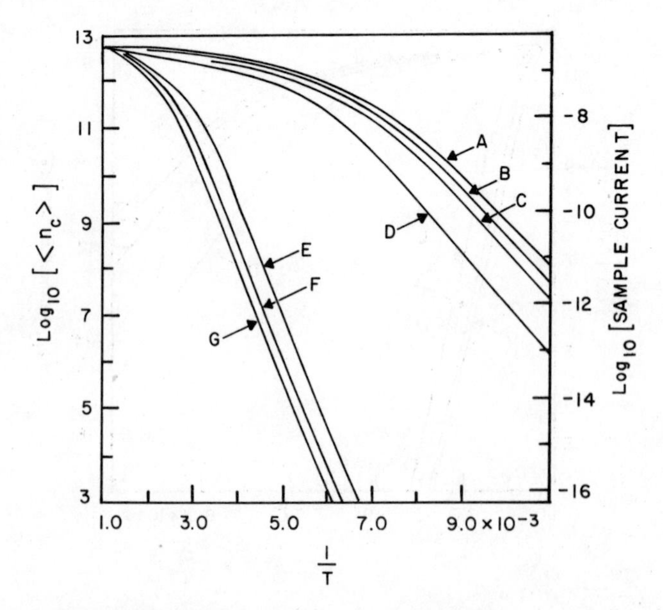


FIG. 8. Logarithm of conductivity vs. $1/T$ for a crystal with an exponential distribution of bulk states and a discrete acceptor-like surface state. [$W=1.0 \times 10^{-3}$ cm, $E_{cs}=0.9$ eV, $N_{d'}=5.4 \times 10^{12}$ per cm³, $N_{bo}=1.0 \times 10^{14}$ per eV-cm³, $T_c=2000$°K.]

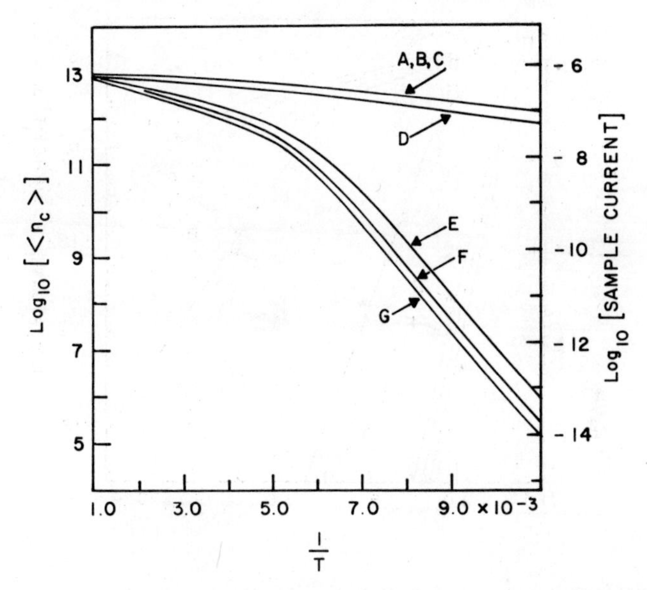


FIG. 9. Logarithm of conductivity vs. $1/T$ for a crystal with an exponential distribution of bulk states and a discrete acceptor-like surface state. [$W = 1.0 \times 10^{-3}$ cm, $E_{cs} = 0.9$ eV, $N_{d'} = 9.6 \times 10^{12}$ per cm³, $N_{bo} = 1.0 \times 10^{14}$ per eV-cm³, $T_c = 2000$°K.]

In the previous three examples $T_c$ has been 2000°K. Following Rose's theory, this $T_c$ produces a variation of photocurrent with the 0.87 power of the illumination intensity at room temperature, while $T_c = 4000$°K corresponds to a 0.93 power law[25]. Since these calculations are of a general nature, the choice of $T_c$ is rather arbitrary, and is not meant to agree with any previous experimental data.

Having gained some feeling for the results of an exponential distribution, one might wish to examine another distribution, such as a uniform distribution of bulk states. The practical problems in making such calculations discourage this. We can, however, go in the direction of the uniform distribution by increasing $T_c$. In Fig. 10, $T_c = 10{,}000$°K, while $N_{bo} = 1.0 \times 10^{12}$ per eV-cm³. As in the first three examples, $N_{d'}$ ($6.8 \times 10^{11}$ per cm³) has been adjusted, so that the bulk is filled to within 0.2 eV of the conduction band at

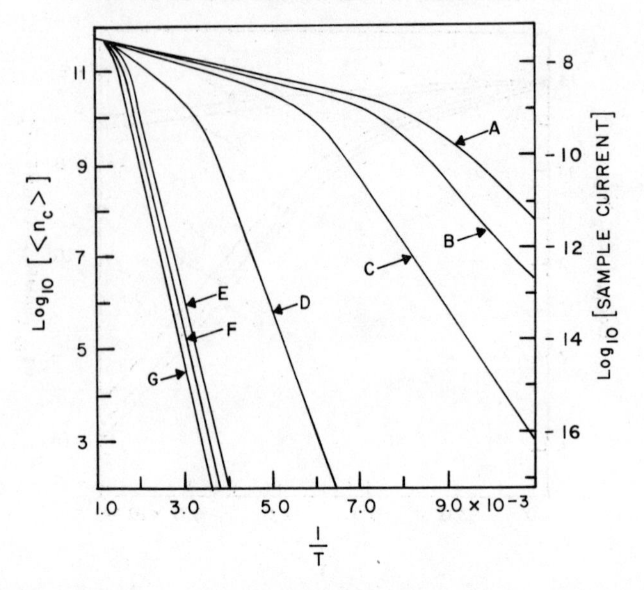


FIG. 10. Logarithm of conductivity vs. $1/T$ for a crystal with an exponential distribution of bulk states and a discrete acceptor-like surface state. [$W = 1.0 \times 10^{-3}$ cm, $E_{cs} = 0.9$ eV, $N_{d'} = 6.8 \times 10^{11}$ per cm$^3$, $N_{bo} = 1.0 \times 10^{12}$ per eV-cm$^3$, $T_c = 10{,}000$°K.]

0°K. Since the doping concentration $N_{d'}$ is very low, the slope of the surface-controlled regime is nearly 0.9 eV. We find that the characteristic behavior of a varying activation energy, as $N_s$ varies between zero and $WN_{d'}$, is even more exaggerated in this example than in the previous examples.

The earlier treatment may be modified very slightly to include a shallow donor level, in addition to the exponential levels. To accomplish this, one merely replaces $N_c$ in eqn. (25) with $N_{c'}$, and then proceeds to solve eqns. (25) and (26) as before. To illustrate the effect of adding the donor level, the example of Fig. 6 was modified by adding two different concentrations of donor states. The results are plotted in Figs. 11 and 12. Comparing Fig. 11 [$N_d = 10^{12}$ per cm$^3$] with Fig. 6 [$N_d = 0$], we find that adding a small density of donor states pulls curves "A" and "B" slightly

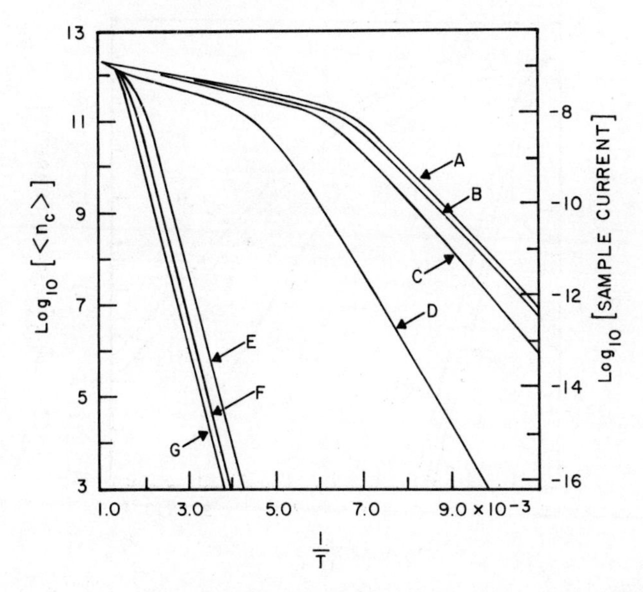


FIG. 11. Logarithm of conductivity vs. $1/T$ for a crystal with a shallow donor level, an exponential distribution of bulk traps, and a discrete, acceptor-like, surface state. [$W = 1.0 \times 10^{-3}$ cm, $E_{cs} = 0.9$ eV, $E_{cd} = 0.2$ eV, $N_{d'} = 1.93 \times 10^{12}$ per cm³, $N_d = 1.0 \times 10^{12}$ per cm³, $N_{bo} = 1.0 \times 10^{13}$, $T_c = 4000$°K.]

downward at the low temperature regions, while curves "C" through "G" remain unchanged. Referring now to Fig. 12 [$N_d = 3.16 \times 10^{13}$ per cm³], we find that curves "A" and "B" are lowered further. In addition, curve "C" is also "pulled down."

These results, and those of other similar calculations, indicate that, as one adds compensated donor states, the surface-controlled regime is unaffected, while the adsorbate free curve is "pulled down" at the low temperature portion of the graph. As the adsorbate free curve is lowered, the curves representing $N_s < WN_{d'}$ become lowered also. The extent of this effect increases as $N_d$ increases.

Whereas the discrete donor model provides curves with only two possible activation energies, the exponential model clearly demonstrates the possibility of observing an array of activation energies. One is limited in making such observations, however, to the region

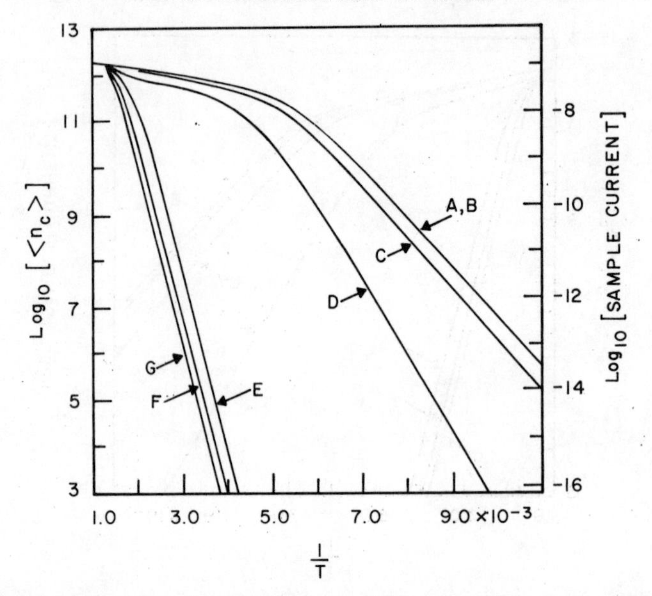


FIG. 12. Logarithm of conductivity vs. $1/T$ for a crystal with a shallow donor level, an exponential distribution of bulk traps, and a discrete acceptor-like surface state. [$W = 1.0 \times 10^{-3}$ cm, $E_{cs} = 0.9$ eV, $E_{cd} = 0.2$ eV, $N_{d'} = 1.93 \times 10^{12}$ per cm$^3$, $N_d = 3.16 \times 10^{13}$ per cm$^3$, $N_{bo} = 1.0 \times 10^{13}$ per eV-cm$^3$, $T_c = 4000$°K.]

where $N_s < WN_{d'}$. Also, the range of activation energies is decreased by increasing the doping concentration $N_{d'}$. One is also limited, in this model, by the rather large curvature of the curves for which $N_s < WN_{d'}$. Any attempt to vary the parameters of the crystal to reduce this curvature also reduces the amount by which the slopes vary. Thus, if one wishes to explain a family of conductivity curves, which are straight lines of varying slope (as $N_s$ varies), one must resort to a significantly different model.

## 4. Effect of a Temperature-dependent Acceptor-like Surface State

The above calculations are for a temperature-independent concentration of surface states. If this constraint is relaxed by letting

$N_s(T)$ increase as $T$ decreases, then the slopes of the surface-controlled curves will be increased. Equation (18) gives the average carrier concentration when a crystal with a shallow donor level is in the surface-controlled regime. This equation was derived with the assumption that $N_s \neq N_s(T)$, but it may be modified quite readily if $N_s = N_s(T)$. Although one may not find an adsorption isobar represented by eqn. (27) over a large temperature range, suppose, for instance, that in some small temperature region we may approximate the concentration of surface states as:

$$N_s(T) = N_0 \exp E_{Ns}/kT. \tag{27}$$

Substituting this into eqn. (18) we obtain:

$$n_c \simeq (N_c N_{d'} W/N_o) G(\Psi_s) \exp[-(\epsilon_{cs} - \Psi_s + \epsilon_{Ns})] \tag{28}$$

where $\epsilon_{Ns} \equiv E_{Ns}/kT$ and $G(\Psi_s)$ is a weakly varying function of temperature compared to the exponential terms. It is clear that the observed activation energy will be increased by the quantity $E_{Ns}$. An experimental test of this effect is discussed later in the paper.

## 5. Case of a Single Donor Level in the Bulk with an Acceptor-like Surface State Between the Donor Level and the Conduction Band

So far, we have considered only the case where the acceptor-like surface state lies deep in the forbidden gap. It is also instructive to determine what effects, if any, are produced by adding an acceptor-like surface state located near the bottom of the conduction band and above the bulk donor level. Since the surface state will not become as heavily populated as in the previous examples, we may, for a simple approximation, ignore the surface potential, since it will be small (the percentage occupation of the surface states must be less than the percentage occupation of the donor states). Assuming that the donor states are partially compensated by deep-lying bulk acceptor states, we may again assume that the Fermi energy is below the energy of the donor state and use Boltzmann statistics. The charge neutrality equation is:

$$n_c + n_d + n_s/W = N_d - N_a \equiv N_{d'}. \tag{29}$$

Since we are representing all of the above occupations by Boltzmann statistics, we may factor $n_c = XN_c$ out of eqn. (29) and obtain:

$$n_c = N_c N_{d'} / [N_c + N_d \exp \epsilon_{cd} + (N_s/W) \exp \epsilon_{cs}]. \tag{30}$$

It is apparent, from eqn. (30), that the surface state will have a negligible effect, unless the third term in the denominator is larger than the other two terms. This means that unless $N_s > WN_{d'}$, there will be no substantial difference in the sample conductivity when the acceptor-like adsorbate is added to the surface. Figure 13 shows a family of curves representing eqn. (30). The break-point temperature for the adsorbate-free case is given by $1/kT_b = \ln (N_c/N_d)/(E_c - E_d)$. If, at this temperature, we equate the second and third terms in the denominator of eqn. (30), we may determine the minimum value of $N_s$ that will have a noticeable effect upon the conductivity. It is

$$N_s(\text{min}) = WN_d[(N_c/N_d)^{(E_s - E_d)/(E_c - E_d)}]. \tag{31}$$

## 6. Case of a Single Donor Level in the Bulk with a Donor-like Surface State Between the Donor Level and the Conduction Band

So far, we have dealt exclusively with acceptor-like surface states. We now complete the picture by showing qualitatively the effect of

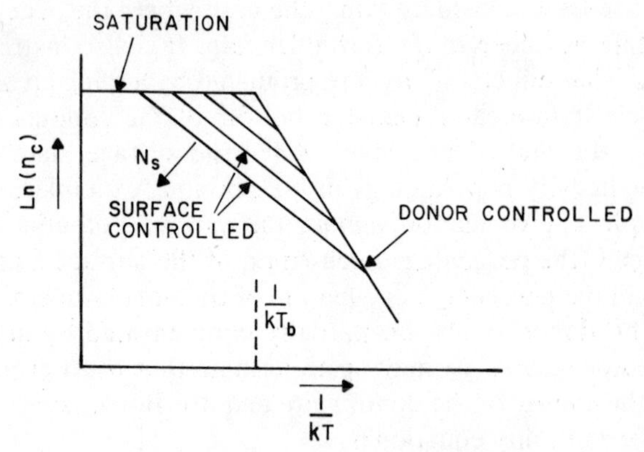


FIG. 13. Plot of $\ln (n_c)$ vs. $1/kT$ for an acceptor-like surface state, which is energetically above the donor level.

a donor-like surface state. If the energy of the surface state is below the bulk donor level, the surface state will remain populated, and there will be little exchange of charge with the bulk. Alternatively, if the surface state energy is between the bottom of the conduction band and the donor level (if more than one bulk level is present, the surface state would be above the lowest partially filled state), then the surface state will give some of its charge to the donor states and affect the conductivity of the crystal.

Unless the crystal is very thin, one cannot ignore the effect of the surface potential due to the transfer of electrons from the surface to the bulk. However, it is instructive to consider first the case where the surface potential can be neglected, and then consider the effects of a surface potential. Since the experimental results reported in this paper do not deal with donor-like adsorbates, this calculation will not be carried as far as that for the acceptor-like surface states, and only the qualitative effects of the surface potential will be considered.

The charge neutrality equation is

$$n_c + n_d + (n_s/W) = N_{d'} + (N_s/W). \tag{32}$$

Since the semiconductor is again partially compensated, the Boltzmann statistics may be applied to every term in eqn. (32). Hence,

$$n_c = N_c[N_{d'} + (N_s/W)]/[N_c + N_d \exp \epsilon_{cd} + (N_s/W) \exp \epsilon_{cs}]. \tag{33}$$

If we plot eqn. (33) as a family of curves, with $N_s$ as the parameter (see Fig. 14), we find that the first effect is merely to raise the saturation regime to $\ln (N_{d'} + N_s/W)$. In general, we will not observe the slope of the surface state energy, until the concentration of surface states is large enough, so that the third term in the denominator of eqn. (33) dominates the other two terms. We may solve for this concentration and obtain the result given by eqn. (31). This means that we may expect to observe an increase in conductivity, without observing the surface regime, when

$$0 < (N_s/W) < N_d[(N_c/N_d)^{(E_s - E_d)/(E_c - E_d)}]. \tag{34}$$

When $N_s$ is larger than this threshold, we may expect to observe regions of reduced activation energy corresponding to the energy of the surface state. We assume above that $(N_{d'} + N_s/W) < N_d$. If

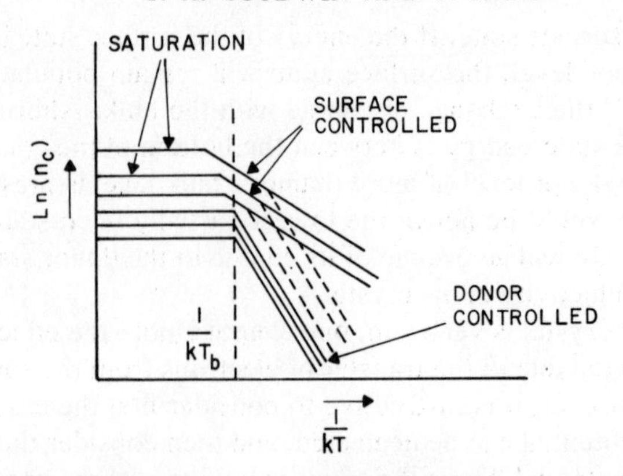


FIG. 14. Plot of ln $(n_c)$ vs. $1/kT$ for a bulk with a single partially filled donor level and a donor-like surface state.

eqn. (34) is violated, this inequality does not hold and we cannot represent the occupation of the donor level by Boltzmann statistics (if $N_s/W$ exceeds the upper limit of eqn. (34), it must exceed $N_d$). This means that eqn. (33) will not hold in the lower temperature regions, when $N_s/W > (N_d - N_{d'})$, and, therefore, the dotted curves of Fig. 14 are incorrect.

Using the correct statistics, and after some rearrangement, we may write eqn. (32) as

$$X^2[N_c + (N_s/W)\exp\epsilon_{cs}] + X\{N_d + [N_c + (N_s/W)\exp\epsilon_{cs}]$$
$$\exp(-\epsilon_{cd}) - [N_{d'} + (N_s/W)]\} = [N_{d'} + (N_s/W)]\exp(-\epsilon_{cd}). \quad (35)$$

Although eqn. (35) may be solved for $X$, the resulting expression will not provide us with the information we want, without plotting the equation point by point. Since we already know how $n_c$ behaves when the donor level is only partly occupied, we need only consider the case where $N_d$ is nearly filled. In this region, the charge neutrality equation may be written as

$$XN_c + (N_s/W)X\exp\epsilon_{cs} = N_{d'} + (N_s/W) - N_d \quad (36)$$

from which $X$ may be factored out to yield $n_c$:

$$n_c = N_c[N_{d'} + (N_s/W) - N_d]/[N_c + (N_s/W)\exp\epsilon_{cs}]. \quad (37)$$

Equation (37) indicates that the donor controlled regime will be absent, and the dotted curves of Fig. 14 must be corrected as shown by the solid curves. Of course, there will be a region where neither Boltzmann statistics nor total occupation will correctly describe $n_d/N_d$ but, since we are interested in qualitative results, it is not necessary to plot the solution of eqn. (35) with great rigor.

The surfact potential, which has so far been ignored or assumed very small, will bend the bands down as one approaches the surface. We must now use the following equation to describe the occupation of the surface states (where $V_s$ is the magnitude of the potential):

$$n_s = N_s X \exp{(E_{cs} + V_s)/kT}. \tag{38}$$

Following a procedure similar to that used in deriving eqn. (18), we find that the slope of the surface state controlled regime is $(E_c - E_s + V_s)$. In other words, the slope indicated by eqn. (37) must be increased by the magnitude of the surface potential.

# PART II EXPERIMENTS

## 7. Description of Samples and Experimental Apparatus

The CdS samples used in this investigation were thin crystal platelets, grown by condensation from the vapor phase. The platelets were approximately 20 microns thick, 2 mm long, and 1 mm wide. Ohmic contacts were prepared by evaporating indium on the ends of the sample.[26] A voltage was applied to the contacts, causing a current to flow parallel to the large surface area [the $(11\bar{2}0)$ surface]. The room temperature dark resistivity of the crystals varied from $10^4$ to $10^5$ ohm-cm, for an ambient pressure of $10^{-8}$ torr. In one atmosphere of air, the crystals had a dark resistivity exceeding $10^{11}$ ohm-cm at room temperature.

The experiments were conducted in a vacuum system equipped with gas-handling apparatus, pressure-measuring devices, a shuttered window, and both electrical and thermal paths to the exterior. The use of a variable leak valve (Varian #951-5100), equipped with a metal-to-sapphire seal, allowed accurate control of gas flow to rates as small as $10^{-10}$ torr-liters/sec. Thus, by leaking and pumping simultaneously, it was possible to obtain pressures from $10^{-8}$ torr to $10^{-4}$ torr. Above $10^{-4}$ torr, the pressure was maintained by closing the pump valve and leaking gas until the required pressure was obtained. The outgassing of the system was not sufficient to perturb pressures above $10^{-4}$ torr, so that continuous pumping was not necessary in that range.

Several pressure gauges were necessary in order to measure pressures from $10^{-8}$ torr to one atmosphere. An ultra-high-vacuum ionization gauge (Varian #971-5005) was used to detect pressures below $10^{-4}$ torr. A millitorr ionization gauge (Varian #971-5009) was employed in the $10^{-4}$ torr to $10^{-2}$ torr range, while a thermocouple gauge (Hastings Model VT-6) was used between $10^{-3}$ torr and 1 torr. A mechanical gauge (Wallace and Tiernan #FA 160) was used to measure pressure from 1 to 800 torr.

In order to provide both thermal contact and electrical insulation, each sample was attached to a sapphire disk (the samples were soldered to the sapphire disks with indium-coated wires, which extended beyond the evaporated contacts, thus sticking to the sapphire), which was then secured to a copper block which had a thermocouple imbedded in the center. The copper block was thermally connected (by a copper rod) to a dewar located at the top of the apparatus. The sample could be cooled by putting refrigerants in the dewar and heated by placing a heater in the dewar. The maximum temperature used in the experiments was 144°C, since the indium contacts melt at 156°C. The electrical leads were brought to the exterior though a metal-to-ceramic seal.

Voltage was supplied by a Hewlett-Packard 6106A d.c. power supply, while sample current was measured with a Keithley 610B electrometer.

In these experiments the surfaces of the crystals were not chemically treated, since it has been observed that chemical etching does not have a consistent effect on the results.[27] Care was taken to use a nitrogen atmosphere, when melting the indium contacts to the sample, for it was observed that heating in air irreversibly alters the characteristics of the sample. It is believed that this is caused by

CO, which reacts with the surface, since some of the resultant characteristics are similar to those produced by heating CdS in CO.[27] A similar effect is observed if the sample is heated before the vacuum chamber has reached a pressure of $10^{-7}$ torr or lower. A mass spectroscopic analysis of the gas in the vacuum chamber shows that a large fraction of the outgassing in the system is composed of CO.

To obtain an adsorbate free surface, it is necessary to remove the adsorbed oxygen and $H_2O$. This is accomplished by placing the sample in a vacuum of $10^{-8}$ torr, and heating the sample to 144°C. To enhance the desorption of oxygen, the sample was illuminated with light passed through a Corning 5-58 filter (blue light). To observe the photodesorption of the oxygen, the process was monitored by measuring the steady state photocurrent[8]. In general, the maximum photocurrent was obtained after about 3 days of desorption.

1715470

## 8. Comments on the Scope of the Data

To compare the experimental data with the calculations developed earlier, the data must be presented in a form which is either equivalent to or analogous to the form in which the calculations were presented. Accordingly, the ideal experimental procedure would be: (1) Place the sample in the best obtainable vacuum ($10^{-8}$ torr). (2) Photodesorb the surfaces of the sample by illumination with bandgap radiation. (3) Heat the sample to the highest temperature used in the experiments (144°C). (4) When the sample is desorbed, remove all light sources and allow the photocurrent to decay to the dark current level. (5) Gradually reduce the temperature to the lowest value of the experiment (110°K) and continuously measure the sample current. The rate of temperature change must be slow enough to insure equilibrium for all measurements. (6) Return temperature to its maximum and repeat the measurements with a higher ambient oxygen pressure.

In this manner one would, in principle, obtain a family of curves which show the sample conductivity vs. temperature with the pressure of ambient oxygen as the parameter. Except for the possibility that the concentration of surface states is temperature dependent, the resultant data should be directly comparable with the computed curves.

One of the most troublesome aspects of performing electrical measurements on insulating materials is to obtain reproducible results.[28,29] Dark conductivity measurements on Zn-doped polycrystalline $SnO_2$ indicate a very strong history dependence and the results are rather difficult to interpret quantitatively.[13] Similar problems were encountered here. When working with low oxygen pressures, the concept of equilibrium is ill-defined. It takes 3 or 4 days for the photocurrent to decay to the dark level. When a change in either temperature or pressure is made one must wait at least 1 day to reach steady state. Even then it is not certain that "equilibrium" is obtained, for the characteristics of the sample vary slowly over a period of weeks. This may be due to diffusion and evaporation processes. To take a family of curves, as described in the preceding section, would be impractical if recording each data point were to take one day, and the results may be inconsistent due to the long-range variation in sample characteristics. In addition to these problems, the measured curves are not very smooth, indicating that there are a large number of interacting bulk and surface states. This makes the data very difficult to analyze.

It is clear that we need a procedure that produces repeatable results and that can also be related to our calculations. If there is an electronic state which interacts very slowly with the conduction band, we can eliminate that state from the experiment by allowing it to come to equilibrium at a fixed starting condition and then taking the data so quickly that the occupation of the state does not change. In addition to allowing one to take the data faster, this procedure simplifies the results by eliminating one state from the model needed to explain the results.

To illustrate this point, refer to Fig. 15, which plots log (sample current, with one volt applied) vs. reciprocal temperature for a desorbed sample of CdS in a vacuum of $10^{-8}$ torr (the left-hand coordinate gives carrier density assuming a constant mobility of 200 $cm^2$/volt-sec). The sample was allowed to come to equilibrium

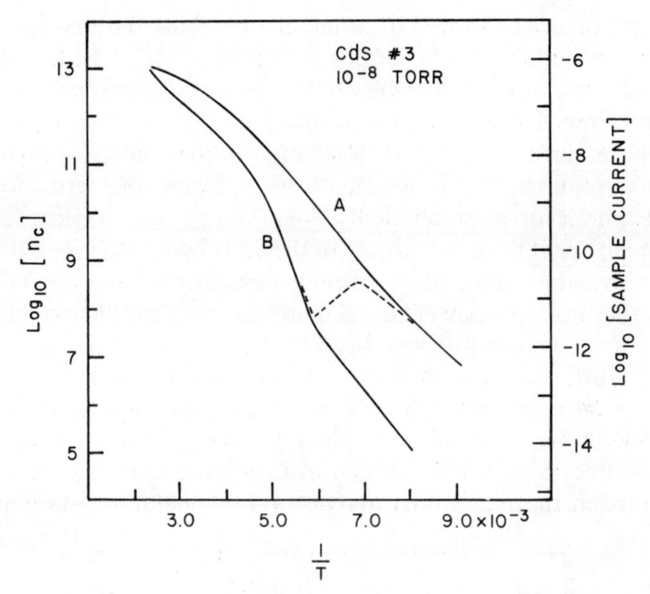


FIG. 15. Sample current vs. reciprocal temperature in $< 10^{-8}$ torr oxygen for two different types of data taking: (A) fast cooling; (B) slow heating.

at 144°C for 18 hours, and was then cooled rapidly. The time for cooling was approximately 1 hour. Curve "A" (fast cooling) shows the resultant sample current. If one waits for several hours, with the sample at the lower temperature, the sample current remains constant (indicating steady state), but if one heats the sample slowly (dotted curve) there is a reversal in the direction of current change at about 150°K to 160°K, which clearly shows that the sample was not in equilibrium. Such a current reversal could be caused by the emptying of a hole trap, which reduces the majority carrier concentration by recombination,[30] or by the filling of an electron trap which, due to a small capture cross-section, did not fill during fast cooling.

If the former is true, one would expect the current to slowly rise, if one holds the temperature at 160°K, because the holes would gradually become depleted. This is not the case, however, for, if one holds the temperature at 160°K, the current remains constant or drops slowly. Further, if one cools the crystal again, the current follows the lower portion of curve "B", which is almost parallel to the fast cooling curve. This is very suggestive of a deep electron trap. It should be noted that, although the trap in question may be a surface trap, it is definitely not associated with oxygen chemisorption in the sense that we are using the term, for the effect is present in an ambient which has far less oxygen than is needed to affect the conductivity of the adsorbent CdS (see below).

If, after the temperature cycling described above, one slowly heats the sample (taking about 5 hours to complete the experiment) the sample current follows curve "B" (slow heating). The higher temperature portion of the curve is devoid of structure.

If one adds an ambient of $3 \times 10^{-6}$ torr and repeats the above experiment, one obtains the results presented in Fig. 16. In addition to the effect just described, one also observes the appearance of structure in the region just above room temperature. As opposed

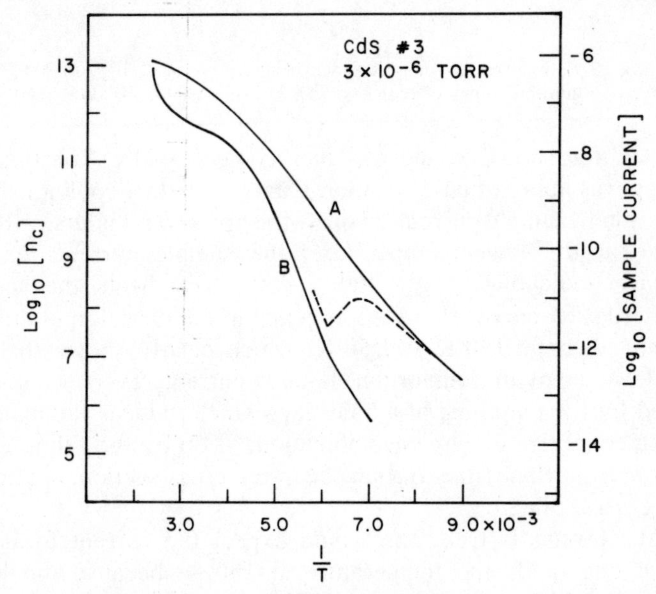


FIG. 16. Sample current vs. reciprocal temperature in $3 \times 10^{-6}$ torr oxygen for two different types of data taking: (A) fast cooling; (B) slow heating.

to the lower temperature effect, this current reversal is due to the presence of oxygen, since the effect was not present in a vacuum of $10^{-8}$ torr. The explanation of these effects (which were present in all platelets studied) is beyond the scope of this paper.

It is quite clear that the data obtained by slow temperature cycling will be very difficult to interpret with the models of Part I. The curves obtained by coming to equilibrium at 144°C and then cooling rapidly, however, have a minimum of structure and can be obtained reproducibly. For this reason, the fast cooling (non-equilibrium) method described above was used in these experiments.

## 9. The Effect of Ambient Oxygen Pressure on the Conductivity of CdS

Figures 17 and 18 show the logarithm of sample current vs. reciprocal temperature with ambient oxygen pressure as a para-

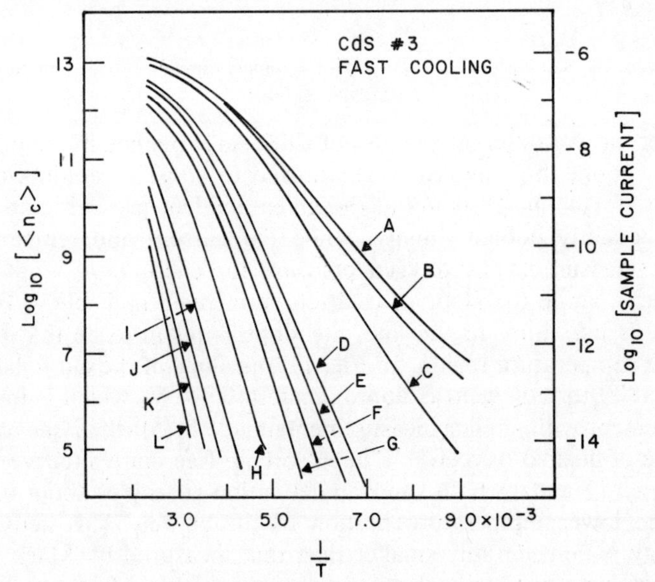


FIG. 17. Sample current vs. reciprocal temperature for various ambient oxygen pressures.

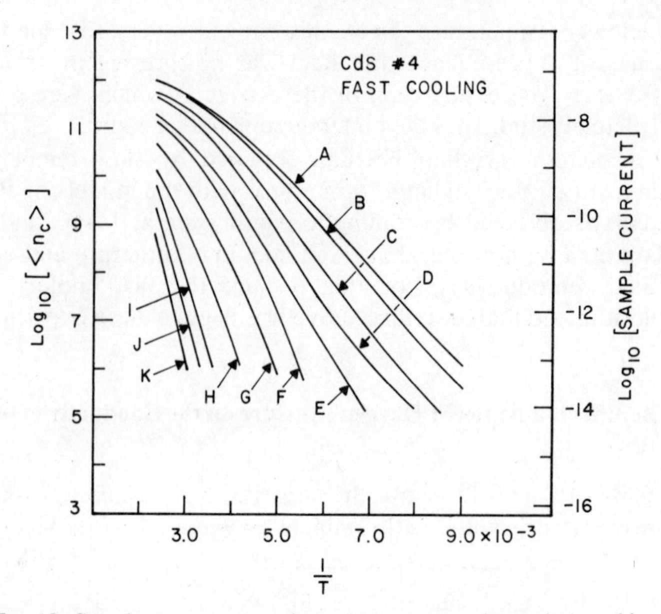


FIG. 18. Sample current vs. reciprocal temperature for various ambient oxygen pressures.

meter for two different samples of CdS. As explained in the preceding section, the samples were allowed to come to equilibrium at 144°C for about 18 hours at the given ambient pressure and were subsequently cooled rapidly, while the current and temperature were measured. The oxygen pressure for each curve is given in Table 3, while the slope of each curve is given in Table 4. Where there is curvature to the line, the slope is taken to be that of the lower temperature region. In Fig. 17, the slope of the curve labeled "L" (800 torr oxygen) is approximately 0.94 eV, which is in good agreement with similar measurements made by Mark in one atmosphere of flushed oxygen.[31] The adsorbate-free curves (curve "A") of Figs. 17 and 18 both show an activation energy of about 0.2 eV in the lower temperature portion of the curves. This activation energy is considerably smaller than that measured by Mark when flushing with one atmosphere of nitrogen, which produced an activation energy of 0.63 eV. At this point, it is helpful to examine the

TABLE 3. KEY TO FIGS. 17, 18, AND 19

| Curve label | Ambient oxygen pressure (torr) |
|---|---|
| A | $10^{-8}$ to $10^{-6}$ |
| B | $3\times10^{-6}$ |
| C | $10^{-5}$ |
| D | $3\times10^{-5}$ |
| E | $10^{-4}$ |
| F | $10^{-3}$ |
| G | $10^{-2}$ |
| H | $10^{-1}$ |
| I | 1 |
| J | 10 |
| K | $10^{2}$ |
| L | $10^{3}$ |

TABLE 4. COMPARISON OF ACTIVATION ENERGIES OBTAINED BY PLOTTING CONDUCTIVITY VS. $1/kT$ UNDER BOTH EQUILIBRIUM AND NON-EQUILIBRIUM CONDITIONS

| | Sample # 3 | Sample # 4 | |
|---|---|---|---|
| Pressure of ambient oxygen (torr) | Slopes of lower temp. portion of fast cooling curves (eV)[Fig. 17] | Slopes of lower temp. portion of fast cooling curves (eV)[Fig. 18] | Slopes between equilibrium points at 144°C and 24°C(eV)[Fig. 19] |
| $< 10^{-6}$ | 0.22 − 0.24 | 0.21 | 0.16 |
| $10^{-6}$ | 0.22 − 0.24 | 0.21 | 0.16 |
| $3\times10^{-6}$ | 0.24 − 0.27 | 0.23 | 0.20 |
| $10^{-5}$ | 0.29 − 0.33 | 0.24 | 0.31 |
| $3\times10^{-5}$ | 0.45 | 0.27 | 0.48 |
| $10^{-4}$ | 0.51 | 0.31 | 0.54 |
| $10^{-3}$ | 0.55 | 0.40 | — |
| $10^{-2}$ | 0.68 | 0.44 | 0.52 |
| $10^{-1}$ | 0.71 | 0.50 | 0.63 |
| 1 | 0.74 | 0.61 | 0.75 |
| 10 | 0.74 | 0.66 | 0.79 |
| $10^{2}$ | 0.82 | 0.78 | 0.78 |
| $10^{3}$ | 0.94[a] | — | 0.96 |
| $2\times10^{3}$ | — | — | 1.04 |

[a]800 torr

assumption, in both the present measurements and in Mark's measurements, that the sample is free of the adsorbate oxygen. In the present case, the sample was photodesorbed in $10^{-8}$ torr.

After the sample was allowed to reach equilibrium in the dark, no measurable change in sample current occurred, until the ambient oxygen pressure was raised to $10^{-6}$ torr. This indicates that the above assumption is correct and that the 0.2 eV activation energy corresponds to a bulk donor level. On the other hand, the purity of the nitrogen in Mark's measurements was about one part per million and may have had from $10^{-4}$ to $10^{-3}$ torr of oxygen present. This pressure range corresponds to curves "E" and "F" of Fig. 17, where the slopes are 0.51 eV and 0.55 eV, respectively. It appears, then, that the slope obtained from Mark's measurement in nitrogen was caused by a small partial pressure of oxygen in the nitrogen gas.

On comparing the data in Fig. 18 with the models of Part I, one finds that the highest partially filled bulk level for the adsorbate-free surface at low temperatures is about 0.2 eV below the bottom of the conduction band. The effective donor concentration $N_{d'}$ must be at least $2 \times 10^{12}$ per $cm^3$. By focusing a microscope on the bottom and top of the sample the thickness was measured to be 20 microns, which gives a half-thickness $W = 10^{-3}$ cm. Substituting these into eqn. (19a), one obtains the height of the Schottky barrier for total depletion of the sample: $V_s = eN_{d'}W^2/2\epsilon \simeq 0.185$ eV.

Assuming that the break temperature for the adsorbate-free curve is above room temperature (we are now assuming that the bulk is described by a single donor level), we can determine that the donor concentration $N_d$ must be at least $10^{16}$ per $cm^3$. All of the above numbers are quite reasonable.

An accurate estimate of the surface potential is not possible for sample #4, because the sample was started with thin ridges running parallel to the direction of current flow. One can only estimate an "effective" sample thickness of $\sim 3 \times 10^{-4}$ cm.

If one compares Fig. 17 with Fig. 18 one observes that, although the families of curves are similar, Fig. 17 displays a definite gap as one goes from $3 \times 10^{-6}$ torr to $10^{-5}$ torr, whereas Fig. 18 does not show this effect (there is only a slight hint of a jump in the same pressure range). This suggests that sample #3 is better described by a single bulk state than is sample #4. One characteristic common to both samples (and to all samples observed) is that there is a continual increase in slope with increasing oxygen pressure. As discussed earlier, there is a limit to the amount of slope increase that can be explained by multiple bulk levels. Note that in Fig. 10 all of the change in activation energies takes place over a two decade change in $N_s$. There is no indication that changes in activation energy can take place at much higher values of $N_s$. Matthews suggests that the increase in activation energies he observed "indicates that it is possible for surface compensation [by an acceptor-like oxygen state] to increase to such a degree that an active donor level can be depleted and a lower-lying one brought into play".[13] This explanation cannot apply to the present experiment. Referring to Fig. 17, as one goes from curve "A" to curve "D", the sample becomes depleted of majority carriers leaving a positive space charge of $(e N_{d'})$, with $N_{d'} \simeq 2 \times 10^{13}$ per $cm^3$. This produces a surface potential of about 0.2 eV. The surface state concentration at this jump is (assuming $W \simeq 3 \times 10^{-4}$ cm): $N_{sl} = W N_{d'} \simeq (3 \times 10^{-4}\ \text{cm}) \times (2 \times 10^{13}\ \text{per cm}^3) \simeq 6 \times 10^9$ per $cm^2$. If we now postulate that a shift from one bulk level to another occurs at a much higher pressure (and therefore a much higher $N_s$), the number of carriers transferred to the surface will be quite large. Suppose, for instance, that the shift in question takes place at a surface-state concentration $N_{s2} = 100 \times N_{s1}$. Then the concentration of bulk carriers $N_2$ transferred to the surface is: $N_2 = N_{s2}/W = 100 \times N_{d'} = 2 \times 10^{15}$ per $cm^3$. This would generate a surface potential of the order of 20 eV, which is impossible. We therefore conclude that the continued increase in slope observed in these experiments cannot be explained solely by the models discussed in Part I.

There are two basic areas one might explore to explain this effect. First, the calculations are for equilibrium conditions, while the experimental curves are not. Second, the calculations assume that the concentration of surface states $N_s$ is temperature independent, while in fact there is no assurance that this is so. We know by the data presented thus far that $N_s$ is a function of oxygen pressure. It is plausible to assume that $N_s$ is related to the total oxygen coverage of the surface of the crystal, although this relation may not be a simple one. As the temperature changes, the concentration of physically adsorbed species changes[32] and, therefore, the total coverage should change.

Before considering this further, we examine the question of equilibrium. If we were to take equilibrium measurements at only two temperatures, it would be possible to obtain some information about the activation energy as a function of pressure, but we would have no information about the detailed structure of the current vs. temperature curve. This experiment is possible, since only two data points are taken, and their equilibrium times are much shorter than for lower temperature points. Figure 19 shows equilibrium

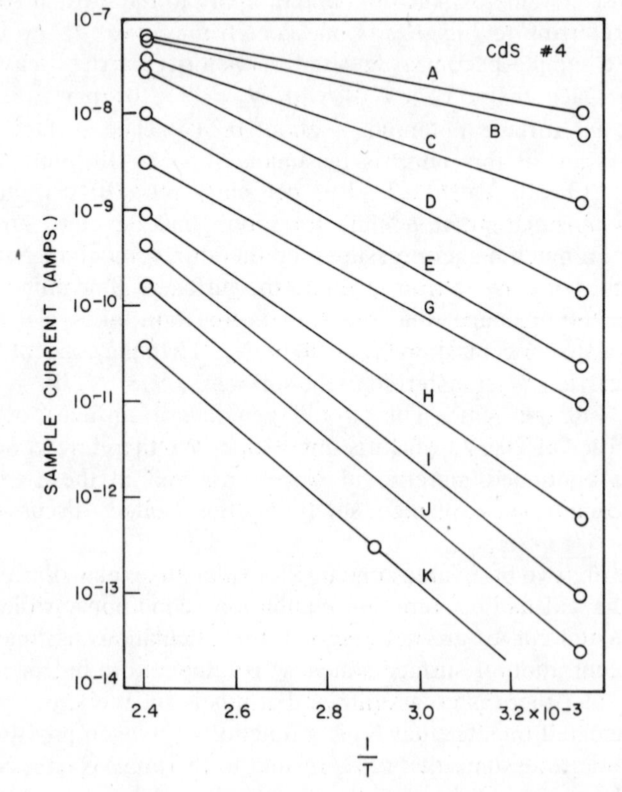


FIG. 19. Equilibrium sample current at 144°C and 24°C for various ambient oxygen pressures.

measurements (equilibration time was over 16 hours) at 144°C and 24°C for various partial pressures of oxygen. For purposes of illustration, a line is drawn between the two equilibrium measurements at each pressure. The slopes of these lines are listed in Table 4. Although an initial jump takes place in the $10^{-5}$ torr range, there is a gradual increase in slope as the ambient pressure is increased. We have exactly the same effect with the equilibrium data as we had with the fast cooling curves. This means that we cannot account for this effect by the method of data taking, and we will have to examine the question of the variation of the concentration of chemisorption states with temperature.

## 10. Concentration of Chemisorption Surface States as a Function of Temperature

Since a temperature-independent concentration of surface states cannot explain the gradual increase in activation energy observed as the oxygen pressure is increased, we examine the effect of a variation of $N_s$ with temperature and look for experimental evidence to confirm or deny this variation. As shown in eqn. (28) an Arrhenius dependence of the surface state concentration $N_s(T)$ produces an increase in the slope of the surface-controlled regime. The observed activation energy is increased by $E_{Ns}$.

If $E_{Ns}(P)$ is an increasing function of pressure, then the data of Figs. 17, 18 and 19 are consistent with the models. The continual increase in observed activation energy would arise from the continual increase in $E_{Ns}$. It is not actually necessary for $E_{Ns}$ to be an increasing function of pressure, in order to explain a continual increase in activation energy. Since the measurements are made rapidly, the concentration of surface states $N_s(T,P)$ may not follow changes in temperature for the lower pressure measurements, while, at higher pressures, $N_s(T,P)$ may come to equilibrium faster and contribute more to the observed activation energy. This is a definite possibility, since we have observed that a change in pressure does not produce immediate changes in sample conductivity when the pressure is below $10^{-4}$ torr, but that immediate effects are observed at pressures above $10^{-3}$ torr. For example, when increasing from $10^{-6}$ to $10^{-5}$ torr, several hours must elapse before the conductivity reaches steady state, while a change from 10 torr to 100 torr produces an immediate reduction in sample conductivity.

Making use of this knowledge, we devise an experiment to test whether or not $N_s(T,P)$ depends on temperature. We allow the sample to come to equilibrium at 144°C and cool quickly to obtain a curve characteristic of $N_s$ (144°C, $P$) [assuming that $N_s$ does not change during the fast cooling]. Then we allow the same sample to come to equilibrium at 24°C, heat quickly to 144°C, and then cool rapidly to obtain a curve characteristic of $N_s$ (24°C, $P$). If $N_s(T,P)$ is temperature dependent, we should observe two distinct curves; specifically, if $N_s(T,P)$ is a decreasing function of temperature, the second curve should lie below the first curve.

When this experiment was conducted at $10^{-8}$ torr, there was no difference between the two curves, which indicates no difference

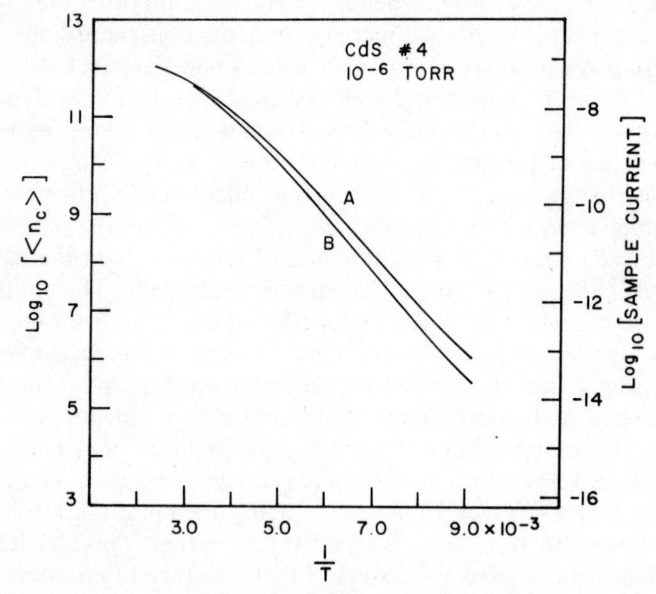


FIG. 20. Fast cooling data for two different starting conditions; (A) after reaching equilibrium at 144°C; (B) after reaching equilibrium at 24°C.

in the concentration of surface states between 24°C and 144°C for the high vacuum or adsorbate-free case. This is expected, since we assume that $N_s$ is zero for the vacuum. Figures 20 through 22 demonstrate the results of this experiment for various oxygen pressures from $10^{-6}$ torr to $10^{-3}$ torr. Clearly, these experiments indicate a definite dependence of the oxygen chemisorption surface state concentration on temperature, and the dependence is in the direction expected from physical reasoning. The effect increases with increasing oxygen pressure until it reaches a maximum at $3 \times 10^{-5}$ torr. An increase in magnitude of this effect indicates that the activation energy associated with the variation of $N_s$ with temperature $E_{Ns}$ is an increasing function of pressure. At $3 \times 10^{-5}$ torr the curve taken after reaching equilibrium at 24°C lies below the fast cooling curve taken after equilibrium at 144°C and $10^{-3}$ torr. This means that the increase in $N_s$ due to lowering the tem-

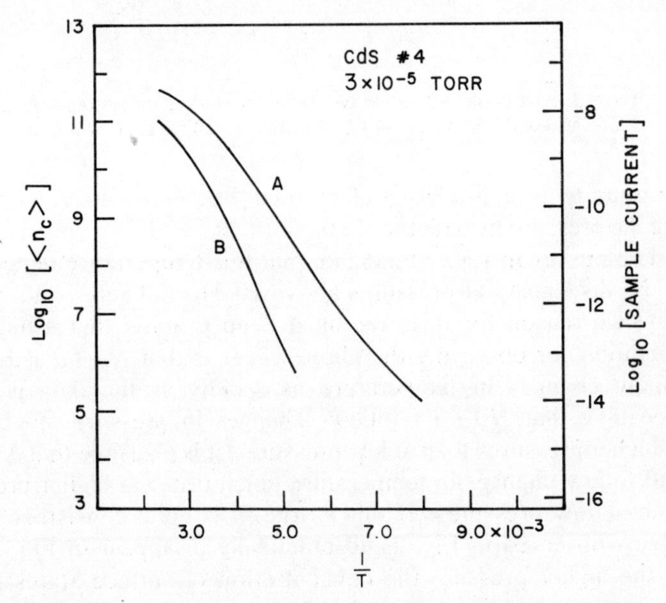


FIG. 21. Fast cooling data for two different starting conditions: (A) after reaching equilibrium at 144°C; (B) after reaching equilibrium at 24°C.

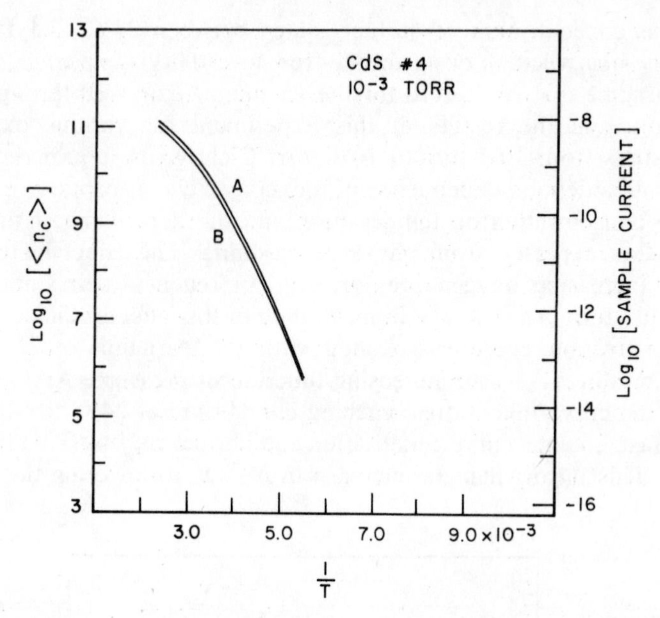


FIG. 22. Fast cooling data for two different starting conditions: (A) after reaching equilibrium at 144°C; (B) after reaching equilibrium at 24°C.

perature to 24°C is a larger effect than the increase in $N_s$ due to raising the pressure by a factor of 30.

The curves in Fig. 20 indicate that this temperature dependence of $N_s$ disappears at pressures above $10^{-3}$ torr. There is no obvious physical reason for this. We must keep in mind that a necessary condition for observing the above effect is that $N_s(T,P)$ does not follow changes in temperature as rapidly as the data is taken. Recalling that $N_s(T,P)$ follows changes in pressure much faster at higher pressures than at low pressures, it is plausible that $N_s(T,P)$ will follow changes in temperature much faster at higher pressures than at low pressures. If this is true, it is quite consistent for the effects observed in Fig. 21 to practically disappear in Fig. 22, for at the higher pressure the concentration of surface states adjusts during the brief heating period.

On comparing the curves in Figs. 18 and 19 for a pressure of

$3 \times 10^{-5}$ torr, we find that the appropriate value of $E_{Ns}$ for this pressure is about 0.2 eV. In other words, the equilibrium curve has a measured activation energy which is 0.2 eV greater than the activation energy of the fast-cooling curve.

## 11. Determination of the Depth of the Chemisorption Surface State $E_{cs}$

In our computational model, the activation energy jumps from $(E_c - E_d)$ to $(E_{cs} - V_s)$, as $N_s$ passes through the value $WN_{d'}$. Figure 23 is a plot of the computed activation energy vs. surface state concentration, for two values of the depth of the surface state $E_{cs}$, as calculated from the exponential bulk state distribution model. The transition from the donor or bulk-controlled regime to the surface-controlled regime is not as abrupt as in the discrete bulk donor model. If we were to convert this plot into one which shows the measured activation energy vs. pressure (where $N_s$ is a

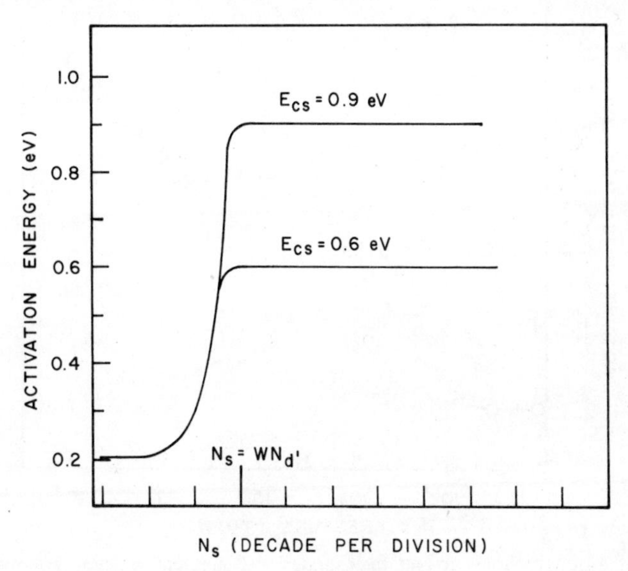


FIG. 23. Calculated activation energies vs. surface state concentration.

function of both temperature and pressure as demonstrated in the previous section), one would expect the same sort of transition of the activation energy to appear, but it would be followed by a gradual rise in the measured activation energy.

By comparing Fig. 23 with a plot of the experimentally determined activation energy for sample #3 (Fig. 24), it appears that there are two possible regions where this transition occurs. Although there is definitely some hint of a leveling off at $10^{-3}$ torr (giving an activation energy of 0.55 eV), it appears more likely that the transition takes place at $10^{-2}$ torr (giving an activation energy of nearly 0.7 eV). The activation energy obtained by this method gives the value for $(E_{cs} - V_s)$, and if $V_s$ were known we would obtain a value of $E_{cs}$. Unfortunately, due to the ridges or striations mentioned earlier, we cannot obtain an accurate value for $V_s$. A reasonable estimate for $V_s$ (0.2 eV ± 0.1 eV) places $E_{cs}$ between 0.8 eV and 1.0 eV. Similar considerations for sample #4 gives a value in the vicinity of 0.65 eV.

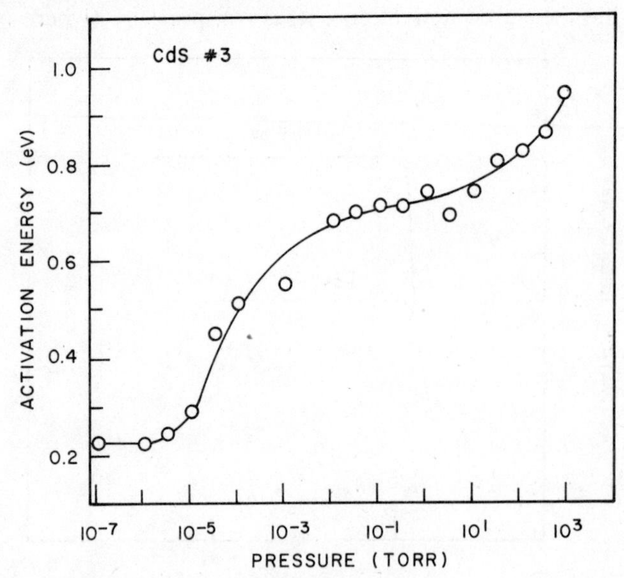


FIG. 24. Measured activation energies vs. ambient oxygen pressure. [Slopes of lower temperature region of fast cooling curves.]

## 12. Comments on the Significance of the Concentration of Surface States

A physically adsorbed electrophilic particle becomes a chemisorbed particle by accepting an electron from the bulk of the adsorbent lattice.[24] We shall represent the concentration of physically adsorbed and chemisorbed particles by $N_X$ and $N_{X^-}$, respectively. The occupation density $n_s$ of the chemisorption surface states is equal to the concentration of chemisorbed particles $N_{X^-}$. Following this reasoning, the concentration of unoccupied surface states should be equal to the concentration of physically adsorbed or uncharged particles $N_X$. The total concentration $N_s$ of surface states can be written as the sum of the occupied plus the unoccupied states:

$$N_s(T,P) = N_X + N_{X^-}.$$

So far, we have not specified the exact nature of the oxygen chemisorption state. The literature does not give a clear argument as to whether the chemisorbed particle is $O_2^-$ or $O^-$. Since the first drafting of this paper, Schubert has published results which indicate that monatomic oxygen can be thermally desorbed from the CdS surface.[33] We show in this section that the experimental evidence suggests that the occupied chemisorption state is $O^-$ and that the surface state concentration is: $N_s(T,P) = N_O + N_{O^-}$.

In the surface-controlled regime the surface state is well above the Fermi energy and will, therefore, be only partially occupied. In this case, $N_s(T,P) \simeq N_X$, and we may be able to determine the nature of $X$ by relating the variation of $N_s(T,P)$ with oxygen pressure to the kinetics of the adsorption process. The concentration of the physically adsorbed particles will adjust (at a given temperature and pressure), so that the rate of adsorption will equal the rate of desorption. If the rate-limiting process in desorption is the release of molecular oxygen from the surface, one would expect the concentration of physically adsorbed oxygen to be approximately proportional to the ambient oxygen pressure. However, if the rate-limiting process is the conversion of atomic oxygen to molecular oxygen [$O + O \rightarrow O_2$], one would expect the rate of desorption to be proportional to the square of the concentration of atomic oxygen or $N_O^2$. This would mean that the concentration of physically adsorbed oxygen would be approximately proportional to the square root of the pressure. Referring to Fig. 17, the concentration of surface states is sublinear with pressure and is almost proportional to the square root of the pressure in the higher-pressure regions (a decade change in pressure produces about one-half decade change in sample current). This suggests that the physically adsorbed species is atomic oxygen.

It was stated earlier that, although the concentration of chemisorption states is a function of both temperature and pressure, $N_s(T, P)$, it does not adjust rapidly to changes in temperature and pressure when the ambient oxygen pressure is below $10^{-4}$ torr. If the physically and chemically adsorbed particles were $O_2$ and $O_2^-$, respectively, then there is no obvious mechanism to account for this slow adjustment, because of the flux of both electrons from the bulk to the surface and $O_2$ particles from the gas to the surface is sufficient to establish rapid equilibrium. But if the conversion from molecular oxygen to atomic oxygen is involved, the process may be slow enough to inhibit the establishment of equilibrium. This does not explain, however, why the establishment of equilibrium occurs rapidly at higher pressures.

A photodesorption experiment was conducted in an effort to settle the question of the nature of the chemisorbed species. If a crystal whose surface is covered with chemisorbed oxygen is placed in a vacuum, the physically adsorbed species are free to leave the surface at their thermal desorption rate. As explained above, a desorption rate proportional to the square of the concentration suggests the process $[O+O \rightarrow O_2]$, while a desorption rate linear with the concentration suggests a monomolecular process.

If a CdS crystal is placed in an oxygen ambient, while under illumination with bandgap light, the photocurrent is determined by a combination of bulk and surface recombination centers.[34] By established theory, the steady state conduction band photoelectron concentration $n$ is determined by:[25]

$$G = n[N_b v \sigma_b + N_s v \sigma_s] \tag{39}$$

where $N_b$, $\sigma_b$, $N_s$, and $\sigma_s$ are the concentration of recombination centers and the electron capture cross-sections of the bulk and surface recombination centers, respectively (normalized to consistent units). Also, $G$ is the carrier generation rate and $\nu$ is the thermal velocity of electrons. If oxygen is removed, lowering $N_s$, $n$, and thus the photocurrent, will rise.

The photocurrent vs. time during desorption at $< 10^{-7}$ torr was obtained as follows. Prior to the start of the experiment, the sample was exposed to 10 torr of oxygen, while under illumination. Next, the illumination was interrupted, while the sample chamber was pumped to below $10^{-7}$ torr. As soon as the illumination was resumed, the current $I(t)$ was recorded vs. time. The photocurrent alone does not give us the information we desire; we need the concentration of surface recombination states as a function of time $N_s(t)$, which may be calculated using eqn. (39), provided that $\partial N_s(t)/\partial t \ll G$. As $t \to \infty$ we have $N_s(\infty) = 0$, so that:

$$G = n(\infty)[N_b \nu \sigma_b]. \tag{40}$$

Inserting eqn. (40) into eqn. (39) and rearranging, we obtain

$$n(\infty)/n(t) = 1 + [N_s(t)\sigma_s/N_b\sigma_b] = I(\infty)/I(t). \tag{41}$$

This may be rearranged to give the function form of $N_s(t)$:

$$N_s(t)\sigma_s/N_b\sigma_s = [I(\infty)/I(t)] - 1. \tag{42}$$

The right-hand side of eqn. (42) was obtained from a measurement of $I(t)$. If $\sigma_s$, $N_b$, $\sigma_b$, and the physical geometry of the problem are constant in time, this function is proportional to $N_s(t)$. Table 5 gives this function and its derivative for various times. For simplicity, the function is labeled $N_s(t)$ in the table, although the constant $\sigma_s/N_b\sigma_b$ has not been included. Looking at the ratio $N_s(t)/[\partial N_s(t)/\partial t]$ we see a steady rise in value as time increases. This indicates that $\partial N_s(t)/\partial t$ is super-linear with $N_s(t)$. However, by looking at $N_s^2(t)/[\partial N_s(t)/\partial t]$ we find that the rate is not proportional to the square of $N_s(t)$ either, because the ratio increases until about 40 hours and then decreases. This indicates a more complicated process than those being considered, and the experiment is inconclusive.

TABLE 5. ANALYSIS OF DESORPTION EXPERIMENT

| Time (hr) | $N_s^a$ | $N_s^2$ | $\partial N_s/\partial t$ | $\dfrac{N_s}{\partial N_s/\partial t}$ | $\dfrac{N_s^2}{\partial N_s/\partial t}$ |
|---|---|---|---|---|---|
| 0.5 | 1.785 | 3.19 | 0.21 | 8.5 | 15.2 |
| 1.0 | 1.70 | 2.89 | 0.155 | 11.0 | 18.7 |
| 2.0 | 1.56 | 2.43 | 0.10 | 15.6 | 24.3 |
| 4.0 | 1.43 | 2.19 | 0.05 | 28.6 | 43.8 |
| 5.0 | 1.39 | 1.93 | 0.04 | 34.8 | 48.25 |
| 10.0 | 1.21 | 1.465 | 0.029 | 41.7 | 50.5 |
| 20.0 | 0.97 | 0.94 | 0.019 | 51.0 | 49.5 |
| 40.0 | 0.71 | 0.504 | 0.00925 | 76.8 | 54.5 |
| 80.0 | 0.44 | 0.194 | 0.0050 | 88.0 | 38.8 |
| 120.0 | 0.29 | 0.084 | 0.0030 | 97.7 | 28.0 |
| 160.0 | 0.185 | 0.0338 | 0.0027 | 68.5 | 12.5 |

[a] Although proportional to $N_s$, this is actually $\{[I(\infty)/I(t)]-1\}$.

## 13. Summary and Conclusions

Several sample computations to explain semi-quantitatively the effects of chemisorption states upon the equilibrium conductivity of a semiconductor adsorbent have been presented. Although the computations are general and applicable to both acceptor-like and donor-like adsorbates on *n*-type and *p*-type adsorbents, the major emphasis has been upon the adsorption of acceptor-like adsorbates on *n*-type adsorbents. The adsorbent semiconductivity has been chosen to arise either from a partially compensated, single, shallow donor, or from a partially compensated distribution of donors and traps. For each case, a set of curves has been generated by computer to illustrate the behavior of the adsorbent bulk carrier concentration, the surface potential and the Fermi energy as functions of the temperature and surface state concentration. In general, the computations predict little or no change in carrier concentration, unless the surface state concentration $N_s$ is greater than $WN_{d'}$ (quantity of available electrons in the bulk). In addition, for a crystal to be sensitive to a chemisorption state, the depth of the state below the conduction band $E_{cs}$ must be greater than the surface potential necessary to completely deplete the bulk $[eW^2N_{d'}/2\epsilon]$. Thus, there are definite bounds on the thickness and conductivity of the adsorbent for the observation of a large chemisorption effect. When the conditions for observing an effect are satisfied, the semiconductivity curves (log conductivity vs. $1/T$) are very sensitive to chemisorbate pressure. In general, both the pre-experimental factor and the activation energy are pressure dependent. Very large conductivity changes can be expected, by as much as seven decades between 1 atm ($\sim 10^3$ torr) and $10^{-6}$ torr, for a properly doped semiconductor. This is an effect which will surely have useful practical applications.

These large chemisorption-induced conductivity changes have been realized with thin, semiconducting, CdS crystals as the adsorbent and oxygen as the adsorbate. To achieve semi-quantitative agreement with the model calculations, the oxygen chemisorption state is associated with an acceptor-like surface state (electron trap) between 0.65 and 1.0 eV below the conduction band. Also, the concentration of oxygen surface states $N_s(P,T)$ is an increasing function of pressure $P$ and a decreasing function temperature $T$. This concentration is associated with total oxygen coverage of the surface (charged and uncharged). The experimental evidence also indicated that atomic, not molecular, oxygen is the adsorbed species. Finally, it was found that $N_s(T,P)$ follows rapid changes in $T$ and $P$, when the oxygen pressure is above $10^{-3}$ torr, but not when the pressure is below $10^{-3}$ torr. In order to avoid erroneous conclusions, this must be taken into account when making electrical measurements at low pressures.

# APPENDIX A

### Numerical Solution of Poisson's Equation for the Model of Fig. 1

The Poisson equation for the model of Fig. 1 is

$$d^2V(Z)/dZ^2 = e[N_{d'} - n_c(Z) - n_d(Z)]/\epsilon. \qquad \text{(A.1)}$$

With the aid of eqns. (3) and (4) and using $E_c(Z) = E_c(0) + V(Z)$, this equation becomes

$$d^2V(Z)/dZ^2 = e\{N_{d'} - N_{c'}X\exp[-V(Z)/kT]\}/\epsilon \qquad (A.2)$$

which may be put into dimensionless form in the following notation: $\Psi \equiv V/kT$, $L_D \equiv \epsilon kT/eN_{d'}$ (the Debye length for ionized donors), $Y \equiv Z/L_D$, $R = XN_{c'}/N_{d'}$. $R$ is the fraction of available electrons in the conduction band and the donor level at the center of the sample. $R$ will be nearly unity when there is little charge on the surface and it will be near zero for complete depletion of the conduction band. Hence, $0 < R \leqslant 1$.

The dimensionless form of Poisson's equation becomes

$$d^2\Psi(Y)/dY^2 = 1 - R\exp[-\Psi(Y)] \qquad (A.3)$$

which may be integrated once to obtain[35]

$$d\Psi(Y)/dY = \sqrt{2}\{\Psi(Y) - R[1 - \exp - \Psi(Y)]\}^{1/2}. \qquad (A.4)$$

If $\Psi(Y)$ is known for some value of $Y$, $\Psi(Y+\Delta)$ can be evaluated by a Taylor expansion about $Y$:

$$\Psi(Y+\Delta) = \Psi(Y) + \Delta\Psi'(Y) + \Delta^2\Psi''(Y)/2 + \cdots \qquad (A.5)$$

where eqns. (A.3) and (A.4) are used, and all terms higher than second order are discarded.

The boundary conditions are established at the center of the sample as follows: $\Psi(0) = 0$, $\Psi'(0) = 0$ and with these substituted into eqn. (A.3), we have $\Psi''(0) = (1-R)$. Using these values, we can obtain the value of $\Psi$ at the point $Y = \Delta$. Since the first two terms in eqn. (A.5) are zero, we have $\Psi(\Delta) = \Delta^2(1-R)/2$, $\Psi'(\Delta) \simeq \Delta(1-R)$, and $\Psi''(\Delta) \simeq (1-R)$. These may, in turn, be used to find $\Psi(Y)$ at the point $Y = 2\Delta$:

$$\Psi(2\Delta) \simeq \Delta^2(1-R)/2 + \Delta^2(1-R) + \Delta^2(1-R)/2 = 2\Delta^2(1-R). \qquad (A.6)$$

Since all higher-order terms are in powers of $\Delta^3$, or higher, we can use the above principle to determine the shape of the potential to any desired degree of accuracy.

A set of curves (with $R$ as a parameter) have been computed by this method and are plotted in Figs. A.1, A.2, and A.3. The value of

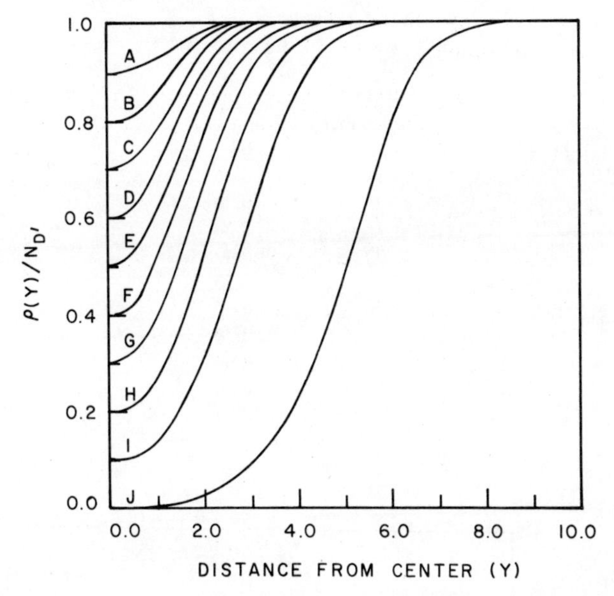


FIG. A.1. Numerical solution of Poisson's equation: space charge divided by effective doping concentration vs. distance from center of specimen in units of Debye lengths.

$R$ for each curve may be found in Table A.1. The results of these calculations may be used to find the space charge and the potential profiles, given the amount of charge on the surface, for any $n$-type material with a shallow donor and whose thickness is less than ten Debye lengths. Suppose, for example, we wish to compute the profile for a flat sample which, at a given temperature, has a Debye length of one-fifth of the distance from the center of the slab to the surface. Then, in Fig. A.3, the surface is located at $Y = 5.0$. If the amount of charge on the surface were, for example, about 3.3 times the available charge in one Debye length (number of electrons on surface per $cm^2 = 3.3 \times L_d \times N_{d'}$), the sample would be described by curve "H" (for this curve $R = 0.80$). In Figs. A.1 and A.2, the space charge and potential profiles would also be described by curves labeled "H". The total height of the potential barrier is, from Fig. A.2, about 6.2 $kT$.

TABLE A.1. VALUES OF $R$ FOR CURVES IN FIGS. A.1, A.2, AND A.3

| Curve label | Value of $R$ |
|---|---|
| A | 0.10 |
| B | 0.20 |
| C | 0.30 |
| D | 0.40 |
| E | 0.50 |
| F | 0.60 |
| G | 0.70 |
| H | 0.80 |
| I | 0.90 |
| J | 0.99 |

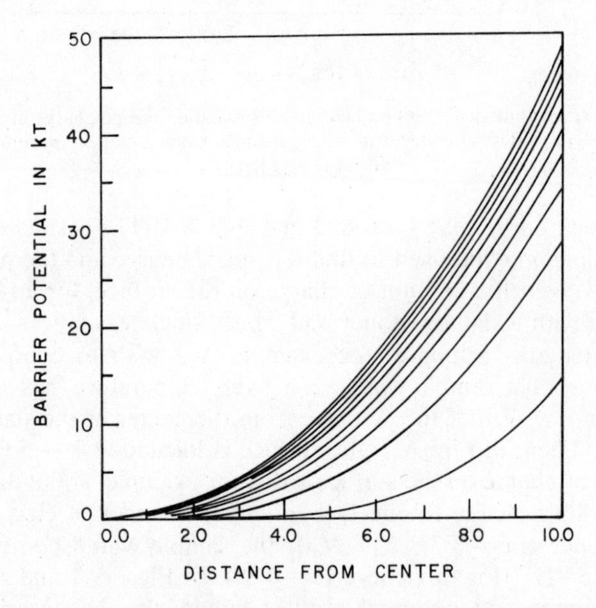


FIG. A.2. Numerical solution of Poisson's equation: Barrier potential in units of $kT$ vs. distance from center of specimen in units of Debye lengths. [Curves from left to right are labeled "A" through "J".]

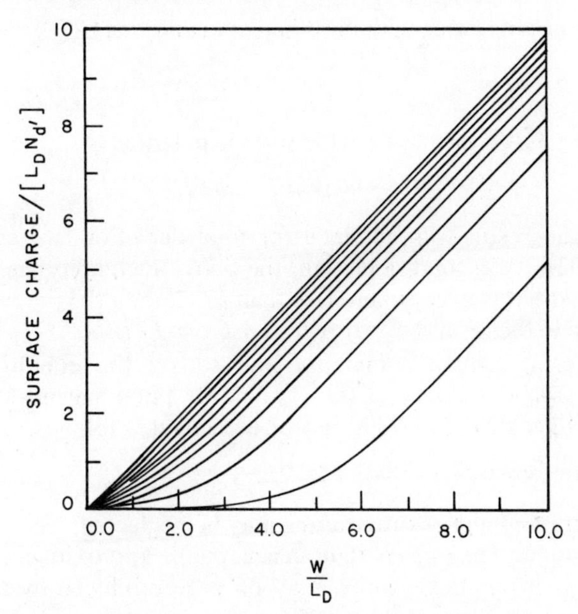


FIG. A.3. Numerical solution of Poisson's equation: Surface charge in units of [$L_D N_{d'}$] vs. half thickness of specimen in units of Debye lengths. [Curves from left to right are labeled "A" through "J".]

The curves calculated above apply to a range of sample half-thickness of about one to ten Debye lengths. The same principle can be applied to other ranges of sample thickness, but the results are rather uninteresting. If one expands the plot of space charge vs. distance to 100 Debye lengths, the curves will look a lot like step functions and the potentials will be parabolic. The surface charge will be the area between the beginning of the step and the surface. Such an approximation is often used for more highly conducting materials.

Let us examine the case where the half-thickness of the same $W$ is less than one Debye length $L_D$. If all of the available charge in the conduction band were moved to the surface, the space charge would be a constant $eN_{d'}$ and we would have:

$$d^2\Psi(Z)/dZ^2 = eN_{d'}/\epsilon kT = 1/L_D^2. \tag{A.7}$$

Upon integrating twice with the boundary conditions $\Psi(0) = \Psi'(0) = 0$,

$$\Psi(Z) = Z^2/2L_D^2. \tag{A.8}$$

Since $W \ll L_D$ by assumption, the surface potential is

$$V_s = \Psi(W)kT < kT/2. \tag{A.9}$$

Since the result was obtained for total depletion, we can state that, for less than total depletion, the above inequality also holds. If the potential is everywhere less than $kT/2$, then the potential can be written in the form $A \exp(-Z/L_D) + B \exp(Z/L_D)$.[36]

In order to obtain symmetry, and satisfy the condition that $d\Psi/dZ = 0$ at the center of the crystal, the potential must be proportional to $\cosh(Z/L_D)$. This may be expanded as follows:

$$\Psi(Z) \sim \cosh(Z/L_D) = 1 + Z^2/2L_D^2 + Z^4/(4! \times L_D^4) + \ldots \tag{A.10}$$

The fourth- and higher-order terms may be neglected, since $Z < L_D$ at every point. This means that an acceptable approximation to the potential is a parabola, which may only be produced by a nearly constant space charge. In view of this, we may assume that the conduction band is uniformly depleted, when the Debye length is greater than the half-thickness of the sample.

## APPENDIX B

### Occupation Statistics for an Exponential Distribution of States

Consider an exponentially distributed set of states extending from the bottom of the conduction band $E_c$ to the top of the valence band $E_v$, described mathematically by eqn. (20). Multiplying this by the Fermi function, produces the occupation density in terms of energy, and the total occupation is obtained by integrating from $E_v$ to $E_c$:

$$n_b = \int_{E_v}^{E_c} N_b(E) f(E - E_f) dE. \tag{B.1}$$

Upon writing $f(E-E_f)$ as

$$f(E-E_f) = [1+\exp(E-E_f)/kT]^{-1}; \quad E \geqslant E_f$$
$$= 1-[1+\exp(E_f-E)/kT]^{-1}; \quad E \leqslant E_f \qquad \text{(B.2)}$$

and using the appropriate substitution of variables, eqn. (B.1) can be reformulated as

$$n_b = N_{bo}\exp[-(E_c-E_f)/kT_c]\Big\{\int_{E_v-E_f}^{0} \exp(E'/kT_c)\,dE$$
$$-\int_{E_v-E_f}^{0} \exp(E'/kT_c)\,dE'/[1+\exp(-E'/kT)]$$
$$+\int_{0}^{E_c-E_f} \exp(E'/kT_c)\,dE'/[1+\exp(E'/kT)]\Big\}. \qquad \text{(B.3)}$$

Upon substituting $Y$ for $-E'/kT_c$, $-E'/kT$, and $E'/kT$ in the first, second and third integrals, respectively, we obtain the following:

$$n_b = N_{bo}\exp[-(E_c-E_f)/kT_c]\Big\{kT_c\int_{0}^{(E_f-E_v)/kT_c} \exp(YT/T_c)\,dY$$
$$-kT\int_{0}^{(E_f-E_v)/kT} \exp(-YT/T_c)\,dY/[1+\exp Y]$$
$$+kT\int_{0}^{(E_c-E_f)/kT} \exp(YT/T_c)\,dY/[1+\exp Y]\Big\}. \qquad \text{(B.4)}$$

If $E_f$ is several $kT_c$ from the conduction band, the upper limit on the first integral can be changed to $\infty$ and the integral reduces to unity. The same procedure is fruitless for the last two integrals, although if $T_c \gg T$, there would be little error in doing so. It is much more helpful to first replace the upper limits by some quantity $Y_1$ (approximately half of the band gap), and then expand the exponents in the numerators in powers of their arguments. Since $Y$ does not go to $\infty$, the expansion will converge more rapidly. The even terms will cancel and the last two integrals reduce to:

$$kT\int_{0}^{Y_1} \frac{[2(T/T_c)Y + (2/3!)(T/T_c)^3Y^3 + (2/5!)(T/T_c)^5Y^5 + \ldots]}{1+\exp Y}\,dY.$$

Since the denominator in the above expression gets large very rapidly, we can now replace the upper limit of integration by $\infty$ and make use of the following well-defined definite integrals.[37]

$$\int_0^\infty Y^n dY/(1+\exp Y) = (1/12)\pi^2,\ n=1$$
$$= (7/120)\pi^4,\ n=3 \quad \text{(B.5)}$$
$$= (31/420)\pi^6,\ n=5$$

Although these integrals increase slowly as $n$ increases, the factor $(T/T_c)^n$ causes the entire expression to converge rapidly and the fifth- and higher-order terms can be dropped. This produces the following expression for the occupation density:

$$n_b \simeq N_{bo} \exp[-(E_c-E_f)/kT_c][kT_c+(\pi^2/6)(T/T_c)kT$$
$$+(7\pi^4/360)(T/T_c)kT], \quad \text{(B.6)}$$

which may be rearranged to yield eqns. (22) and (23).

## APPENDIX C

### Numerical Solution of Poisson's Equation in the Space Charge Region For a Thin Specimen with an Exponential Distribution of Bulk States

The present derivation is similar to that of Appendix A except that the occupation of the bulk states (donor and/or traps) is not proportional to the occupation of the conduction band as it was in Appendix A. Since the bulk states cannot be combined with the conduction band states, the mathematics will be slightly more complex.

The Poisson's equation is the same as eqn. (A.1) except that $n_d$ is replaced by $n_b$. Using eqn. (22) we obtain:

$$d^2V(Z)/dZ^2 = (e/\epsilon)\{N_{d'}-N_cX\exp[-\Psi(Z)]$$
$$-N_{ef}X^{(T/T_c)}\exp[-\Psi(Z)T/T_c]\}. \quad \text{(C.1)}$$

Following the method of Appendix A, but with $R_1 \equiv XN_c/N_{d'}$ and $R_2 \equiv X^{(T/T_c)}N_{ef}/N_{d'}$, eqn. (C.1) becomes

$$d^2\Psi(Y)/dY^2 = 1 - R_1 \exp[-\Psi(Y)] - R_2 \exp[-\Psi(Y)T/T_c]. \quad \text{(C.2)}$$

which may be integrated directly to obtain:

$$(1/2)[d\Psi(Y)/dY]^2 = \Psi(Y) - R_1\{1 - \exp[-\Psi(Y)]\} - (R_2T_c/T) \times \{1 - \exp[-\Psi(Y)T/T_c]\}. \quad \text{(C.3)}$$

The significances of $R_1$ and $R_2$ are similar to that of $R$ in Appendix A. $R_1$ is the fraction of available electrons in the conduction band at the center of the crystal, while $R_2$ is the fraction of available electrons in the distributed bulk states at the center of the crystal. The $R$ in Appendix A corresponds to the sum $(R_1 + R_2)$ in the present development.

Equations (C.2) and (C.3) can be solved on a computer in the same manner as eqns. (A.3) and (A.4), but in this case there are three parameters ($R_1$, $R_2$, and $T/T_c$) to consider. This means that it will not be possible to represent the result with a single family of curves.

## REFERENCES


1. K. Hauffe, *Adv. Catalysis* **7**, 213 (1955); F. F. Volkenstein, *Adv. Catalysis* **12**, 189 (1957).
2. C. E. K. Mees and T. H. James, *The Theory of the Photographic Process*, Macmillan, New York, 1966, 3rd ed., chap. 12.
3. D. B. Medved, *J. Phys. Chem. Solids* **20**, 255 (1961).
4. F. S. Stone, *Adv. Catalysis* **13**, 1 (1962).
5. M. Sakamoto, S. Kobayashis and S. Ishii, *Phys. Rev.* **98**, 552 (1955).
6. R. H. Bube, *J. Chem. Phys.* **27**, 496 (1957).
7. G. A. Somorjai, *J. Phys. Chem. Solids* **24**, 175 (1963).
8. P. Mark, *J. Phys. Chem. Solids* **25**, 911 (1964).
9. J. Bardeen and S. R. Morrison, *Physica* **20**, 873 (1954).
10. R. H. Jones, *Proc. Phys. Soc.* (London) **70**B, 704 (1957).
11. H. T. Minden, *J. Chem. Phys.* **23**, 1948 (1955).
12. D. E. Bode and H. Levinstein, *Phys. Rev.* **96**, 259 (1954).
13. H. E. Matthews and E. E. Kohnke, *J. Phys. Chem. Solids* **29**, 653 (1968).
14. A. Kobayashi and J. Kawaji, *J. Phys. Soc. Japan* **10**, 270 (1955).
15. P. M. Heyman and G. H. Heilmeier, *Proc. IEEE* **54**, 842 (1966).
16. K. K. Reinhartz and V. A. Russel, *Solid State Electronics* **9**, 911 (1966).

17. R. Williams, *J. Phys. Chem. Solids* **23**, 1057 (1962).
18. K. J. Haas, D. C. Fox and M. J. Katz, *J. Phys. Chem. Solids* **26**, 1779 (1965).
19. K. W. Böer and R. Schubert, *Phys. Stat. Sol. Letters* **16**, K5 (1966).
20. R. S. Crandel, *Phys. Rev.* **169**, 577 (1968).
21. A. Many, Y. Goldstein and N. B. Grover, *Semiconductor Surfaces*, North-Holland, Amsterdam, 1965, chap. 4.
22. D. R. Frankel, *Electrical Properties of Semiconductor Surfaces*, Pergamon Press, New York, 1967, chap. 2.
23. P. Handler, *J. Phys. Chem. Solids* **14**, 1 (1960).
24. P. Mark, *J. Phys. Chem. Solids* **29**, 689 (1968).
25. A. Rose, *Concepts in Photoconductivity and Allied Problems*, Interscience, New York, 1963.
26. R. W. Smith, *Phys. Rev.* **97**, 1525 (1955).
27. P. Mark, *RCA Review* **26**, 461 (1965).
28. R. W. Smith and A. Rose, *Phys. Rev.* **97**, 1531 (1955).
29. R. H. Bube, *Photoconductivity of Solids*, Wiley, New York, 1960.
30. P. Braunlich, *J. Appl. Phys.* **39**, 2953 (1968).
31. P. Mark, *J. Phys. Chem. Solids* **26**, 1767 (1965).
32. S. Ross and J. P. Oliver, *On Physical Adsorption*, Interscience, New York, 1964.
33. R. Schubert, Ph.D. Thesis, University of Delaware, June 1969.
34. P. Mark, *J. Phys. Chem. Solids* **26**, 959 (1965).
35. R. Baker, Private communication (1968).
36. J. M. Ziman, *Principles of the Theory of Solids*, Cambridge University Press, Cambridge, 1964, p. 131.
37. H. B. Dwight, *Table of Integrals and Other Mathematical Data*, MacMillan, New York, 1961, chap. 5.

# THE QUANTUM THEORY OF ADSORPTION ON METAL SURFACES*

KAZIMIERZ F. WOJCIECHOWSKI
*Institute of Experimental Physics, University of Wrocław, Wrocław, Poland*

## CONTENTS



## 1. Introduction

The adsorption phenomenon is connected with forces arising when a molecule or an atom approaches a metal surface. These forces are attractive at great distances from the surface, and repulsive at small distances. Denoting by $W(l)$ the interaction energy between an atom ($A$) and a crystal surface, we can write the following conditions, which the function $W(l)$ must fulfil, regardless of its shape: $W(\infty)=0$, $W(0)=\infty$. As a result of attractive forces (curve $a$ in Fig. 1) and repulsive forces (curve $r$ in Fig. 1) the function $W(l)$ has a minimum $W(l_0)$ at a distance $l_0$ from the surface, and is represented by the full line in Fig. 1. The value of $W(l_0)$ is a measure of the so-called adsorption energy $-E(l_0)$[1].

An adsorbed atom we shall call, from now on, an adatom or adsorbate, and crystal on which adsorption can take place adsorbent.

Calculation of the adsorption energy between adatom and adsorbent and the description of the electron processes, which are

*This is an enlarged English version of the article which appeared in *Postepy Fizyki Journal* **20**, 580 (1969).

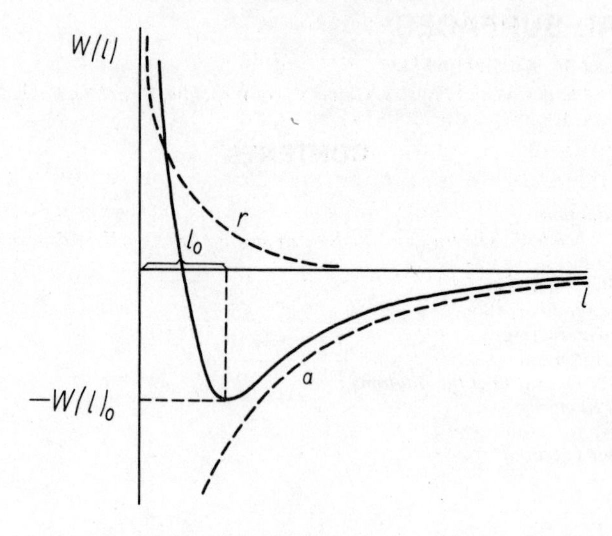


FIG. 1. The adsorbate–adsorbent interaction energy vs. the distance between them.

accompanying the adsorption, are the basic tasks of the adsorption theory.

In experimental work on adsorption, it has become customary to speak of physical and chemical adsorption or chemisorption, the distinction being on the basis of binding energy: the former has low binding energy, of the order of $10^{-2}$ eV, the latter a greater one, of the order of several eV.[2, 3] An up-to-date account of the research on physical adsorption has been given in a monograph by Young and Crowell.[4] Data on chemisorption on semiconductors are collected in a monograph by Wolkenstein,[2] and on chemisorption on metals in a monograph by Trapnell[5] and review articles by Tompkins and co-workers.[6, 7] From a theoretical point of view the distinction between chemisorption and physical adsorption is not sharp;[8] therefore, we shall not adhere to it in what follows.

In this article we will try to present the state of the quantum theory of adsorption on metals. The phenomenon of adsorption is of fundamental importance in understanding the catalysis process, and finds great application in electronic tubes technology and vacuum techniques; it also plays an essential role in such processes as corrosion. No wonder, then, that the theory of this phenomenon still constitutes a central problem of surface physics. However, unlike the theory of adsorption on ionic crystals and semiconductors, the theory of adsorption on metals is less developed and still lacks a review treatment. This is a consequence of the fact that the theoretical description of adsorption on metals is far more complicated than in the other two cases. Nevertheless, a number of works on the quantum theory of adsorption on metals have appeared; and the number is still growing. So it seems desirable to review the most important of them.

## 2. The "Chemical" Theories and the Role of Surface States in the Adsorption Process

Before discussing work which takes into account the interaction between the adsorbate and the metal as a whole, we will discuss briefly the work which treats adsorption as an act of developing surface chemical bond, created by an adsorbate atom or molecule with an atom or several adsorbent surface atoms.

The first such attempt was made by Eley.[9] He considered the adsorption of $H_2$ on tungsten, treating it as a chemical reaction between H and W, $H_2 + 2W \rightarrow 2HW$, and determined the adsorption energy as the following difference:

$$Q = 2E\,(\mathrm{W{-}H}) - E\,(\mathrm{H{-}H})$$

where $E(\mathrm{W{-}H})$ is the binding energy of the tungsten and hydrogen atom, and $E(\mathrm{H{-}H})$ is the $H_2$ molecule binding energy. Eley determined the binding energy using the Pauling equation[10] for single bond energy:

$$E(\mathrm{W{-}H}) = \tfrac{1}{2}[E(\mathrm{W{-}W}) + E(\mathrm{H{-}H})] + 23.06(x_{\mathrm{W}} - x_{\mathrm{H}})^2$$

where $x_{\mathrm{W}}$ and $x_{\mathrm{H}}$ denote the electronegativity of W and H. The W—W binding energy is found from the metal sublimation heat $L$, taking

$2L/12$ for face-centred structures and $2L/8$ for body-centred structures. A similar attempt to calculate the binding energy for adsorption on metals was made by other authors.[11]

Slightly different attempts were based on the so-called "charge-transfer-no bond" theory, first used by Mulliken[12] for calculating the interactions between organic compounds. Application of this theory to adsorption is discussed in the paper by Gundry and Tompkins,[7] where references to earlier works are also given. The method is based on the well-known quantum chemistry method of linear combination of orbitals.

All the above-mentioned works have a common feature: they omit the interaction with conduction electrons. Another group of works, in a sense connected with the chemical approach, attributes a decisive role in the adsorption process on metals to the surface states. The first such work was by Pollard.[13] According to his suggestion, the adsorption may be explained by creation of a one-electron bond between the surface-state orbital, localized on the adsorbent surface, and the adsorbate valence orbitals. This interpretation seems, however, very controversial. First, because the existence of surface states having adequate energy for creating such bonds is doubtful.[5] Second, the interaction between the bonding electron and the metal electrons may be of the same order as the bonding interaction.

A similar view of adsorption theory is given by Grimley[14] and Koutecky.[15] However, they neglect the interaction of the binding electrons with the metal electrons. It seems, therefore, that the calculation of the adsorption energy, without assuming surface states, by consideration of the interaction between an atom and metal as a whole is the more general and direct method. The following paragraphs are devoted to a discussion of treatments which adhere to such a method.

### 3. Van der Waals Forces

We will now consider the Van der Waals attractive forces, sometimes called dispersional forces. The quantum theory of this problem is the subject of several papers.[16–19] Treating atom $A$ as a virtual dipole, and using the free-electron approximation in respect to a surface $M$, Bardeen[17] evaluated the dispersion interaction energy $E_D(l)$ between $A$ and $M$ as

$$E_D^B(l) = -\frac{\gamma\bar{\mu}^2}{12(1+\gamma)}\frac{1}{l^3} \tag{1}$$

where $\bar{\mu}^2$ denotes the mean-square dipole moment of $A$, and $\gamma$ a constant approximately equal to unity, regardless of the kind of $A$. According to Bardeen's suggestion, one can assume that

$$\bar{\mu}^2 = \tfrac{3}{2}\alpha I \tag{2}$$

where $\alpha$ is the polarizability of $A$, and $I$ its ionization energy.

Margenau and Pollard[18] treated the dispersional interaction between $A$ and $M$ as an interaction of induced dipoles – that of a molecule and a small volume of metal. The existence of the conduction electrons was completely ignored in this work. For this reason, the above approach is to be treated as a rough approximation, which may, eventually, be used for distances $l \gg l_0$. But for these distances one can hardly speak of adsorption.[20] The authors obtained the following result:

$$E_D^{MP}(l) = -\frac{\epsilon^2\alpha}{8l^3}\left(\frac{1.25}{r_s} - \frac{h^2\rho_0}{2\pi m I}\right) \tag{3}$$

where $\epsilon$ = charge of electron, $r_s$ = radius of a sphere which contains one conduction electron, $h$ = the Planck constant and $m$ = mass of electron.

The $E_D^B$ and $E_D^{MP}$ calculated for helium, argon, hydrogen and nitrogen, which interact with a series of monovalent and divalent metals, are given in ref. 18. In general, formula (3) gives considerably bigger values than formula (1).

Prosen and Sachs[19] considered the interaction between $A$ and $M$ as an interaction of a field created by a virtual dipole of $A$ with the conduction electrons and metal ions. They expressed the dispersional interaction energy by the following quantities:

$$E_D = -\tfrac{1}{2}\alpha \sum_{\xi,i} H_{0i}^{\xi} H_{i0}^{\xi} \tag{4}$$

where $\alpha$ is the adatom or molecule polarizability, and

$$H_{0i}^{\xi} = \int \Phi_0^* H^{\xi} \Phi_i d\tau, \quad \xi = x, y, z. \tag{5}$$

In the formula $\Phi_0$ is the metal ground state wave function, and $\Phi_i$ = the excited state wave function;

$$H^{\xi} = \sum_e H^{\xi}(e) + \sum_n H^{\xi}(n)$$

where

$$H^{\xi}(e) = \frac{\epsilon \xi_e}{r_e^3}, \quad H^{\xi}(n) = -\frac{Z\epsilon \xi_n}{r_n^3} \tag{6}$$

and $r_e = (x_e, y_e, z_e)$ is the distance between a molecule and the $e$th metal electron; $r_n = (x_n, y_n, z_n)$ is the distance between a molecule and the $n$th metal ion; $Z$ is the metal ion valency.

The authors, assuming that during the interaction between $A$ and $M$ the metal ions do not undergo transitions to an excited state (compare ref. 23), obtained for the sum in (4) the following expression:

$$\sum_{\xi, i} \left[\sum_e H_{0i}^{\xi}(e) + \sum_n H_{00}^{\xi}(n)\delta_{0i}\right]\left[\sum_e H_{i0}^{\xi}(e) + \sum_n H_{i0}^{\xi}(n)\delta_{i0}\right] \tag{7}$$

which they simplified to:

$$\sum_{i \neq 0} \left[\sum_e H_{0i}^{\xi}(e)\right]\left[\sum_e H_{0i}^{\xi}(e)\right] \tag{7a}$$

assuming that the charge distribution of the positive ions may be smeared out over the whole metal. In such a case, the sums of the diagonal elements in (7) cancel with the expressions $\sum_n H_{00}^{\xi}(n)$ and (7) becomes (7a).

Further calculations are based on the assumption that the wave function $\Phi$ which describes the metal electrons is a simple product of plane waves. Taking into account the Pauli exclusion principle, by choosing proper integration limits in the momentum space, Prosen and Sachs[19] obtained the following formula for $E_D$.

$$E_D^{PS}(l) = -\frac{\alpha \epsilon^2 k_F^2}{8\pi^2} \frac{\ln(2k_F l)}{l^2} \tag{8}$$

where $k_F$ is the wave vector value corresponding to the Fermi level; for free electrons we have:

$$k_F = \left(\frac{3\rho_0}{\pi}\right)^{1/3}. \tag{9}$$

In Fig. 2, a graphic comparison of the functions $E_D(l)$, calculated from formulae (1), (3) and (8) for the interaction between benzene and mercury, are given after Kemball.[21] It can be seen that $E_D^B$ and $E_D^{MP}$ lie close to one another, whereas $E_D^{PS}$, although of the same order, increases much more slowly with increasing $l$.

It should be noted that in the works[17, 19] plane waves were used to describe the electron gas in the metal. It seems that this assumption in the case of the adsorption phenomenon contains an

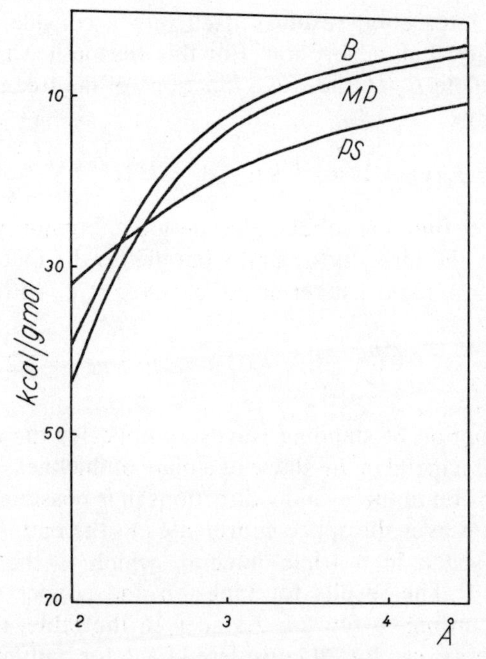


FIG. 2. The comparison of the dispersional energy vs. benzene–mercury distance calculated after Bardeen[17] (B), Margenau and Pollard[18] (MP) and Prosen and Sachs[19] (PS).

inconsistency. Namely, it is well known that the plane wave function $\varphi_k(\boldsymbol{r}) = e^{i\,kr}$, fulfills the so-called Born–Kármán's periodic conditions, which require that the function has the same values on the boundaries of cube of side $L$. Thus, it describes the bulk properties of the metal, which may be imagined as a cube with side $L$ embedded in a very large, theoretically infinite, metal block. Hence the plane wave does not allow for the surface phenomena[22] and may be used only for describing the bulk properties which, when the sample is big enough, do not depend on these effects, e.g. conductivity. Adsorption is a purely surface phenomenon; in this case we have, roughly speaking, an atom outside a metal which occupies the whole half-space. The Born–Kármán's conditions require the $\varphi_k(\boldsymbol{r})$ to be periodic, so they do not allow such a possibility. Therefore, the application of plane waves to the description of the atom–metal interaction resolves itself into a consideration of impurities, rather than adsorption. For this reason it was assumed[23] that, to calculate $E_D(l)$, the wave functions of the free electrons are determined by

$$\varphi_k(\boldsymbol{r}) = 2^{1/2}L^{-3/2}\sin\left[k_3(z-l)\right]e^{i(k_1x+k_2y)}. \tag{10}$$

The above function obeys the periodic boundary conditions (period $L$) in the directions $x$ and $y$ parallel to the metal surface; it vanishes for $z = 1$ and is determined for $0 \leqslant z \leqslant L$, with

$$k_{1,2} = \frac{2\pi n_{1,2}}{L}, \quad k_3 = \frac{\pi n_3}{L}, \quad n_{1,2} = 0, \pm 1, \ldots; \quad n_3 = 1, 2, \ldots \tag{11}$$

The assumption of standing waves complicates the calculations. However, for a metal in the shape of a plate of thickness $L$ and of unlimited extension in the $x$- and $y$-directions, it is possible to carry out the integration over the space coordinates to the end, and over the momentum space in a triple integral, which is then calculated numerically.[23] The results for tungsten and copper are given in Table 1, according to ref. 24. As seen in the table, the obtained values of energy are, for all considered $l$ and for both metals, about 10 times smaller than the $E_D^{PS}(l)$ which were calculated according to ref. 19. For example, for the Ba–W system, with $l = 2.6$ Å and $\alpha_{Ba} =$

TABLE 1. THE VALUES OF $-E_D/\alpha$ in eV COMPUTED AFTER REFS. 19 AND 23

| $l$ Å | Cu | | W | |
|---|---|---|---|---|
| | ref. 19 | ref. 23 | ref. 19 | ref. 23 |
| 0.8 | 0.866 | 0.083 | 1.304 | 0.119 |
| 1.0 | 0.677 | 0.060 | 0.997 | 0.086 |
| 1.2 | 0.537 | 0.046 | 0.782 | 0.065 |
| 1.4 | 0.437 | 0.035 | 0.629 | 0.051 |
| 1.6 | 0.363 | 0.029 | 0.520 | 0.041 |
| 1.8 | 0.306 | 0.024 | 0.437 | 0.034 |
| 2.0 | 0.263 | 0.020 | 0.374 | 0.028 |
| 2.2 | 0.229 | 0.017 | 0.338 | 0.023 |
| 2.4 | 0.199 | 0.015 | 0.280 | 0.020 |
| 2.6 | 0.170 | 0.013 | 0.246 | 0.017 |

63 Å$^3$,[25] the Prosen and Sachs formula gives $-15.25$ eV, while from Table 1 we read $-1.05$ eV. These numbers prove that formula (8) gives the wrong results.

Let us return now to the assumption of the continuous metal ions positive charge distribution, which permitted us to simplify (7) to (7a). Thus, if we dismiss this assumption, the diagonal elements which are dependent on the metal surface atomic structure will appear in (4) via (7). So, in order to make allowance for the dependence of $E_D$ on the crystallographic direction, the following integrals have to be calculated

$$H^{\xi}_{kk} = \int \varphi^*_k(\boldsymbol{r}) H^{\xi}_e(\boldsymbol{r}) \varphi_k(\boldsymbol{r})\, d\boldsymbol{r}, \quad \xi = x, y, z \tag{12}$$

which, in the case of $\varphi_k(\boldsymbol{r})$ being a plane or standing wave and $H^{\xi}_e(\boldsymbol{r})$ being given by (6), become divergent. (It is well known that similar difficulties arise in metal theories when Coulomb's interaction is taken into account.[22, 26]) Thus, in the case of $H^{\xi}(e)$ having the form of (6), only the assumption that

$$\sum_e H^{\xi}_{00}(e) - \sum_n H^{\xi}_{00}(n) = 0$$

made by Prosen and Sachs[19] gives reasonable results. However, assuming a screened Coulomb interaction instead of (6), the integrals (12) are not divergent and the dependence of $E_D$ on the crystallographic direction is given by;[27]

$$E^D_{(hkl)}(l) = -\frac{1}{2}\alpha\epsilon^2\left[\frac{1}{(2\pi)^6} J^2(l) - \frac{2Z}{(2\pi)^3} J(l) \sum_{\xi n} \frac{\xi_n e^{-\lambda r_n}}{r_n^3} + \right.$$

$$\left. + Z^2 \sum_{\xi} \sum_{n,n'} \frac{\xi_n \xi_{n'} e^{-\lambda(r_n + r_{n'})}}{r_n^3 r_{n'}^3}\right] + E_D(l) e^{-2\lambda l}$$

where

$$J(l) = 2\pi^2 k_F^3 \lambda \int_l^{\infty} z f(z) \Gamma(-1, \lambda z) dz$$

$$f(z) = \frac{1}{3} - \frac{\sin[2k_F(z-l)] - 2k_F(z-l)\cos[2k_F(z-l)]}{[2k_F(z-l)]^3}$$

$\Gamma(a, x)$ is the incomplete gamma function, $\lambda^{-1}$ screening radius, and $E_D(l)$ dispersional energy computed in refs. 23 and 24 (the values of this function for copper and tungsten, computed for several $l$, are given in Table 1).

It is to be noted that in earlier works on the adsorption theory of metals, the surface atomic structure, which depends on the kind of crystallographic plane, was not taken into account.

## 4. Exchange Interaction

When the distance between two molecules or atoms becomes comparable with their sizes, the exchange of electrons between the interacting systems begins to play a considerable role. The first work on adsorption which considered the exchange forces was made by Pollard.[28] In this work, the atom is represented by the valence electron and an ion located at a point $r_A$, at a distance $l$ from the metal surface. The valence electron has space coordinates $\boldsymbol{r}_a$ and spin $s_a$, and is described by the wave function $[c_+\alpha(s_a) + c_-\beta(s_a)]\psi_{1s}(\boldsymbol{r}_a)$, where $c_+$ and $c_-$ are constant coefficients, and $\alpha$ and $\beta$ spin functions. The metal ions have charge $Z\epsilon$ and are located at points $\boldsymbol{r}_j$, while the metal electrons at points $\boldsymbol{r}_i$ are described by one-electron wave functions $\varphi_i(\boldsymbol{r}_j)$ and by spin functions $\alpha(s_i)$ and $\beta(s_i)$.

The (metal + atom) system, having $(N+1)$ free electrons, can be described as one unit by an antisymmetrical function of $(N+1)$ coordinates. Since the spin function of an atom is assumed to be a linear combination, the $\psi_E$ function, which describes the system exchange phenomenon, may be written as

$$\psi_E = c_+\psi_1 + c_-\psi_2 \tag{13}$$

where

$$\psi_1 = \frac{1}{\sqrt{(N+1)!}} \left| \begin{array}{c:c} \psi_{1s}(\boldsymbol{r}_1)\alpha(s_1) & \\ \cdot\quad\cdot\quad\cdot & \\ \cdot\quad\cdot\quad\cdot & \Phi \\ & \\ & a \\ \psi_{1s}(\boldsymbol{r}_N)\alpha(s_N) & \\ \hdashline \multicolumn{2}{c}{\psi_{1s}(\boldsymbol{r}_a)\alpha(s_a), \varphi_1(\boldsymbol{r}_a)\alpha(s_a), \ldots, \varphi_N(\boldsymbol{r}_a)\beta(s_a)} \end{array} \right| \tag{14}$$

and function $\psi_2$ is of the same form save for the difference that it has a column made up of $\psi(\boldsymbol{r}_i)\beta(s_i)$ on its left side.

In (14) $\Phi$ is the Slater's determinant of the $N$th order, constructed from functions $\varphi_i(\boldsymbol{r}_j)$:

$$\Phi = \det\,[\varphi_i(\boldsymbol{r}_j)]. \tag{15}$$

The energy $E_E$ is found from the formula

$$E_E = \frac{\int \psi_E^* H \psi_E d\tau}{\int \psi_E^* \psi_E d\tau} \tag{16}$$

where $H$ is the Hamiltonian of the system.

Pollard made allowance in $H$ for the Coulomb interaction between the adatom electron and the atom ion, the metal electrons and the ions and also the interaction between them. He showed that the individual terms in the exchange interaction refer to the electrons which have the same spin as the electron of the visiting atom.

Neglecting terms of order (crystal surface area)$^{-1}$, the exchange interaction energy may be expressed as follows:[29]

$$E_E(l) = \frac{J_1(l) - J_2(l) + 2J_3(l)}{1 - D^2(l)} \tag{17}$$

where

$$\left.\begin{aligned} D^2 &= \int\int \psi_{1s}^*(1)\psi_{1s}(2) f(1,2)\, d\tau_1 d\tau_2 \\ J_1 &= \epsilon^2 \int\int \frac{\psi_{1s}^*(1)\psi_{1s}(2) f(1,2)}{r_{a1}}\, d\tau_1 d\tau_2 \\ J_2 &= \epsilon^2 \int\int \frac{\psi_{1s}^*(1)\psi_{1s}(2) f(1,2)}{r_{12}}\, d\tau_1 d\tau_2 \\ J_3 &= \epsilon^2 \int\int\int \frac{\psi_{1s}^*(1)\psi_{1s}(2) f(3,2) f(1,3)}{r_{13}}\, d\tau_1 d\tau_2 d\tau_3 \end{aligned}\right\} \tag{18}$$

$$f(1,2) = \sum_{i=1}^{n} \varphi_i^*(\mathbf{r}_2)\varphi_i(\mathbf{r}_1). \tag{19}$$

Taking plane waves for $\varphi_i(\mathbf{r}_j)$, Pollard calculated the integrals (18) assuming that

$$f(1,2) = \begin{cases} N/2V, & \text{for } r_{12} \leqslant R = \left(\dfrac{8\pi\rho_0}{3}\right)^{1/3} \\ 0, & \text{for } r_{12} > R \end{cases} \tag{20}$$

where $\rho_0$ is the electron density. This assumption permitted further simplifications

$$r_{A2} = r_{A1} - r_{12}\cos\theta \tag{21}$$

where $\theta$ is the angle between $\mathbf{r}_{A1}$ and $\mathbf{r}_{12}$, and

$$\psi_{1s}(2) = \psi_{1s}(1)e^{ar_{12}\cos\theta}$$

$$\int\int d\tau_1 d\tau_2 = \left(\int_M d\tau_1\right)\left(\int_S d\tau_{12}\right) \tag{22}$$

where $a$ is constant, $M$ = space occupied by metal, and $S$ = sphere of radius $R$ centred at $\mathbf{r}_1$, inside of which $f(1,2)$, according to (20), has constant value and equals zero outside.

Pollard calculated $E_E$ for the helium atom and hydrogen molecule, assuming that the whole exchange interaction of both the electrons is twice as great as the interaction of one of them with a metal, which has the same number of free electrons with the same spin direction. In both cases, $E_E(l)$ was a positive function, which shows that the exchange interaction, in the case of an atom with one valence electron with the metal, gives rise to a repulsion. The Pollard calculations were repeated for $H_2$ in the work done by Toya.[30]

Calculations of $E_E(l)$ for alkaline atoms Sr and Ba interacting with tungsten and copper have been made.[23] In all these cases, like in those described above, $E_E$ is connected with repulsive forces. The valence electron was described by Slater's orbital, and the metal electrons by functions of the type (10), which formed the function $\psi_E$ in the same way as above. As we have mentioned in § 3, the wave functions of type (10) are more suitable for describing the adsorption phenomena than the plane waves are, but they complicate the calculations considerably. In the case of plane waves, $f(1,2)$ (the so-called Wigner–Seitz density) could be approximated according to (20), because then

$$f_B(i,j) = 3\left(\frac{N}{2V}\right)\frac{\sin\xi_{ij}-\xi_{ij}\cos\xi_{ij}}{\xi_{ij}^3}$$

where $\xi_{ij} = k_F r_{ij}$, $k_F =$ the Fermi momentum; while in the case of function (10)

$$f(i,j) = f_B(i,j) - 3\left(\frac{N}{2V}\right)\frac{\sin\eta_{ij}-\eta_{ij}\cos\eta_{ij}}{\eta_{ij}^3} \tag{23}$$

where

$$\eta_{ij} = k_F[(x_i-x_j)^2+(y_i-y_j)^2+(z_i+z_j-2l)^2]^{1/2}.$$

Thus, for $\xi_{ij}\to 0$ we do not have $f_{ij}\to N/2V$, but

$$3\left(\frac{N}{2V}\right)\left[\frac{1}{3}-\frac{\sin x - x\cos x}{x^3}\right] \equiv f_{\xi\to 0}\,(i,j)$$

where $x = 2k_F(z_i - l)$, so we can approximate $f(i,j)$ by

$$f(i,j) = \begin{cases} f_{\xi\to 0}(i,j), & \text{for} \quad r_{ij} \leq R \\ 0, & \text{for} \quad r_{ij} > R. \end{cases} \tag{24}$$

The above fact brings about an additional complication when transforming (18) to double and triple integrals. We shall see it in the example of the $D^2(l)$ integral. In the case of plane waves, one can make the simplifying assumptions (20), (21) and (22), and carry out the integration, obtaining

$$D^2(l) = \frac{3}{2} \frac{x \cosh x - \sinh x}{x^3} e^{-2l} \tag{25}$$

where $x = R/a_0$, $a_0 =$ the Bohr radius.

In the case of the function of type (10), one can approximate $f(i,j)$ according to (24), and assuming instead of (21) the more precise expression:

$$r_2^2 = r_1^2 + r_{12}^2 - 2r_1 r_2 \cos\theta \tag{26}$$

and using the polar coordinates $r = r_{12}$, $\theta$, $\varphi$ the integral $D^2$ may be written as

$$D^2(l) = \frac{2\pi}{\gamma^{\beta+2}} \int_l^\infty g(z_1)\,dz_1 \int\!\!\int_{-\infty}^{+\infty} r_1^{\beta-1} e^{-\gamma r_2} h_{\beta+1}(r_1)\,dx_1 dx_2.$$

where $\gamma$ and $\beta$ are the Slater constants in the atom wave function (in place of fractional values of $\beta$, the nearest integers were used), and

$$g(z_1) \equiv f_{\xi\to 0}(1,2)$$

$$h_{\beta+1}(r_1) = \int_0^R rR_{\beta+1}(r, r_1)\,dr$$

$$R_{\beta+1}(r, r_1) = \int_{\gamma(r_1-r)}^{\gamma(r_1+r)} v^{\beta+1} e^{-v}\,dv.$$

Expressing $x_1$, $y_1$, $z_1$ by polar coordinates $r$, $\theta$, $\varphi$ gives a double integral over $r_1$ and $\theta$ which can be calculated numerically. The result of the $D^2$ integral calculation is given in ref. 24, under assumption (20), but using (26) instead of (21). It turns out that the error caused by the assumption (21) is considerable, and increases with decreasing $l$.

## 5. Ionic Interaction

In the case of dispersional and exchange interactions, we had to deal with the neutral electron configurations of the system, that is, with configurations which were not accompanied by a charge transfer from atom to metal and vice versa. There exists also an interaction during which charge is partly transferred from adsorbate to adsorbent. Using chemical language, we can denote the first two configurations by $A-M$, and the last by $A^+-M^-$, or $A^--M^+$.

We shall deal with the $A^+-M^-$ configuration when considering, for example, the adsorption of alkaline atoms on metals (compare with the experimental works[5,6,31,32]). Such a configuration corresponds to the following situation: an atomic electron goes into the free level $\alpha$ in the metal and we have a state which can be described by a wave function $\psi_I$ of $(N+1)$ electrons. The ionic state wave function may be obtained in the same way as function $\psi_E$ (compare (13)), save for substituting $\psi_{ns}(\boldsymbol{r})$ for $\varphi(\boldsymbol{k}_\alpha, \boldsymbol{r})$ as the valence electron goes into the metal

$$\psi_I = \frac{1}{\sqrt{(N+1)!}} \begin{vmatrix} \varphi(\boldsymbol{k}_\alpha, \boldsymbol{r}_1) & \\ \cdots & \Phi \\ \cdots & \\ \varphi(\boldsymbol{k}_\alpha, \boldsymbol{r}_a), \varphi(\boldsymbol{k}_1, \boldsymbol{r}_a), \ldots, \varphi(\boldsymbol{k}_N, \boldsymbol{r}_a) & \end{vmatrix} \tag{27}$$

where the spin functions are already included in $\varphi$ anf $\Phi$ has the same form as in (15).

The energy of this configuration can be expressed as:

$$E_I(\boldsymbol{k}_\alpha, l) = I - \epsilon(\boldsymbol{k}_\alpha) - Q(l) \tag{28}$$

where $I$ is the atom ionization energy, $\epsilon(k_\alpha)$ denotes the valence electron energy in level $\alpha$ in the metal, and $Q(l)$ the electrostatic interaction energy between the adion and the metal.

In case we have to deal with a reversed charge transfer direction, or when such a configuration is also to be considered, a function $\psi_I^-$ must be constructed, which will differ from (27) by having in the delimiting column the atomic wave function of an electron taken from a level inside metal.

The energy of the A – M configuration is usually taken in the form[5, 6, 30]

$$E_I = I - \varphi + \frac{\epsilon^2}{4l} \tag{29}$$

where $\varphi$ is the work function and the last term arises from the classical image force. The image force was invented at the beginning of the development of adsorption theory.[33] The application of the classical image force to adsorption was discussed by Bardeen.[17] He found that, for distances comparable with atomic dimensions, a considerable deviation from the classical formula exists, because of the approximate form of the last term in (29). The correction to this term has a quantum mechanical character[17, 34] and may be neglected only at distances of the order of $10^{-4}$ cm from the metal surface.[35] In view of the fact that adsorption is a surface phenomenon, publications[23, 24] allow for the quantum correction to the classical image force in formula (28) according to ref. 36 and, by taking into account the electronic interaction between the adions and the metal surface ions, a dependence of $E_I$ on the crystallographic direction is obtained.

## 6. Adsorption Energy

As it was mentioned in § 1, the adsorption energy is one of the more important parameters determinating the adsorption process. Since this quantity depends closely on the degree of coverage $\theta$,[5, 6, 37] and it is difficult to allow theoretically for the effect of $\theta$ on the adsorption energy, it is the so-called initial energy that is calculated. The initial adsorption energy $E(l_0)$ is the measure of the binding energy between a clean metal surface and a single adatom;[1] thus, its experimental determination is made by extrapolating the curve $E(\theta)$ to the value at $\theta = 0$. There are several methods of estimating $E(l_0)$ which are described in other papers. One of them, the so-called chemical method, is described in § 2, where a criticism of this approach is also given. Another method consists of summing directly the energy of the attractive and repulsive forces, which were calculated independently. In such a way, the bonding energy between $H_2$ or He and Ni was calculated by Pollard.[28] He added the exchange energy and the dispersional energy calculated from (3) obtaining

$$E(l) = E_E(l) - E_D(l) \tag{30}$$

from which he found $E(l_0)$.

The diagram of $E(l)$, calculated in such a way for a benzene–mercury system, can be seen in Fig. 3. $E_D(l)$ is found from (3) and $E_E(l)$ as in ref. 28, using the results obtained there for hydrogen. The calculated value of $E_E(l_0)$ is $-10.2$ kcal/gmol, whereas the experimental value is $-15.4$ kcal/gmol. In the case of adsorption of

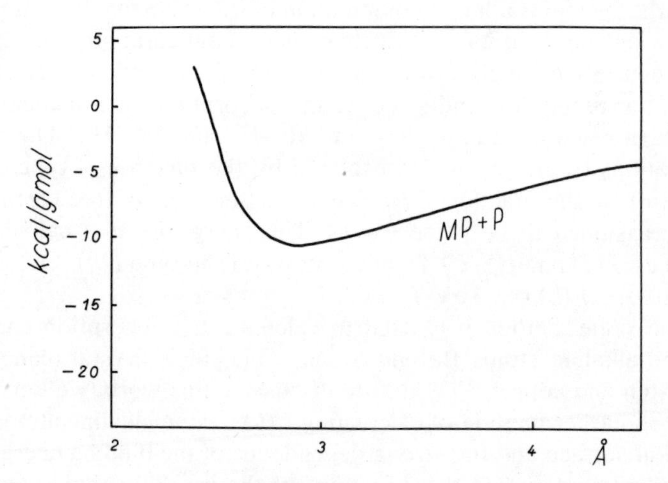


FIG. 3. The adsorption energy for a benzene–mercury system, calculated in ref. 21.

the inert gases, such a way of calculating the binding energy is the least controversial, because one can think of the electron cloud of the adsorbate as being only slightly deformed when an atom approaches the metal surface. But in other cases the deformation of the electron cloud of an adsorbate is significant, because of interaction between the cloud and the metal. This will be particularly strong in the case of the alkaline atoms, which have a low first ionization potential and because of this the valence electron can break off easily from the atom and go into the metal during the adsorption process. In such a case, the adsorbate ionizes at some distance from a metal surface. Thus, it becomes a physical system which is characterized by quite different polarizability and charge, in comparison with a free atom.

To include effects following the electron cloud deformation of the adsorbates in the adsorption process, and their influence on the binding energy, a more accurate method must be used than the above. It seems that for this purpose a particularly useful method is that of a linear combination of atomic orbitals based on the Ritz method.[38] This method was applied by Toya[30] to calculate the adsorption energy of $H_2$ on Pt. He assumed that the wave function $\psi$ of the system is a linear combination of the wave functions for the following configurations: (i) exchange configuration (see §4); (ii) neutral configuration, which corresponds to an electron transition under the influence of the adsorption on a higher free level; (iii) two ionic configurations $M^+$–$H^-$ and $M^-$–$H^+$. The configurations (i) and (ii) were described by the function of type (14), and (iii) by the function (27). The wave functions of free electrons were assumed to be plane waves. The energy $E_E$ was calculated from eq. (17) and $E_I^+$, $E_I^-$ from equations of the type (29). Obtained results are: $E(l_0) = -5$ eV, $l_0 = 0.8$ Å.

The same method was used to calculate the adsorption energy of the alkaline atoms Ba and Sr on some single crystal planes of tungsten and copper.[23,24] Before discussing this work, we will discuss an earlier method of calculating $E(l_0)$ for the alkaline atoms on a metal surface and the strong dependence of the binding energy of the alkaline atoms on the directions of the crystallographic planes of the metal monocrystals. For the systems considered, the following method is very often applied to calculate the $E(l_0)$;[5,6,37] at a small distance from the metal surface the weakly bound valence electron goes into the metal. During this process, energy is released equal to $\varphi - I$, where $\varphi$ is the work function and $I$ is the ionization energy of the atom. The ion generated in such a way remains at a distance $r_0$ from the metal surface and, as a result of bringing the adion into the equilibrium position, the energy $Q$ is released, so: $E(r_0) = \varphi - I + Q$, where $Q$ is treated as the image force energy. The distance between the ion centre and the metal surface is assumed to be equal to the ion radius $r_0$, thus, $Q = \epsilon^2/4r_0$. Hence, it is the ion configuration energy, which was calculated for $l = r_0$, that is identified with the adsorption energy. It is evident that such an approach cannot be treated seriously, because it is assumed, in is equal to the ion radius and the repulsive forces are quite negligible. The adsorption energy of the alkaline atoms Sr and Ba depends significantly on the crystallographic direction. Table 2 contains the experimental values of $E(l_0)$ for $\theta \to 0$. It can be seen that addition to the assumption that the metal surface is an ideal plane and perfectly conductive, that the equilibrium position of the adion $l_0$ there exist many discrepancies between the results reported by the various workers concerning the same system. These discrepancies may be due to the different techniques, to surface contamination and experimental conditions under which the results were obtained. Extrapolation of the results obtained for $\theta \neq 0$ can also cause some discrepancies. On the other hand, the mean values of $E(l_0)$, corresponding to small coverages, cannot give any real information, because in this case many crystal planes are not quite covered and the others are covered with $\theta_{hkl} > \theta$, where $\theta_{hkl}$ is the degree of coverage of the crystal plane having Miller indices $hkl$ and $\theta$ is the mean degree of coverage of all the crystal planes. This problem now attracts more and more attention.[45,53] The comparison of the values $E(l_0)$, measured by the same author and for the same crystal plane, can give some information about the adsorption process, in spite of the above-mentioned remarks. For the crystal planes (100) and (112), one can write the following sequence of the increasing adsorption energies of the alkaline atoms on tungsten, on the basis of the data in ref. 40.

TABLE 2. SUMMARY OF EXPERIMENTAL VALUES OF ADSORPTION HEAT ON TUNGSTEN IN eV

| Adsorbate | $-E(l_0)$ | Plane | Author | Experimental technique |
|---|---|---|---|---|
| Na | 1.39 | — | Bosworth[39] | Thermal desorption TD |
| | 3.09 | (112) | Kühl[40] | * |
| | 2.51 | (110) | Kühl[40] | * |
| | 2.73 | — | Starodubtsev[41] | Pulsed-beam techn. PB |
| K | 2.64 | — | Evans[42] | TD |
| | 2.55 | — | Starodubtsev[41] | PB |
| | 2.90 | — | Knauer[43] | PB |
| | 3.36 | (112) | Kühl[40] | * |
| | 2.75 | (110) | Kühl[40] | * |
| | 2.6† | — | Schmidt and Gomer[44] | TD |
| | 2.9 | (110) | Schmidt and Gomer[32] | TD |
| | 2.5 | (111) | Schmidt and Gomer[32] | TD |
| Rb | 2.54 | — | Evans[42] | TD |
| | 2.61 | (110) | Hughes *et al.*[46] | PB |
| Cs | 2.46 | — | Evans[42] | TD |
| | 2.83 | — | Taylor and Langmuir[47] | TD |
| | 3.6 | — | Knauer[43] | PB |
| | 3.58 | (112) | Kühl[40] | * |
| | 3.08 | (110) | Kühl[40] | * |
| | 2.04 | — | Scheer and Fine[48] | PB |
| Sr | 3.26 | — | Moore and Allison[40] | TD |
| | 4.21 | — | Madey *et al.*[50] | TD |
| Ba | 3.48 | — | Moore and Allison[49] | TD |
| | 4.68 | — | Zingerman *et al.*[31] | TD |
| | 3.77 | — | Utsugi and Gomer[51] | Field desorption |

*The experimental technique of Kühl is unknown to the author, and Popp[40] did not give the description of this technique.

†For $\theta = 0.2$.

$$E_{Na} < E_K < E^*_{Rb} < E_{Cs}. \tag{31}$$

In this sequence we have distinguished $E_{Rb}$, because it was measured by another author. For all the investigated systems the following inequality holds

$$E_{(110)} > E_{(100)}. \tag{32}$$

In the case of Sr and Ba, it can be seen in Table 2 that:

$$E_{Sr} < E_{Ba}. \tag{33}$$

In some works,[23,24] the binding energy of the alkaline atoms Sr and Ba on the crystal planes (100), (110) and (111) of W and on the crystal planes (100) and (110) of Cu is calculated using the linear combination of the atomic orbitals technique. The system atom–metal was described by means of the wave function $\psi_{12}$, which is a combination of wave functions $\psi_1$ and $\psi_2$ of the configurations: (1) exchange and (2) ionic $A^+$–$M^-$: $\psi_{12} = c_1\psi_1 + c_2\psi_2$. It is known that $|c_i|^2$ can be interpreted as the probability of a particular configuration. Thus, calculation of $c_2^2(l_0)$ can also determine the degree of ionicity on a particular crystal plane.

The wave functions (10) were taken for the conduction electrons and the calculations described in § 3 and § 4 were carried out. The surface atomic structure, represented by the additional term $Q^j(l)$ in equation (29), was taken into account in the calculations. For this purpose, the work functions used were obtained experimentally for the various crystal planes. The calculated values of $E(l_0)$, for the combination of the configurations 1 and 2, correspond to the cases when the atom approaches the metal surface along the perpendiculars through the points marked by the cross on each of the crystal planes considered (see Fig. 4). $E(l_0)$ was obtained by minimizing $E(l)$ with respect to $l$. The curves in Figs. 5 and 6 represent a typical shape of $E(l)$ for the Na–W and Sr–W systems. The comparison of the calculated and experimental values, obtained for tungsten, is given in Table 3. It can be seen that the ionization probability $c = c_2^2$ of the alkaline atom near its equilibrium position $l_0$ is almost 1. From this fact, we can conclude that these atoms are adsorbed on the tungsten surface as ions; thus, we can treat the calculated resonance energy $E(l_0)$, between the configurations A–M and $A^+$–$M^-$, as the binding energy ($E_D$ is very low, because of the small polarizability of the alkaline ions). Calculated energy values are in good agreement with the experimental values. For each system considered, the dependence of the following sequence of the adsorption energies on crystallographic direction was obtained

$$E_{(110)} > E_{(100)} > E_{(111)}.$$

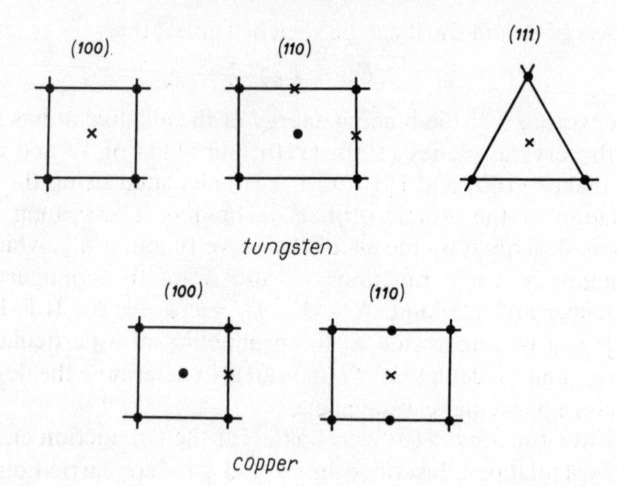


FIG. 4. The adsorption centres on different crystal planes of tungsten and copper.

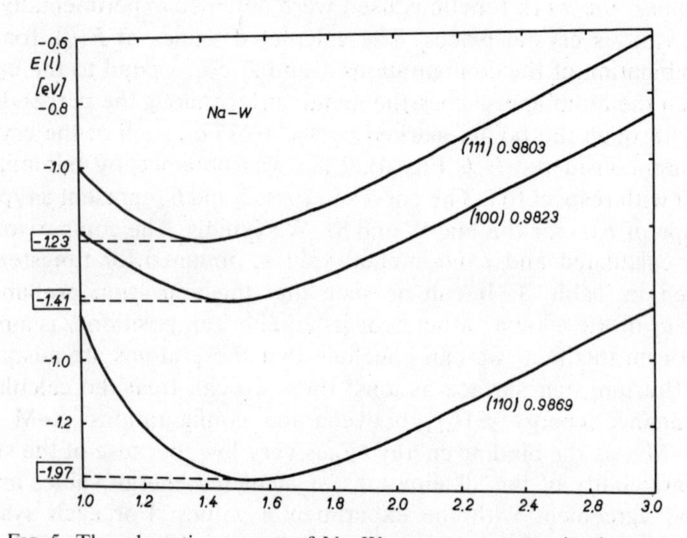


FIG. 5. The adsorption energy of Na–W system vs. separation between the atom and the metal, calculated for some single planes of tungsten in ref. 24. The numbers in brackets indicate the Miller indices and the other the ionization probabilities.

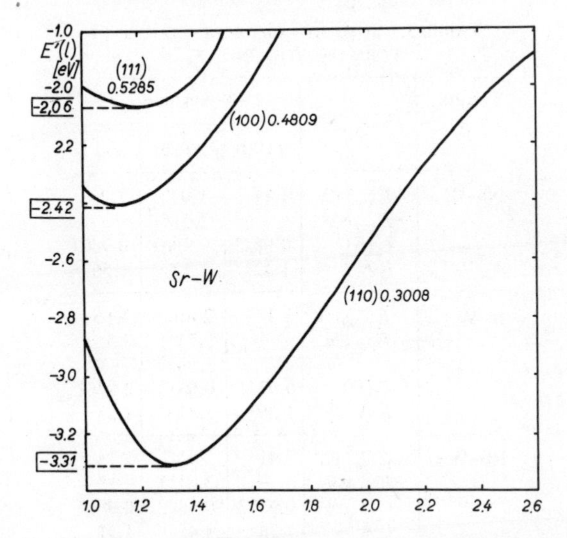


FIG. 6. The adsorption energy of Sr–W system vs. separation between the atom and the metal, calculated for some single planes of tungsten in ref. 24.

It was shown[52] that, in the case of the adsorption of $K$ on a single crystal, deposition of the adsorbate on the crystal planes occurs, in the first stages, on the planes of relatively high Miller indices as, for example, (124) and next on (112), (110) and (100) in the sequence. A higher value of the adsorption energy on the crystal plane (110) than on the plane (100) was obtained also in ref. 53: $E_{(110)} = 3.2$ eV, $E_{(100)} = 2.8$ eV. Thus, it can be assumed that the theoretical results obtained are in good agreement with the experimental data.

In the case of Sr or Ba adsorption, for all the crystal planes considered, relatively small ionization probabilities were obtained. Due to this, and to the fact that the resonance energy $V = \int \psi_1 (H-E)\psi_2 d\tau$ is considerable (see Fig. 7), the conclusion can be drawn that the adsorption of Sr and Ba on W has only partially ionic character. Since the polarizability of the free ion is several orders of magnitude smaller than the polarizability of the free atom of Sr or Ba, we can conclude that the adatoms have a greater polarizability when a larger negative charge is near them. In ref. 31, the experimental value of the polarizability of Ba adsorbed on W

TABLE 3.* SUMMARY OF OBTAINED RESULTS FOR TUNGSTEN IN REF. 24

| System | | Surface | | |
|---|---|---|---|---|
| | | (100) | (110) | (111) |
| Na–W | $-E(l_0)$ eV | 1.41 | 1.97 | 1.23 |
| | $-E_{exp}$ eV | — | 2.51*1 | — |
| | $C(l_0)$ | 0.9823 | 0.9869 | 0.9803 |
| | $l_0$ Å | 1.42 | 1.52 | 1.56 |
| K–W | $-E(l_0)$ eV | 2.34 | 2.86 | 2.16 |
| | $-E_{exp}$ eV | — | 2.75*1<br>2.90*2 | 2.5*2 |
| | $C(l_0)$ | 0.9771 | 0.9813 | 0.9762 |
| | $l_0$ Å | 1.26 | 1.42 | 1.23 |
| Rb–W | $-E(l_0)$ eV | 2.41 | 2.95 | 2.24 |
| | $-E_{exp}$ eV | — | 2,61*3 | — |
| | $C(l_0)$ | 0.9877 | 0.9892 | 0.9865 |
| | $l_0$ Å | 1.33 | 1.45 | 1.31 |
| Cs-W | $-E(l_0)$ eV | 2.67 | 3.23 | 2.51 |
| | $-E_{exp}$ eV | — | 3.08*1 | — |
| | $C(l_0)$ | 0.9914 | 0.9924 | 0.9910 |
| | $l_0$ Å | 1.31 | 1.41 | 1.31 |
| Sr–W | $-E(l_0)$ eV | 2.42 | 3.31 | 2.06 |
| | $-E_{exp}$ eV | | 3.26*4 | 4.21*5 |
| | $C(l_0)$ | 0.4809 | 0.3008 | 0.5285 |
| | $l_0$ Å | 1.13 | 1.32 | 1.22 |
| Ba–W | $-E(l_0)$ eV | 3.23 | 4.17 | 2.89 |
| | $-E_{exp}$ eV | 3.48*4 | 4.68*5 | 3.77*7 |
| | $C(l_0)$ | 0.3801 | 0.2311 | 0.4192 |
| | $l_0$ Å | 1.20 | 1.14 | 1.35 |

*Mean values obtained in (1) ref. 40; (2) ref. 32; (3) ref. 46; (4) ref. 49; (5) ref. 50; (6) ref. 31; (7) ref. 51.

is quoted as 7.22 Å; on assuming this value and taking $l_0 = 1.20$ Å we can calculate, with help from Table 1, that the corresponding dispersional energy is $-0.448$ eV. Therefore, we can conclude that the dispersional energy increases the adsorption energy by about

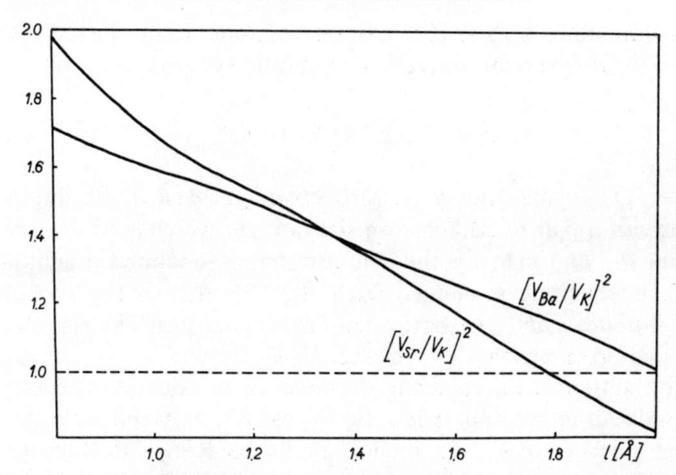


FIG. 7. The plots vs. M–A distance, of the squared ratio of resonance energies of Sr–W and Ba–W to the resonance energy of K–W, which has the greatest value of the energies of the systems: alkaline adsorbate–tungsten, after ref. 24.

0.5 eV and at the same time decreases the difference between the calculated and measured values.

Recently, Newns[55] published a paper devoted to the calculation of the adsorption energy of hydrogen on Ti, Cr, Ni and Cu on the basis of a self-consistent model. To evaluate the adsorption energy, he adopted the Anderson Hamiltonian, which was first used for describing magnetic impurities in alloys.[56] The author obtained the adsorption energies of the same order as the experimental ones and showed that, in agreement with experiment, $|E_{Ni}| > |E_{Cu}|$. He also found a correlation between the observations of catalytic ortho–para hydrogen interconversion in Pd–Au alloys and the rigid-band calculation of $E$. In the paper, the calculation of the charge on the adatom for adsorption on the above metals is also quoted; this is reviewed in the next paragraph.

So far we have dealt with the initial adsorption energy, but as experiment shows (see, for instance, ref. 32), the adsorption energy depends strongly on surface coverage $\theta$. The explanation of the initial dependence $E = E(\theta)$ was given by Grimley and Walker.[57] The authors considered a uniform surface with $M$ adsorption sites and $N$ adatoms as a sum of pair interactions. They started with the following formula for differential adsorption energy

$$Q = Q_\infty - \tfrac{1}{2} \sum_s{}' g(\boldsymbol{R}_{st})\phi(\boldsymbol{R}_{st})$$

where $Q_\infty$ is the energy at zero coverage, $\phi(\boldsymbol{R}_{st})$ the interaction energy of a pair of adatoms on sites $s$ and $t$ which are joined by the vector $\boldsymbol{R}_{st}$, and $g(\boldsymbol{R}_{st})$ is the pair correlation function for adatoms at surface coverage $\theta$ defined such that $\theta^2 g(\boldsymbol{R}_{st})$ is the probability that the sites $s$ and $t$ are both occupied by adatoms. The prime on the summation means that $s = t$ is excluded.

The authors examined the dependence of adsorption energy on the indirect interaction (see refs. 64 and 65) between adatoms on a metal surface, using the model similar to Kim and Nagaoka's[58] generalization of the Anderson[56] model of a dilute alloy. Their final result is

$$Q = Q_\infty - K\theta + O(\theta^2)$$

where $K$ is a function of the indirect interaction energy of a pair of adatoms, and is equal to 0.171 eV for immobile adsorption, and −1.84 eV for mobile adsorption. According to this result, the initial slope of the $Q$ versus $\theta$ curve is small and negative for immobile adsorption, but large and positive for mobile adsorption, so that a rising adsorption energy with surface coverage is predicted in the latter case. As was shown by the authors, the indirect interaction energy can determine the initial slope of $E = E(\theta)$ only for systems where the dipole moment of the surface bond is rather small. The authors themselves write; "This is certainly not the case for the alkali metal atoms on tungsten, but it may be realized for transition metal adatoms, where the Coulomb repulsion energy of a pair of electrons in the same atomic orbital is large. The repulsion suppresses the transfer of electrons either to, or from, the adatom and so prevents the occurrence of a large surface dipole moment. We do not know of any experimental determinations of, either the dipole moment of the surface bond, or the initial slope of the $E$ versus $\theta$ curve for a transition metal adatom on tungsten."

### 7. Charge Density near the Adatoms

The theoretical determination of the electronic density near the adsorbed atom or molecule is as important as the calculation of the binding energy. This problem, because of its difficulty, has only been partially resolved in some works dealing with alkaline atom adsorption. Bennett and Falicov[59] gave one of the solutions. Neglecting the possible influence of the surface states and assuming that the wave function $\psi$ of the atom–metal system can be represented as a product of the one-electron functions, they calculated the charge $q$ in the volume $V$ around the adatom. Assuming that the effective charge density per unit energy can be represented by a Lorentzian distribution

$$\rho(E) = \frac{1}{\pi}\frac{\Delta}{(E-E_{\varphi})^2+\Delta^2} \tag{34}$$

they obtained

$$q = |\epsilon|(1-q^-)$$

where

$$q^- = \sum_{E,l,\nu}\int_V |\psi(E,l,\nu,\mathbf{r})|^2 d^3\mathbf{r} = \frac{1}{\pi}\left[\tan^{-1}\frac{E_F-E_{\varphi}}{\Delta}+\frac{\pi}{2}\right]. \tag{34a}$$

In this formula $E_F$ is the Fermi energy, $E_{\varphi}$ and $\Delta$ are quantities calculated on the basis of the following simplifying assumptions: the Hamiltonian $H$ describing the system in the presence of the external electric field $F$ is

$$H = T+V_M+V_A+V_F+V_{ee} \tag{35}$$

where $T$ is the kinetic energy operator, $V_M$ is the potential of the metal ions, $V_A$ is the potential of the adsorbed ions, $V_F$ is the potential due to an external field and $V_{ee}$ is the self-consistent potential acting on the electron due to the average charge distribution of all the other electrons. Since accurate solution of the Hartree equation is impossible, the authors simplified (35) to the form

$$H = T+UP_L+(V'_A+V_F+V_{\text{im}})P_R \tag{36}$$

where $U$ is a constant potential inside the metal, $V'_A$ is the adion potential, modified by the effective charge $q^-$ of the neighbour electrons, and $P_L$, $P_R$ are the following functions

$$P_L = 1, \quad P_R = 0, \quad x < 0, \quad \text{metal}$$
$$P_L = 0, \quad P_R = 1, \quad x > 0, \quad \text{vacuum.}$$

The assumptions which allow us to modify the Hamiltonian (35) into the form (36) can be summarized as follows: (1) the metal electrons screen the external field so it cannot penetrate into the metal, (2) the potential inside the metal is assumed constant, (3) the double layer on the metal surface has a negligible thickness, (4) $V_{ee}$ can be approximately replaced by the image potential $V_{\text{im}}$. Next, the authors calculated $E_\varphi$ and $\Delta$ by means of the method given in refs. 56 and 60

$$E_\varphi = \langle \varphi | H | \varphi \rangle \tag{37}$$

$$\Delta = \pi \int |\langle \chi | H | \varphi \rangle|^2 N_0(\mu, E)\, d\mu, \tag{38}$$

where $\varphi$ are the atomic $s$-like wave functions, $\chi$ are the wave functions of the metal, $\mu$ is the radial linear momentum and $N_0$ is the metallic states density for a single spin.

It was also assumed that: (1) the considerations are limited to such distances $l$ of the atoms from the metal surface that less than 10% of the charge related to $|\varphi|^2$ is inside of the metal, (2) $V_F + V_{\text{im}}$ is calculated for a fixed distance $l$; (3) the functions $\varphi$ have the Slater form

$$\varphi(r) = \left[ \frac{(2\gamma)^{2n^*+1}}{4\pi\Gamma(2n^*+1)} \right]^{1/2} r^{n^*-1} e^{-\gamma r} \tag{39}$$

where $\gamma$ is the coefficient of screening and $n^*$ is the effective quantum number, (4) to account for the change of $E_\varphi$, caused by the presence of the effective charge $q^-$, the coefficient of screening is replaced by $\gamma^* = \gamma - \alpha q^-$, where the polarizability $\alpha$ is constant, (5) $V_{\text{im}}$ is partially screened by $q^-$, thus

$$V_{\text{im}} = \frac{(1 - \beta q^-)\epsilon^2}{4l}, \quad \text{where } \beta \sim 1.$$

Finally, the shifting of the atom level $E_\varphi$ (see Fig. 8) is given by $F$, $l$ and $q^-$

$$E_\varphi = -R\left(\frac{\gamma^*}{n^*}\right)^2 - |\epsilon|Fl + \frac{(1-\beta q^-)\epsilon^2}{4l} \tag{40}$$

where $R$ is the Rydberg constant.

In the calculations, which were carried out by means of a computer, the authors also assumed that the zeroth order Bessel function $J_0$ appearing in matrix element $\langle\chi|H|\varphi\rangle$ can be replaced by one. This is a rough assumption and changes remarkably the final value of the matrix element and, therefore, also the value of $\Delta$. The $\Delta$ computed with $J \neq 1$ will be smaller than the value computed under assumption that $J_0 = 1$. Thus, as follows from Fig. 9, the final value of $q^-(\Delta)$ will be smaller than that computed in the absence of the external electric field. The results obtained are given in Table 4[(61)] and Figs. 10 and 11, where the computed points are marked and joined by dotted lines to reveal the dependence of $q^-$ on the metal–atom distance and the external field.

It can be seen in Table 4 and Fig. 9, that during the adsorption in the absence of the external field ($F = 0$), the alkaline atoms are already ionized at considerable distances ($l = 8$ Å) from the metal surface. On the other hand, at smaller distances the degree of ionization becomes less than 1 and reaches the value about 0.8 when $l =$ 2.8 Å. In spite of the simplifying assumptions, these results lead to the conclusion that the alkaline atom becomes the adion. The authors

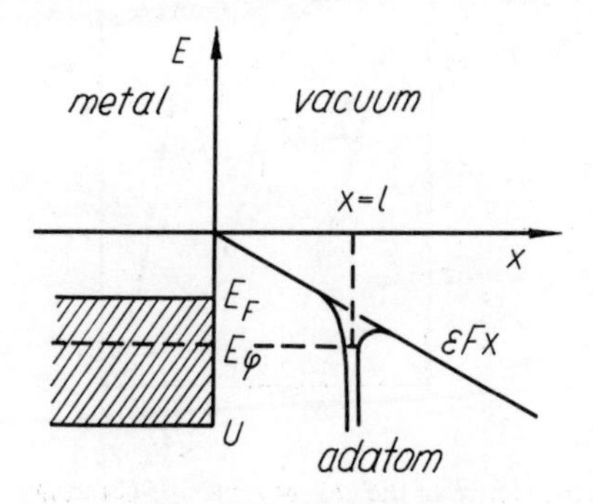


FIG. 8. The potential model adopted in ref. 59.

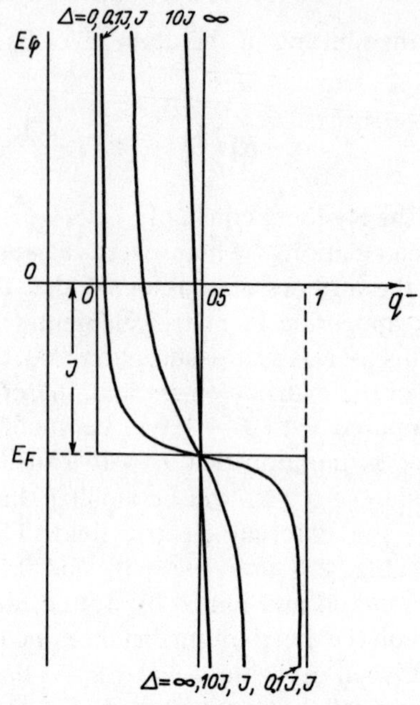


FIG. 9. The effective charge $q^-$ as a function of the position of $E_\varphi$ with respect to the Fermi level $E_F$.

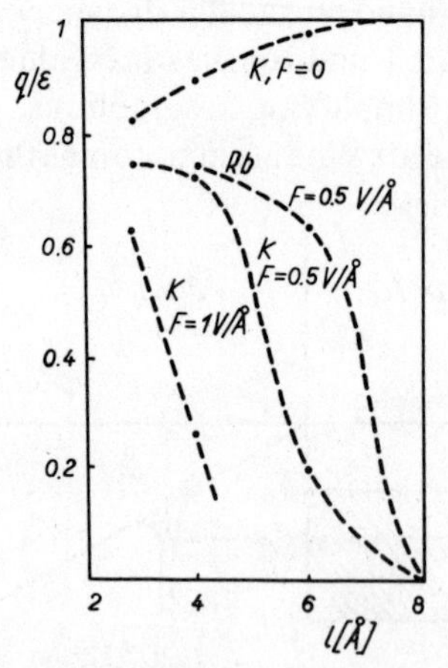


FIG. 10. The dependence of the charge on K and Cs adsorbed on tungsten, at M–A distance calculated in ref. 59 at various external fields.

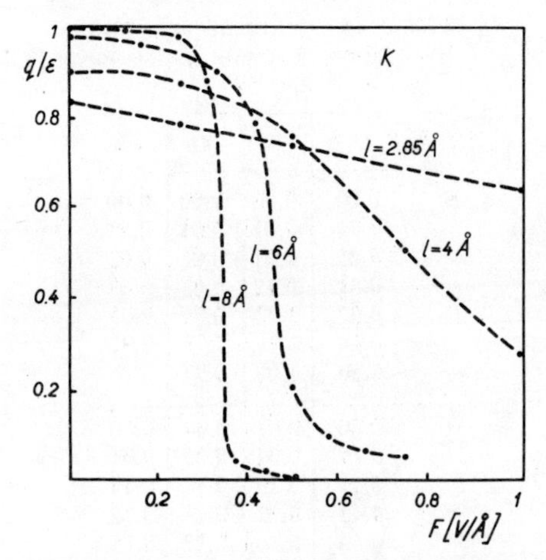


FIG. 11. The dependence of the adatom charges on the external field calculated in ref. 59 at different A–M distances.

make the reservation that more accurate comparison of their theory with the experiment is not possible, because of the lack of satisfactory data. This matter will be discussed later.

Bennett[62] has generalized the above paper to include the low-coverage limit and two types of adsorbate–adsorbate interactions. As a numerical example, he calculated the effective charge on potassium adsorbates 2.85 Å from a (100) tungsten surface. The adsorbate–metal separation (assumed to be independent of $\theta$) 2.85 Å is one postulated in ref. 32. The tungsten metal is represented by a Sommerfeld well of depth 12.0 eV, which corresponds to one free $s$-electron (see ref. 54). The adsorption sites are taken to be in the centre of the squares formed by four tungsten atoms (Fig. 12), and it was assumed that the ions of the dimer occupy nearest-neighbor sites and that the direct overlap interaction is not important between next-nearest-neighbour sites. The author considers two cases. The first is when the ions are always adsorbed in pairs; therefore, only dimer species are considered. In the second case, the individual ions are regarded as randomly occupying sites on the surface. The numerical results (based on the assumption $J_0 = 1$) concerning values of the electronic charge per ion at various surface coverages are given in Table 5.

TABLE 4. VALUES OF THE ELECTRONIC CHARGE $q^-$ AFTER REF. 59

| $l$, Å | $F$, V/Å | K | Rb | Cs |
|---|---|---|---|---|
| 8 | 0.00 | 0.00 | 0.00 | 0.00 |
| | 0.12 | 0.01 | 0.01 | 0.01 |
| | 0.25 | 0.02 | 0.02 | 0.02 |
| | 0.31 | 0.12 | 0.03 | 0.05 |
| | 0.37 | 0.97 | 0.75 | 0.25 |
| | 0.44 | 0.98 | 0.98 | 0.92 |
| | 0.50 | 1.00 | 0.99 | 0.96 |
| 6 | 0.00 | 0.02 | 0.04 | 0.06 |
| | 0.16 | 0.04 | 0.05 | 0.07 |
| | 0.33 | 0.10 | 0.10 | 0.14 |
| | 0.42 | 0.22 | 0.17 | 0.22 |
| | 0.50 | 0.80 | 0.37 | 0.35 |
| | 0.58 | 0.91 | 0.77 | 0.65 |
| | 0.66 | 0.94 | 0.86 | 0.77 |
| | 0.75 | 0.96 | 0.91 | 0.85 |
| 4 | 0.00 | 0.10 | 0.12 | 0.14 |
| | 0.25 | 0.13 | 0.17 | 0.18 |
| | 0.50 | 0.27 | 0.26 | 0.26 |
| | 0.75 | 0.50 | 0.44 | 0.38 |
| | 1.00 | 0.74 | 0.67 | 0.57 |
| | 1.25 | 0.85 | 0.78 | 0.70 |
| 2.85 | 0.00 | 0.17 | | |
| | 0.25 | 0.22 | | |
| | 0.50 | 0.25 | | |
| | 1.00 | 0.37 | | |
| | 1.50 | 0.62 | | |
| | 2.00 | 0.77 | | |

The work of Gadzuk[63] confirms the presumptions that the alkaline atoms are adsorbed as ions, but it is based on very simplified assumptions. We will describe the basic one of these assumptions concerning the Hamiltonian used in the work. The author assumes that the interaction between the alkaline atom and the metal can be approximated by the interaction between the adion and the valence

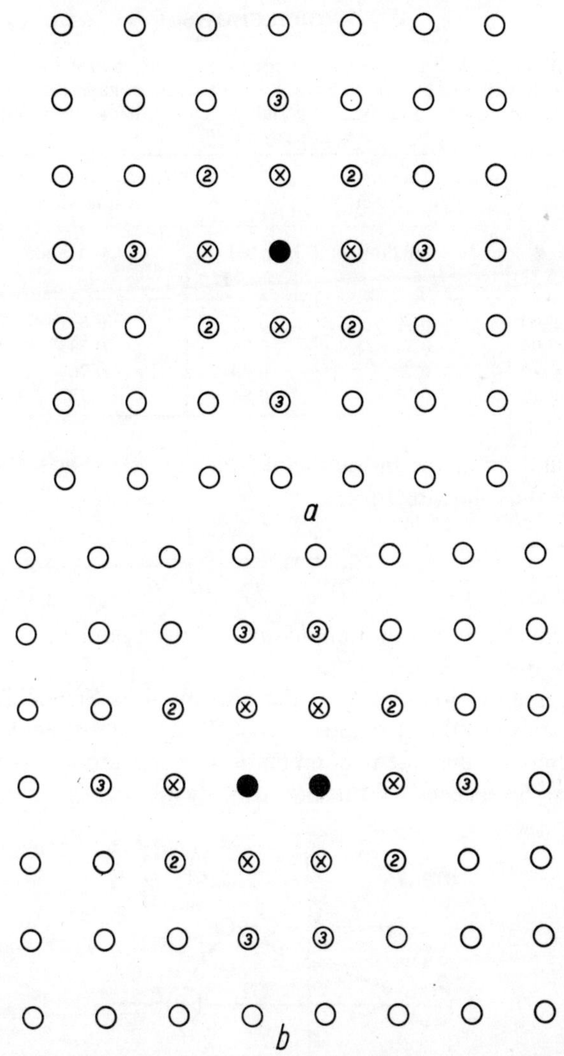

FIG. 12. (a) The lattice sites available for adsorption in the vicinity of an isolated adsorbed ion (solid circle) on a (100) surface. (b) The lattice sites available for adsorption in the vicinity of an isolated ion dimer (solid circles) on a (100) surface. The crossed circles represent empty sites, the circles enclosing the number 2 (3) are sites which can be occupied by a member of two (three) distinct dimers, and the empty circles are sites which can be occupied by a member of four distinct dimers.

TABLE 5. VALUES OF THE ELECTRONIC CHARGE PER ION (IN UNITS OF $e$) AT VARIOUS $\theta$ FOR RANDOMLY ADSORBED SINGLE K IONS AND RANDOMLY ADSORBED K ION DIMERS ON (100) W SUBSTRATE[62]

| | Random adsorption of ion dimers | Random adsorption of single ions | |
|---|---|---|---|
| $\theta$ | Isolated dimer $q_D^-$ | Isolated ion $2q_A^-$ | Isolated dimer $q_D^-$ |
| 0.00 | 0.70 | 0.65 | 0.69 |
| 0.05 | 0.74 | 0.73 | 0.73 |
| 0.10 | 0.76 | 0.79 | 0.76 |
| 0.15 | 0.78 | 0.84 | 0.77 |

electron and its image in the metal (see Fig. 13). Thus, the Hamiltonian has the following form

$$H = -\frac{\hbar^2}{2m}\nabla^2 - \frac{\epsilon^2}{r} - \frac{\epsilon^2}{4d} + \frac{\epsilon^2}{R}\cdot$$

It is evident that the author could not obtain any quantitative results with such simplifications.

The degrees of ionicity $C$ of the alkaline ions Sr and Ba on the crystal planes (100), (110) and (111) of W have been calculated.[23, 24] The method of the linear combination of the atomic orbitals described in § 6 was used for this purpose. Results are given in Table 3.

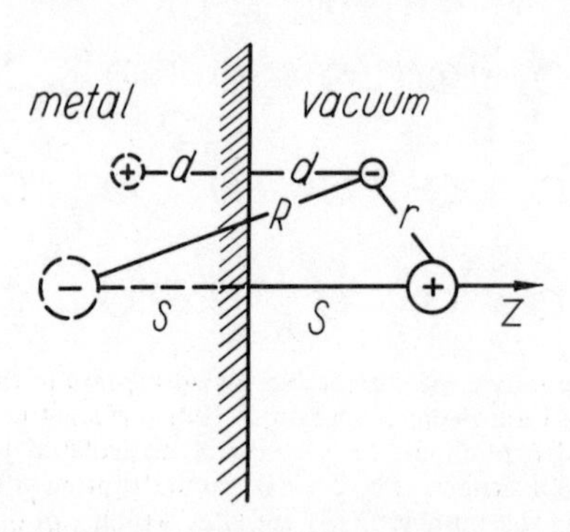


FIG. 13. The model for the interaction between the electropositive atom and metal surface adopted in ref. 63.

For the alkaline atoms, the sequence of increasing values of the degree of ionization and the dependence on the crystallographic direction is

$$C_{(110)} > C_{(100)} > C_{(111)}.$$

For Sr and Ba this sequence is

$$C_{(111)} > C_{(100)} > C_{(110)}.$$

These results are in a good agreement with experiment.[32, 44, 45]

Since the one-dimensional model is often employed in the literature concerning surface phenomena, such a model was used to calculate the probability of the ionization of the atom visiting the metal surface[24] (see Fig. 14).

As a result of the tunnelling through the rectangular barrier between the metal and the adatom, the valence electron goes into the metal. The calculated probability of such a transition is

$$P(l) = nT(l)$$

where $n$ is the percentage of the unoccupied states in the metal and $T(l)$ is the transmission factor of the barrier. In Table 6 are the results obtained for a K–W system. The sequence of the values of $P(l)$ for other alkaline atoms Sr and Ba on W is the same as in Table 6. It can be seen that the one-dimensional model gives higher values of $P$ for the (100) crystal plane than for (110) plane, which

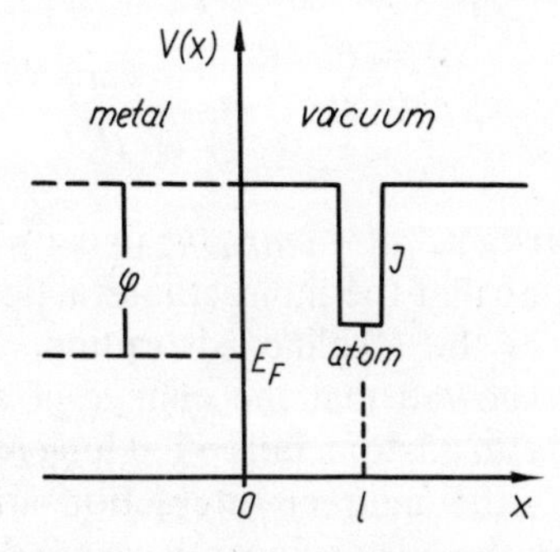


FIG. 14. Schematic diagram of the one-dimensional square potential barrier between A and M.

is not in agreement with experiment. Hence, it may be concluded that the application of the one-dimensional models to surface problems should be used with caution.

TABLE 6. VALUES OF $P(l)$ FOR THE K–W SYSTEM

| Temperature | 300°K | | 1000°K | |
|---|---|---|---|---|
| Surface | (100) | (110) | (100) | (110) |
| $l$, Å 1.96 | $5.10 \times 10^{-2}$ | $4.88 \times 10^{-2}$ | $5.01 \times 10^{-2}$ | $4.82 \times 10^{-2}$ |
| 2.16 | 3.35 | 3.20 | 3.29 | 3.16 |
| 2.36 | 2.19 | 2.10 | 2.16 | 2.07 |
| 2.56 | 1.43 | 1.37 | 1.41 | 1.35 |
| 2.76 | 0.94 | 0.89 | 0.92 | 0.88 |

The adsorbing of an atom or molecule onto a metal causes a disturbance of the electron density in the metal, which can be so big that it will bc perceptible at great distances from the adsorbate. Thus, it can be concluded that the indirect interaction between adatoms on a metal surface arises, even if they were at such a large distance apart that direct interaction between them is negligible. Grimley[(64)] considered this curious problem in the case of the metal–two adsorbed atoms system. The calculation[(65)] was carried out by means of a Green function and established that the energy of this indirect interaction between the two atoms is an oscillating function of $R$, the distance between the adatoms, i.e.

$$\Delta W(R) = C\frac{\sin(2p_F R)}{R^3}$$

where $C$ is a constant, $p_F = 2mE_F/\hbar^2$ and $E_F$ is the Fermi energy. Grimley also states that the indirect interaction may be very essential in the case of the alkaline adsorption. Using a very similar technique,[(66)] he showed that the change of the electronic density $\Delta N(R)$, caused by an adatom, falls off at long range like $R^{-3}$.

In both cases – the indirect interaction and the change of the density caused by the adsorption – there are directions along which both $\Delta W(R)$ and $\Delta N(R)$ decrease with $R$ more rapidly than indicated by the above formulae. The author finds that, up to now, there is no experimental confirmation of such effects. Actually, experimental information about the electronic density near an adatom, and its change with the coverage increase, is very scant and in some cases questionable. Such as, for instance, the diagrams in the works,[32 44,45] showing the dependence of the dipole moment of the alkaline adatoms on the adsorbate density; to obtain them it was assumed that the adatom polarizabilities $\alpha$ are not dependent on $\theta$. Taking into account that the alkaline ion polarizabilities are several orders of magnitude smaller than the atom polarizabilities, and that the adsorption energy decreases considerably with increasing coverage, the above assumption must be treated as a rough approximation.

Newns[55] predicted that the total charge on the hydrogen atom decreases from $q = 1.4\,e$ on Ti to $q = 1.2\,e$ on Ni. The author states: "These values of $q$ are greater than unity, as is indeed observed, but are too large to be in quantitative agreement with the observed surface potentials, unless partial burying of the adsorbate in concave surface sites is assumed. This is in general plausible, though it is questionable whether closely packed planes such as (110) of the bcc and (111) of the fcc structures can offer such sites.... More detailed comparison of the present theory with experiment would require measurements of $E$ and charge distribution for adsorption on single crystal surfaces."

## 8. Final Remarks

Many problems in the field of absorption theory on metals have been resolved in an approximate way, as was shown in this review. Although we cannot view this knowledge too optimistically, it cannot be treated too sceptically either. From this paradoxical conclusion comes the analysis of the adsorption process, which is a very complex and complicated physical phenomenon. It is well known that adsorption is a many-electron phenomenon, but a theoretical description in the field of many-electron theory is not yet possible, although some steps have been taken for that purpose.[55,57,59,62–64,66–68] To improve the approach to the theoretical problems related to adsorption, one needs to obtain more accurate information about the electronic structure of the metal surface and its influence on the bulk characteristics of the metal, for example, on the Fermi energy and shape of the Fermi surface. Unfortunately, these problems still await solution, because of their difficulty. Actually, we do not even know the positions of the metal surface atoms; such investigations were started a short time ago. The problem is additionally complicated when one wants to consider the adsorption of the many-electron atoms, but in spite of the lack of quantum calculations related to this problem, it appears that one can at least obtain qualitative information by means of the one-electron methods. Hence, the one-electron treatments referred to here must not be treated too sceptically, but as a first approximation in acquiring knowledge about the adsorption process on metals. In fact, one can hazard a statement that such treatments have already given much information; in particular, it seems certain that the alkaline atoms are adsorbed as ions.

The author wishes to thank Professor J. Nikliborc for reading the manuscript and for his critical remarks.

## REFERENCES

1. The adsorption energy depends strongly on the number of atoms adsorbed on a unit area, thus $-E(l_0)$ will be understood as the so-called initial adsorption energy (heat), which is a measure of the binding energy between adatom and a clean surface.
2. S. Brunauer, *The Adsorption of Gases and Vapours*, vol. I. *Physical Adsorption*, Clarendon Press, Oxford and Princeton University Press, Princeton, 1945.
3. F. F. Wolkenstein, *Electronic Theory of Catalysis on Semiconductors*, Pergamon Press, Oxford, 1963.
4. D. M. Young and A. D. Crowell, *Physical Adsorption of Gases*, Butterworths, London, 1962.
5. B. M. W. Trapnell, *Chemisorption* (edited by W. E. Garner), Butterworths, London, 1957.
6. R. V. Culver and F. C. Tompkins, *Adv. Catalysis* **11**, 67 (1957).
7. P. M. Gundry and F. C. Tompkins, *Quarterly Rev. London* **1**, 257 (1960).
8. M. Boudart, Adsorption and chemisorption, in: *Surface Chemistry of Metals and Semiconductors*, John Wiley and Sons Inc., New York and London, 1960.

9. D. D. Eley, *Disc. Faraday Soc.* **8**, 34 (1950); *Trans. Faraday Soc.* **49**, 643 (1953).
10. L. Pauling, *Nature of Chemical Bond*, Cornell Univ. Press, Ithaca, New York, 1948.
11. J. Higuchi, T. Ree and H. Eyring, *J. Amer. Chem. Soc.* **79**, 1330 (1957).
12. R. S. Mulliken, *J. Amer. Chem. Soc.* **74**, 811 (1953).
13. W. G. Pollard, *Phys. Rev.* **56**, 324 (1939).
14. T. B. Grimley, *Chemisorption* (edited by W. E. Garner), Butterworths, London, 1957, p. 17.
15. J. Koutecky, *Trans. Faraday Soc.* **54**, 1038 (1958).
16. J. E. Djaloshynsky, E. M. Lifshyc and L. P. Pytaevsky, *Uspekhi Fiz. Nauk* (in Russ.) **73**, 381 (1961).
17. J. Bardeen, *Phys. Rev.* **58**, 727 (1940).
18. H. Margenau and W. G. Pollard, *Phys. Rev.* **60**, 128 (1941).
19. E. J. R. Prosen and R. G. Sachs, *Phys. Rev.* **61**, 65 (1942).
20. V. L. Bonch-Bruevich, *Uspekhi Fiz. Nauk* (in Russ.) **56**, 55 (1955).
21. C. Kemball, *Proc. Roy. Soc.* A **187**, 73 (1946).
22. S. Raimes, *The Wave Mechanics of Electrons in Metals*, North Holland, Amsterdam, 1949.
23. K. F. Wojciechowski, *Acta Phys. Polon.* **29**, 119 (1966).
24. K. F. Wojciechowski, *Acta Phys. Polon.* **33**, 363 (1968).
25. A. Dalgarno, *Adv. Phys.* **11**, 281 (1962).
26. D. Pines, *Phys. Rev.* **92**, 626 (1953); *The Many-Body Problem*, Benjamin, New York, 1962.
27. K. F. Wojciechowski, *Surface Science* **17**, 490 (1969).
28. W. G. Pollard, *Phys. Rev.* **60**, 578 (1941).
29. In Pollard's paper the coefficient 2 before $I_3$ was mistakenly omitted (see ref. 30).
30. T. Toya, *J. Res. Inst. Catalysis*, Hokkaido Univ. **6**, 308 (1958).
31. J. Zingerman, V. A. Ischuk and V. A. Morozovsky, *Solid State Phys. USSR* **2**, 2276 (1960).
32. L. D. Schmidt and R. Gomer, *J. Chem. Phys.* **42**, 3573 (1965); **45**, 1605 (1966).
33. J. E. Lennard-Jones, *Trans. Faraday Soc.* **28**, 334 (1932).
34. R. G. Sachs and D. L. Dexter, *J. Appl. Phys.* **21**, 1304 (1950).
35. R. Parsons, *Modern Aspects of Electrochemistry* (edited by J. O. M. Bockirs), Butterworths, London, 1954, § 3.
36. P. H. Cutler and J. C. Davis, *Surface Science* **1**, 194 (1964).
37. M. Kaminsky, *Atomic and Ionic Phenomena on Metal Surfaces*, Springer-Verlag, Berlin, 1965.
38. H. Eyring, J. Walter and G. E. Kimball, *Quantum Chemistry*, Wiley, New York and London, 1960.
39. R. C. L. Bosworth, *Proc. Roy. Soc.* A **162**, 32 (1937).
40. W. Kühl – after G. Popp, *Ann. Phys.* **13**, 115 (1964).
41. S. V. Starodubtsev, *J. Exp. Theor. Phys.* (*USSR*) **19**, 215 (1949).
42. R. C. Evans, *Proc. Roy. Soc.* A **139**, 604 (1933).
43. F. Knauer, *Z. Phys.* **125**, 278 (1949).
44. L. Schmidt and R. Gomer, *J. Chem. Phys.* **42**, 3573 (1965).
45. Z. Sidorski, I. Pelly and R. Gomer, *J. Chem. Phys.* **50**, 2382 (1969).
46. F. L. Hughes, H. Levinstein and R. Kaplan, *Phys. Rev.* **113**, 1023 (1959); F. L. Hughes and H. Levinstein, *Phys. Rev.* **113**, 1029 (1959).

47. J. B. TAYLOR and I. LANGMUIR, *Phys. Rev.* **44**, 423 (1933).
48. M. D. SCHEER and J. FINE, *J. Chem. Phys.* **37**, 107 (1962).
49. G. E. MOORE and H. W. ALLISON, *J. Chem. Phys.* **23**, 109 (1955).
50. T. E. MADEY, A. A. PETRAUSKAS and E. A. COOMES, *J. Chem. Phys.* **42**, 479 (1965).
51. H. UTSUGI and R. GOMER, *J. Chem. Phys.* **37**, 1706 (1962).
52. R. BLASZCZYSZYN, M. MASLANKÓWNA, R. MECLEWSKI and J. NIKLIBORC, *Acta Phys. Polon.* **29**, 403 (1966).
53. V. M. GAVRILUK, J. S. VEDULA, A. G. NAUMOVETS and A. G. FEDORUS, *Solid State Phys. (USSR)* **9**, 1127 (1967).
54. It is to be noted that, in these works, the two *s*-electrons per W-atom were taken as free, giving a Fermi energy $E_F = 9.173$ eV. Measurements of B. ROZENFELD, *Acta Phys. Polon.* **31**, 197 (1967), gave 2–2.7 *s*-electrons per W-atom, which agrees with Ziman's suggestion (J. M. ZIMAN, *Electrons and Phonons*, Oxford, Clarendon Press, 1960, pp. 168–70). In the earlier papers it was assumed, by virtue of Manning and Chodorow's result, that there is only one free electron per atom of tungsten (M. F. MANNING and M. I. CHODOROW, *Phys. Rev.* **56**, 787 (1939)).
55. D. M. NEWNS, *Phys. Rev.* **178**, 1123 (1969).
56. P. W. ANDERSON, *Phys. Rev.* **124**, 41 (1961).
57. T. B. GRIMLEY and S. M. WALKER, *Surface Science* **14**, 395 (1969).
58. D. J. KIM and T. NAGAOKA, *Prog. Theor. Phys.* **30**, 743 (1963).
59. A. J. BENNETT and L. M. FALICOV, *Phys. Rev.* **151**, 512 (1966).
60. V. FANO, *Phys. Rev.* **124**, 1866 (1961).
61. As was pointed by Bennett,[62] some numerical mistakes are present in Table 4. They increase the value of $q^-$ given, but do not change the general dependence of the effective charge on the type of adsorbate, external field, and adsorbate–metal separation.
62. A. J. BENNETT, *J. Chem. Phys.* **49**, 1340 (1968).
63. J. W. GADZUK, *Surface Science* **6**, 133 (1967); **6**, 159 (1967).
64. T. B. GRIMLEY, *Proc. Phys. Soc.* **90**, 751 (1967).
65. In the original paper,[64] this function is mistakenly given as falling off at long range like $R^{-2}$ (see ref. 57).
66. T. B. GRIMLEY, *Proc. Phys. Soc.* **92**, 776 (1967).
67. V. L. BONCH-BRUEVICH and V. B. HLASKO, *Vestnik MGU (USSR)* **5**, 91 (1958).
68. V. L. BONCH-BRUEVICH and S. V. TIABLIKOV, *Green Function Method in Statistical Mechanics*, North Holland, Amsterdam, 1962.

## NOTE ADDED IN PROOF

The author is very indebted to Professor S. G. Davison for reading through the manuscript of the paper and for pointing out to the author the interesting review article by J. Koutecky in *Advances in Chemical Physics*, **9**, 85 (1965); this article contains a lot of information pertaining to the quantum theory of adsorption. The author is also grateful to The Royal Society for permission to reproduce Figs. 2 and 3, and to A. J. Bennett for permission to reproduce Figs. 9 and 12, and Tables 4 and 5.

# SURFACE PHENOMENA ASSOCIATED WITH THE SEMICONDUCTOR/ELECTROLYTE INTERFACE

S. Roy Morrison

*Stanford Research Institute*
*Menlo Park, California 94025*

## CONTENTS

# 1. INTRODUCTION

The use of semiconductor electrochemistry to study the exchange of mobile carriers (holes or electrons) between a solid and a surface molecule is leading to new understanding of interactions at an interface.

Of interest to the chemist are the chemical reactions occurring on a solid surface. Using semiconductor electrochemistry, the chemist has been able to elucidate, to some extent, how the electrons in a solid become involved in such catalytic reactions. The electrochemist is interested in the details of the reduction (electron capture) or oxidation (electron loss) of an ion in solution. With semiconductors, the electrochemist has been able to control the availability of electrons and holes, and thus gain a better understanding of these processes.

The surface physicist is interested in the determination of the behavior and energy of surface states† arising from the adsorption of foreign species on the semiconductor surface. In order to control the surface properties of a semiconductor, one must know how foreign adsorbates will behave, how they will interact with the semiconductor and with each other (the chemical characteristics), and how the resulting species will interact with the holes and electrons in the semiconductor. Only when such behavior is understood will surface control be possible (e.g., surface passivation, control of surface resistance and recombination velocity, control of field emission or thermionic emission, etc.).

This review will be devoted primarily to electrochemical studies as applied to the elucidation of the exchange of electrons between the energy bands and chemically active foreign molecules at the semiconductor surface. Our knowledge is minimal regarding (a) whether a given electron or hole exchange will be favorable, (b) the laws limiting the rate, and (c) the various alternating chemical and electronic steps involved. We will attempt to evaluate what we can learn about (a), (b), and (c) from semiconductor electrochemistry.

† The term "surface state" in this discussion will be used in its broadest sense: any localized electronic level at the surface capable of exchanging electrons with the conduction and valence bands.

From an experimental point of view, we hope to show that for clarifying such electron exchange reactions, studies at the liquid/semiconductor interface may be substantially simpler and more powerful than those presently available for studying the gas/semiconductor interface.

We shall restrict the discussion almost exclusively to work done on germanium and zinc oxide. There is a solid foundation of studies of these materials in an inert electrolyte, initiated for germanium by Brattain and Garrett[1] and for ZnO by Dewald.[2] The two materials have since been studied extensively with active species in the solution and the results illustrate well the types of interactions that have been observed in general, and also the techniques available to study such interactions.

Thus, we will examine reduction of (electron capture by or hole injection from) surface species and oxidation of (electron injection from or hole capture by) surface species. We will discuss cases, for example, where hole capture by a surface species is spontaneously followed by electron injection from the product, and where hole injection by one species is followed by hole capture by another. We will discuss the energetic considerations of electron capture, and examine evidence that a weak oxidizing agent tends to capture electrons from the conduction band, whereas more powerful oxidizing agents strip electrons from the valence band (injecting holes). We will examine attempts to relate the surface state energy level of the species to the standard redox potential. Finally, we will examine the generality of the behavior; whether we can expect the characteristics of a given electron/surface species reaction, which are observed in electrochemical measurements, to be similar at the semiconductor/gas interface.

## 2. BACKGROUND

The advantages of the semiconductor/liquid interface over the semiconductor/gas interface for studies of carrier/molecule interactions are threefold: (a) the flow of electrons and holes to the surface can be followed directly, (b) the measurement and control of surface potentials at the surface can be directly and easily done, and (c) the chemical reactants can be dissolved in the liquid, so that the variety of materials easily studied is much greater.

The motion of electrons and holes to the surface cannot be directly followed at the gas/solid interface, because such motion is not part of a completed circuit. However, at the liquid/solid interface when an electron moves to the surface, an ion moves through the solution completing the circuit, and we can follow capture of electrons at the surface by a simple current measurement.

At the gas/solid interface, the surface potential is very hard to measure and control. Although many measurement techniques have been developed, they are difficult to use and interpret. At the liquid/semiconductor interface, on the other hand, the surface potentials are controlled directly by the voltage applied in the external circuit. They are measured in part by this voltage, and in more detail by a reasonably simple measurement of capacity.

## A. Model of the Semiconductor/Electrolyte Interface

In discussing the actual measurement techniques and the interpretation of the measurements, it is sometimes useful to refer to a model of the semiconductor/electrolyte interface. We will use the model illustrated in Fig. 1.

On the ordinate of Fig. 1 is the energy of an electron; on the abscissa is distance. Thus, we are illustrating allowed energy levels for electrons in three adjacent phases: the semiconductor, the aqueous solution, and another electrode (a reference electrode) immersed in the solution. The symbols shown in Fig. 1, which we will use throughout the article, are as follows:

$E_F$ = Fermi energy in the semiconductor;

$E_F(Hg)$ = Fermi energy in the mercury contact to a calomel reference electrode, used to test the "solution potential";

$E_{CB}$ = conduction band energy deep in the semiconductor;

$E_{VB}$ = valence band energy deep in the semiconductor;

$E_{CS}$ = conduction band energy at the semiconductor surface;

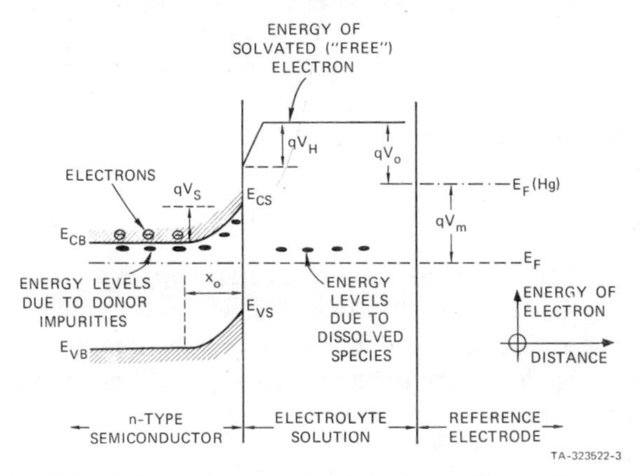


FIG. 1. Allowed electronic energy levels in three adjacent phases: *n*-type semiconductor, electrolyte, and reference electrode.

$E_{VS}$ = valence band energy at the semiconductor surface;
$q$ = electronic charge;
$qV_0$ = energy difference between an electron at the Fermi energy of the reference electrode and a solvated electron in solution;
$V_H$ = Helmholz double-layer potential difference;
$V_s$ = surface barrier potential difference;
$V_m$ = potential difference measured on the voltmeter.

In such a representation we have, of course, the normal band diagram in the region occupied by the semiconductor. Various electronic energy levels could be indicated in the electrolyte region. For illustration we indicate the energy of a solvated ("free") electron in the solution by the solid line. The concept of a solvated electron is not useful in this discussion, except to provide a link with the solid/gas diagram—the free electron in vacuum is analogous to the solvated electron in a liquid. Of course, any electron in an aqueous solution, including a solvated electron, will have a localized wave function. Finally, we indicate the position of the Fermi level in the metal (mercury) contact to a calomel reference electrode, to illustrate a potential difference $V_m$, which can be directly measured.

Energy levels due to dissolved chemical species are indicated.

In experiments of interest here, ions are added to the aqueous solution, which can exchange electrons or holes with the semiconductor. Reducing agents may be added, which tend to give up electrons, or oxidizing agents may be added, which tend to capture electrons. Thus, in the reaction

$$Fe^{2+} \rightleftharpoons Fe^{3+} + e$$

the iron may be added as $FeCl_2$, the reducing agent, and $FeCl_3$, the oxidizing agent. If both are added, we are studying a "redox couple." The reducing agent would be represented in Fig. 1 by an occupied energy level, the oxidizing agent by an unoccupied energy level. In an energy level diagram, such as Fig. 1, we are representing an electronic energy level associated with the iron molecule, *not* the iron molecule itself. Thus, it is conceivable that, under certain simplifying conditions, a single energy level could represent both forms of iron; if the level is occupied, the ion is $Fe^{2+}$; if the level is unoccupied, the ion is $Fe^{3+}$.

In the experiments to be discussed, there is always a high concentration of inert salt, such as KCl, dissolved in the solution to lower, by ion movement, the resistance through the solution, and thus maintain a constant potential through the solution. The energy levels associated with these ions do not need to be indicated, as they cannot exchange electrons with the semiconductor bands.

We choose this form (Fig. 1) of a diagram on which to base our discussion for several reasons. First is simplicity. As we will be dealing primarily with phenomenological concepts, we prefer to avoid the complexity, both in concept and notation, of the hybrid model which is often used,[(3)] where the solid phase is described in terms of electronic energy levels and where the liquid phase is described in terms of the electrochemical potential of ions. Second is consistency with the present objective. The objective here is to describe electron and hole capture or release by ions in solution, but at the semiconductor surface. Attempts have been made to describe the rate of such processes in terms of energy levels for the electron on the ion, following the lead of surface physicists in studies of electron capture by surface states. Thus, it is convenient to use electronic energy levels throughout the diagram.

The disadvantage of a representation strictly in terms of electronic energy levels is that it cannot show changes in the chemical form of a species, but only the resulting energy level. If a given species is highly complexed, by some chemical complexing agent in solution, the associated electrons will have a different energy level than they would have on the uncomplexed form. This problem is discussed by Morrison.[4] The omission is not a serious shortcoming for the present discussion of phenomena.

The various potential differences, shown in Fig. 1, are always important at the solid/liquid interface. The voltage measured is the potential difference $V_m$, where $qV_m$ is the difference between $E_F$ and $E_F(Hg)$.

The Helmholz double-layer potential difference $V_H$ arises due to charge exchange between the solid surface and the electrolyte. In the case of metals, the charge exchange is often in the form of electrons; if an excess of ions of a strong oxidizing agent is in the solution, the ions will capture electrons, building up a negative charge in the solution until the electron-capture rate is balanced by a reverse electron-injection rate. The potential difference associated with this charge will be $V_H$. In the case of semiconductors and insulators, on the other hand, the charge exchange usually results from strong adsorption of ions of one sign on the surface, with the resulting surface charge balanced by mobile ions of opposite charge in the solution. (See, for example, Parks.[5]) This adsorption occurs until a sufficient double-layer potential is built up so that the adsorption rate is balanced by the desorption rate.

There is evidence that ZnO generally behaves as an insulator, in this respect, with the value of $V_H$ determined by the pH of the solution[6] and not by redox reactions.[4,7] Germanium has the interesting characteristic (as will be discussed) that it exhibits either of two insulator-like Helmholz layers, which can be interchanged reversibly by passing anodic or cathodic currents. As with ZnO, the value of $V_H$ is pH sensitive for germanium.[3,8] In both materials of interest here, then, the Helmholz potential $V_H$ is generally determined by an insulator-like Helmholz layer.

The surface barrier potential $V_s$ in Fig. 1 is a double-layer potential arising between a charge within the semiconductor (a space charge or a free carrier charge) and a counter charge due to mobile ions in the solution. In most of the discussions to follow, a change in applied voltage $\Delta V_m$ appears entirely as a change in the surface barrier potential $\Delta V_s$, so that the surface barrier potential $V_s$ can be controlled by the experimenter. In an *n*-type semiconductor, biased as shown in Fig. 1, with the solution negative and semiconductor positive, the donor ions supply a positive space charge. Conduction electrons are repelled from the surface region by the applied voltage, leading to an insulating layer, where the positive ions are uncompensated by conduction band electrons. As the voltage on the semiconductor becomes less positive, the thickness $x_0$ of the "space-charge layer" becomes smaller and the barrier $V_s$ becomes smaller. With increasing negative voltage applied to the semiconductor, the bands become flat (no charge on the semiconductor) and then bend down ($V_s$ negative) leading to a potential well for electrons at the surface. In this "accumulation layer", there is excess negative charge, due to the electrons in the well. As will be discussed later, the relation between $V_s$ and $x_0$ can be analyzed by applying Poisson's equation.

The last parameter in Fig. 1 that needs discussion is the energy difference $qV_0$, which represents the change of energy when an electron moves from a particular energy level in the solution to the Fermi energy of the reference electrode. Reference electrodes are chosen such that $qV_0$ can be assumed independent of solution variables, such as pH, and chemical species present. When a change in measured voltage $V_m$ occurs, the change is due to effects occurring at, or near, the semiconductor electrode.

In the present discussion, mathematical analysis of electron and hole capture will not be considered in detail. Thus, the simplified diagram of Fig. 1 should provide a satisfactory basis for the discussions below. However, there are three approaches to a more detailed mathematical description of electrode processes at a semiconductor electrode, which we list for reference.

The first approach,[3] used by most authors, simply utilizes the thermodynamic model of classical electrochemistry to describe the solution side of the interface. The band model is used in describing the semiconductor, but emphasis is placed also on thermodynamic potentials within the solid.

The second approach, due to Gerischer and his coworkers,[9–11] describes the system in terms of electronic energy levels throughout, and thereby simplifies the use of concepts based on the band diagram to describe carrier transfer. He describes the system in terms of equilibrium electron transfer and suggests that a spectrum of energy levels, arising from statistical fluctuations, is associated with each ionic species in solution. Gerischer then describes the electron exchange rate near thermodynamic equilibrium in terms of these energy level spectra for the competing oxidizing and reducing agents.

A third approach, due to Morrison,[4] also describes the system throughout in terms of electronic energy levels. There are two major differences between this approach and that of Gerischer. First is the assumption by Morrison that the semiconductor/electrolyte interface is not generally at thermodynamic equilibrium with respect to charge transfer. Equilibrium occurs when the rate of any electron-capture process is balanced by the rate of the reverse electron-injection process. The rate in the two directions at equilibrium can be called the "exchange current." With semiconductors, the energy level of the species in solution can be in the band-gap region, so the exchange current can be extremely low. Flaws in the crystal surface will then permit currents to pass far in excess of the exchange current of a redox couple. Thus, equilibrium conditions can seldom be assumed, and electron exchange must be described in terms of irreversible capture or injection. A second difference is the explicit separation of the overall electron transfer process into: (a) electronic steps, which can be represented on the band diagram (such as electrons being captured by protons to form H atoms), and (b) chemical transformations, which cannot be represented (such as two H atoms forming an $H_2$ molecule). A third difference is that, in common with usual energy level representations, statistical fluctuations of the energy of allowed states are not explicitly represented, so that each form of a dissolved ion in a dilute solution is represented by one energy level only. With these differences, the mathematical analysis of Morrison becomes quite different from that of Gerischer.

Although Fig. 1 is essentially a simplification of the model proposed by Morrison, it is not inconsistent with the other approaches.

## B. Experimental Methods and Interpretation

To familiarize the reader with the techniques used, a brief discussion of the basic measurements and their interpretation will be presented.

A typical electrochemical cell is shown schematically in Fig. 2.

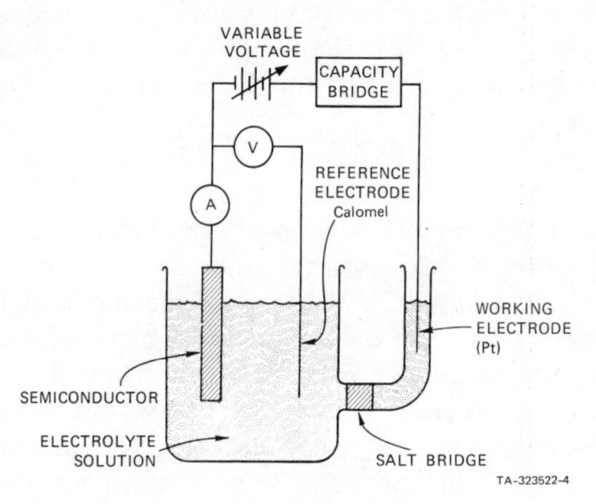


FIG. 2. Schematic of an electrochemical cell. $V$ represents a voltmeter, $A$ an ammeter.

The ammeter $A$ measures the current across the semiconductor/electrolyte interface. The current flows through the electrolyte, through a conducting "salt bridge," and through a platinum "working electrode." Due to the current flow, chemical changes occur both at the semiconductor/electrolyte interface (desired) and at the platinum/electrolyte interface (not desired). The purpose of the salt bridge is to isolate the working electrode, so that the ions formed cannot diffuse to the semiconductor electrode under study.

The voltmeter $V$ measures the potential difference between the semiconductor and a "reference electrode." Unfortunately, the reference electrode potential is not the potential in the solution (a "solution potential" can never be measured directly), but differs from it by a constant and unknown amount (see the discussion of $V_0$ above). Thus, the potential difference $V_m$ measured is always termed the semiconductor potential "with respect to a saturated calomel electrode," or "with respect to a standard hydrogen electrode," or other stable reference electrode.

The third parameter measured by the apparatus in Fig. 2 is the capacity. This measurement is designed to monitor the thickness of the insulating region at the semiconductor surface. When it can be shown that such an insulating "space-charge layer" actually dominates the capacity measurement, the data provide information about $V_s$, as discussed below.

### (*i*) *Current and Voltage Measurement*

The steady-state current equals the net rate at which electrons (or holes) cross the surface barrier to the semiconductor surface from the bulk; this rate, in turn, equals the net rate at which chemical species are reduced and oxidized at the surface. (We exclude the possibility, in these studies, that solvated electrons will exist as such in the solution.) Also these rates must equal the net rate of ion flow through the electrolyte solution to the working electrode. Thus, the simple current measurement provides considerable information about processes at steady state. Transient measurements are more difficult to interpret, due to capacity charging effects.

With $V_0$ constant, changes in the measured voltage $V_m$ reflect changes in $V_s$ and $V_H$, and these therefore can be followed as we change parameters, such as the current, the reactant ions in solution, or the light intensity, etc. From the potential changes only, we cannot separate changes in $V_s$ and $V_H$, but the capacity measurement monitors $V_s$, in general, thus permitting us to monitor the variables separately.

### (*ii*) *Capacity Measurement*

In order to show how $V_s$ is obtained from the capacity, we must examine the model in more detail. The capacity measurement measures the thickness $x_0$ of the insulating space-charge region at the semiconductor surface (Fig. 1). Such an insulating space-charge region arises when the solution (the working electrode) is biased to repel the majority carriers from the semiconductor surface. The capacity measurement is not so easily interpreted, if the surface layer is not insulating. The parallel plate capacity formula applies:

$$C = A\epsilon\epsilon_0/x_0 \tag{1}$$

with $\epsilon$ the dielectric constant, as the configuration simulates a parallel plate capacitor of area $A$. (The conducting region of the semiconductor bulk is one "plate," separated by the space-charge region from the other "plate," the high-conductivity electrolyte.)

In the case of ZnO, a wide band-gap, $n$-type material, there are no equilibrium minority carrier effects. Then the solution of Poisson's equation in the space-charge region yields the Schottky relation:

$$V_s \cong qN_D x_0^2/2\epsilon\epsilon_0 \tag{2}$$

with the approximation that the density of charge in the space-charge region equals $N_D$, the density of donor ions. From these two equations we have $V_s$ as a function of the capacity

$$V_s \cong qN_D A^2\epsilon\epsilon_0/2C^2. \tag{3}$$

As $V_s$ goes to zero, the capacity obviously goes to infinity (because the space-charge region disappears). In practice, of course, one of two events occurs: either the simple theory becomes a poor approximation (Dewald[2,12] developed the complete theory) or other impedances in the circuit begin to dominate the measurement. In the work to be discussed on ZnO, the capacity measurement, where applicable, is interpreted with eq. (3) and provides a quantitative measure of $V_s$, the surface barrier.

In the case of germanium, on the other hand, minority carrier effects are important. If conducting minority carriers are induced at the surface, the conducting region extends to the surface, and the apparent $x_0$ goes to zero. This happens on $n$-type material, when $V_s$ becomes very high. Thus, in germanium, eq. (3) applies only for a limited range – it breaks down if $V_s$ approaches zero (just as in the case of ZnO), and also if $V_s$ becomes too large. For $V_s$ large, a complex theory is required and the capacity increases with increasing $V_s$. (See, for example, the book by Myamlin and Pleskov[13] for a detailed analysis.)

The overall effect with germanium is a minimum in the capacity vs. $V_s$ curve, which appears at some surface barrier $V_{sm}$. The value of $V_{sm}$ depends only on the bulk properties of germanium. The observation of this capacity minimum has been used to measure variations of $V_H$ on germanium. The measured voltage at the capacity minimum is given by

$$V_m = V_{sm} + V_H + \text{const.} \tag{4}$$

With $V_{sm}$ a constant for a given sample, any observed variation in $V_m$ can immediately be related to variations in the surface double-layer potential $V_H$.

Alternatively, the value of $V_{sm}$ can be calculated from theory; then measurement of the position of the minimum "calibrates" the surface barrier at one value of $V_m$. Then, if there is reason to believe $V_H$ is a constant, any variation in $V_m$ can be interpreted as a variation in $V_s$, according to $V_m = V_s + \text{constant}$.

Thus, the measurements indicated in Fig. 1, the current, voltage, and capacity, can provide information about both reaction rates and the steady-state potential differences $V_H$ and $V_s$.

### (*iii*) *Transistor Method*

The study of semiconductor electrochemistry is facilitated by the ability to control the behavior of the electrode by the methods developed in semiconductor physics. To study the reactions of holes and electrons with surface species, we are interested in methods to provide carriers at the surface for the reaction. For example, the electrode can be doped; for some semiconductors either *n*-type (electrons the majority carrier) or *p*-type (holes the majority carrier) material can be used. With very heavy doping, the concentration of minority carriers is diminished. If, on the other hand, minority carriers are desired, they can be produced by illumination, as is done in studies of ZnO (where only $n$-type material is available, so photoproduced holes are used to study hole reactions).

The transistor method available for germanium is a powerful way of studying minority carrier effects, one which can sense whether a given reaction involves minority carriers or majority carriers or both. The measurement is made using the "transistor" configuration, as shown in Fig. 3. A thin wafer of $n$-type germanium is the electrode; measurements involving the $n$-type region only are completely analogous to those of Fig. 2. On the face of the wafer opposite to the electrolyte, however, a $p$-type region is provided.

The $p$-type region can be utilized in two ways. First, we can provide holes, by passing positive current through the $p$-region, causing the holes to flow into the $n$-region. If the $n$-region is thin enough, the holes will flow through it to the surface. Thus, we can force hole reactions at the surface and measure the flux of holes by the current $A_p$. Simultaneously, of course, we can provide electrons in the normal way using the $n$-region and measure the flux with $A_n$.

A second utilization of this configuration is the measurement of holes produced by the surface reaction. Hole injection occurs when a reactant in solution captures an electron from the valence band. The injected hole can pass across the $n$-region and into the $p$-region,

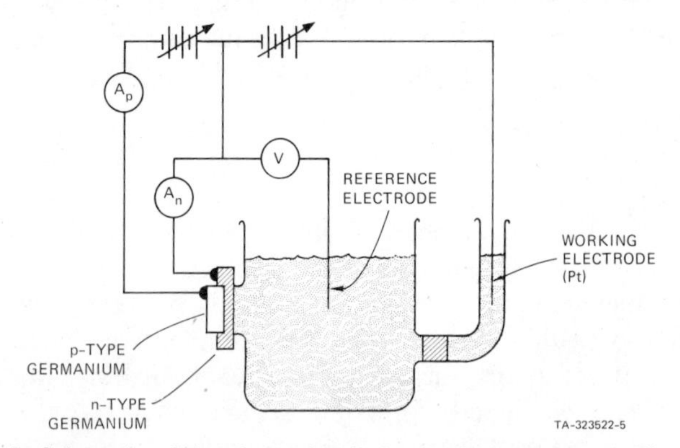


FIG. 3. Schematic of the electrochemical cell adapted to use the "Transistor Method." The meter $A_p$ measures hole flow to and from the electrolyte interface, the meter $A_n$ measures electron flow.

and will do so if the $p$-region is biased very negative (reverse bias of the $p$-$n$ junction). If there were no other source of holes except the hole injection process, then the current measured by $A_p$ would correspond to the hole injection rate. In practice it is the increase in hole current that is related to the hole injection rate. Simultaneously, of course, the electron flow to or from the surface is measured with the ammeter $A_n$.

## 3. HOLE CAPTURE AND HOLE INJECTION

### A. Hole Capture on Germanium, Current Multiplication

One of the first important studies of semiconductor electrochemistry was that of Brattain and Garrett.[1] They used the transistor configuration described under Experimental Methods (Fig. 3) biasing the $p$-region to supply holes to the surface, and biasing the $n$-region anodically, measuring electron flow from the solution into the germanium.

For moderate voltages they found that, when the hole current to the surface is zero, there is negligible electron injection into the conduction band. When, however, holes are caused to flow, they found that about one electron is injected into the germanium conduction band for each two holes reaching the surface. This injection is measured as current through $A_n$ in the circuit Fig. 3. The effect is termed current multiplication. The overall chemical reaction occurring is presumably

$$\mathrm{Ge} + (4x)p + 6\mathrm{OH}^- \longrightarrow (\mathrm{GeO_3})^{2-} + 4(1-x)e + 3\mathrm{H_2O} \quad (5)$$

where $p$ represents a hole and $e$ an electron (in this case an electron injected into the conduction band). Here $x$ is the fraction of total interface current carried by the holes. The value of $x$ has been found to range from 0.6, for very low hole currents, to 0.8 for higher hole currents.[14]

The process occurring may be visualized as follows. As holes become localized at the surface, the germanium molecule must go through four stages of oxidation, going from $Ge^0$ to $Ge^{4+}$ before it is dissolved as $(GeO_3)^{2-}$. Each step of oxidation can occur by hole capture or by electron injection. Pleskov's result simply indicates that the higher the hole concentration, the greater the probability that each oxidation step occurs by hole capture. Unfortunately, the complexity of the reaction obviously precludes detailed analysis in terms of energy levels. If there were only one or two oxidation states, the fraction $x$ might indicate the position of the surface state associated with the species. (The higher the surface-state energy, the more probable is electron injection as the oxidation route.) But, with four oxidation states, an analysis of the behavior of the corresponding surface states to determine their energy levels would be difficult.

However, the current multiplication effect has been utilized as a tool to study other reactions, in particular, some reactions with foreign ions involving hole injection, as discussed in the following section.

### B. Hole Injection into Germanium, Hole Injection vs. Electron Capture

Gerischer and Beck[15] studied the influence of ferricyanide ions, $Fe(CN)_6^{3-}$, on the current/voltage characteristics of an interface between $n$-type germanium and an aqueous solution; their results are shown in Fig. 4.

Since the ferricyanide ion is an oxidizing agent, it can be expected to capture electrons from the germanium. Its ability to capture electrons is observed at negative voltages, where there are large negative (cathodic) currents. The curve shape in this region, with the ferricyanide ion present, is typical of the current/voltage characteristics observed when electrolytic current is diffusion-limited. The current (between $-1$ and $-2$ V) is limited by the rate at which ferricyanide ions can diffuse to the surface to become reduced, i.e. capture electrons.

The unique feature in Fig. 4 is that there is also an increase in the

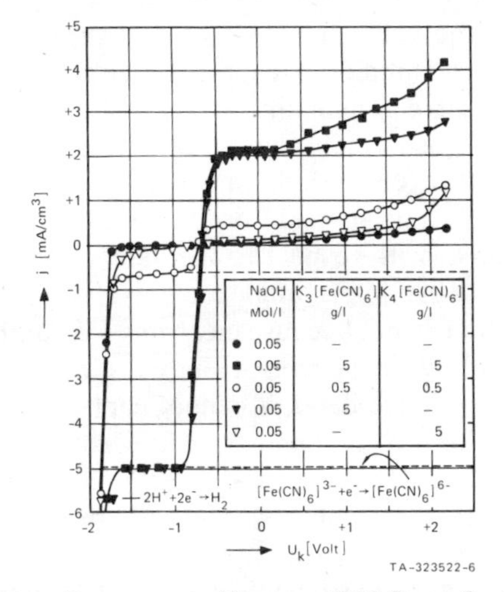


FIG. 4. Current–voltage curves with *n*-type (0.08 Ω cm) Ge. The potential is relative to a calomel reference electrode. The excess current due to $Fe^{3+}$ both in negative and positive directions is shown. From Gerischer and Beck.[15]

current in the *positive* direction, when the ferricyanide ion is present. Electrons must flow into the semiconductor from the solution to account for this current. However, it is inconceivable under the experimental conditions that the ferricyanide ion could be oxidized, that is, could inject electrons into the germanium, and thus account for the excess positive current. The explanation is that the ferricyanide ion captures electrons from the valence band, i.e. injects holes. These holes are then available to oxidize the germanium, and the positive (anodic) current observed arises from the $4(1-x)$ electrons injected, as in eq. (5), per germanium oxidized.

If we assume that the rate of hole injection by the ferricyanide ion is limited only by the rate at which the ferricyanide ion can diffuse to the surface, then the hole injection rate can be calculated from the observed negative cathodic diffusion-limited current. It can easily be shown, using this model, that the anodic current of Fig. 4 should be lower than the cathodic current by a factor $(1-x)/x$. The experimentally observed anodic current leads to $x = 0.7$, which is consistent with the values of Pleskov, and supports the interpretation that the ferricyanide ion captures electrons from the valence band (injects holes) in germanium.

This type of measurement can be used to show whether a given species will be reduced by hole injection (electrons captured from the valence band), thus giving information about the energy level of the oxidizing agent with respect to the valence band. For example, the energy level of the ferricyanide ion must be close to or below the edge of the valence band, because electrons are captured freely from the valence band by the species. If, however, a species has a level much above the valence band, it cannot capture electrons from the valence band, although it could from the conduction band. Then no current would flow in the anodic direction (compare Fig. 4) as no holes are injected, but current would perhaps flow in the cathodic direction, because of electron capture from the conduction band. Finally, if the level is extremely high, no electrons could be captured, so no current would flow in either direction.

Beck and Gerischer[10] employed this criterion in estimating the energy level of a series of oxidizing agents, and employed the standard oxidation-reduction potential of the corresponding redox couple as a measure of the chemical activity of the species. They reasoned that, as a species with a large negative redox potential is a strong oxidizing agent (tends to capture electrons), it should have a low energy level. Species with a large positive redox potential, where it is difficult to reduce them, should have a high energy level. The former should be able to capture electrons from the valence band. In their measurements, they found that several oxidizing agents, that are stronger than ferricyanide, showed equivalent behavior (i.e. electron capture from the valence band): $O_2$, $Fe^{3+}$, $MnO_4^-$, and $Ce^{4+}$. Weaker oxidizing agents, such as $TiO^{2+}$, $V^{3+}$, and $Cr^{3+}$, showed no reduction (electron capture). These results are consistent with the model. However, one strong oxidizing agent, $Cr_2O_7^{2-}$, was not reduced at a detectable rate. Thus, there is some indication that species that are strong oxidizing agents can capture valence band electrons, but the model may be oversimplified.

The ratio of reduction by valence electron capture to total reduction was measured more quantitatively by Pleskov[16,17] using the transistor method. The results are discussed by Myamlin and Pleskov[13] and by Turner.[18] In this application of the method, the *n*-germanium is made cathodic (negative with respect to the solution) to cause electrons to flow from the conduction band to the solution, and the junction is reverse biased, so that any holes injected during the reduction are collected at the *p-n* junction. Thus, the junction current (as measured on $A_p$) is the rate of electron capture from the valence band, and the current in the *n*-region is the rate of electron capture from the conduction band. The fraction captured from the valence band, *y*, for various reducible surface species is given in Table 1. The remaining reduction is with conduction band electrons. The right-hand column lists the standard oxidation-reduction potential $E^0$ in acid medium (obtained from Latimer.[19]) One would perhaps expect the strongest oxidizing agents (most negative $E^0$) to extract electrons from the valence band, but this is not borne out. In fact, $H_2O_2$ and $Cr_2O_7^=$, apparently the strongest oxidizing agents, were reduced almost entirely by electrons from the conduction band.

From the value of *y* in Table 1, some information about the surface-state energy should be obtained. The high value of *y* for $KMnO_4$ suggests that the energy level associated with $MnO_4^-$ is close to the valence band, as it is reduced primarily by valence band electrons. Moving up the table to lower *y*, it might be concluded that the energy level for the corresponding ion is higher. The peroxide level, showing no hole injection, must be near the conduction band and captures electrons only from this source. This simple interpretation

TABLE 1. FRACTION (*y*) OF ELECTRONS CAPTURED FROM VALENCE BAND DURING REDUCTION OF VARIOUS SPECIES

| Surface species | *y* | $E^0$ |
|---|---|---|
| $H_2O_2$ | 0 | −1.77 |
| $K_2Cr_2O_7$ | 0.03–0.08 | −1.33 |
| Quinone | 0.38 | −0.70 |
| $KI_3$ | 0.42 | −0.54 |
| $K_3Fe(CN)_6$ | 0.66–0.80 | −0.36 |
| $KMnO_4$ | 0.78–0.88 | −0.56 |

would permit some analysis of the energy levels associated with the various chemical species. However, this simple interpretation is much too naive for several reasons:

(a) The influence of the Helmholz double-layer potential $V_H$ has been neglected, and this could move the energy levels up or down. (Variations of $V_H$ on germanium are discussed in Section 5B.)

(b) With most of the species, more than one energy level must be involved, for two or more electrons are captured on a single molecule.

(c) The density of electrons in the conduction band is variable in these experiments, and the fraction of electrons captured from this band is expected to depend on the electron density.

(d) Under certain conditions during these measurements (see Section 5B) there is electron-hole recombination or generation at the Ge species on the germanium surface.[20] Also, hole reactions such as eq. (5) are difficult to monitor.

(e) The degree of adsorption of the species is not known; adsorption at various sites on the surface could presumably cause variations in energy levels.

In summary, the approaches of Beck and Gerischer, Pleskov, and Harvey, yielding information on relative rates of electron capture from the valence and conduction bands; (a) provide important qualitative understanding of the various steps possible in hole-capture reactions, and (b) suggest a method to obtain information about the surface energy level of chemical species, but the method will be difficult to make quantitative.

## C. Hole Capture from ZnO, Current Doubling

Zinc oxide is an *n*-type semiconductor with a band gap of about 3.2 eV. Because of the wide band gap and the high density of conduction electrons in undoped monocrystalline zinc oxide (the ionized donor density $N_D$ is normally greater than $10^{16}$ cm$^{-3}$) the hole concentration is usually negligible. Holes can, however, be produced by light at $\lambda \lesssim 4000$ Å. The reaction of such photoproduced holes with surface species is the topic of this section.

Measurements of hole capture are performed with the ZnO

positively biased with respect to the solution, so that the photoproduced holes move to the surface and provide the current flow through the space-charge region. In the dark, the current flow is negligible for a well-etched sample ($< 10^{-10}$ amps/cm$^2$), indicating no electron injection. Upon illumination, anodic current is observed proportional to the light intensity.

In an aqueous KCl solution, it is generally assumed that the photoproduced holes reaching the surface decompose the ZnO; the current is carried into the solution by zinc ions[21] and the lattice oxygen is oxidized, leading to observable generation of $H_2O_2$:

$$ZnO + p + H^+ \longrightarrow \tfrac{1}{2}H_2O_2 + Zn^{++}.$$

If a sufficient concentration of reducing agent (formic acid, HCOOH, for example) is added to the solution, however, the holes oxidize the reducing agent. One indication that the hole reaction is changed is that the decomposition product, $H_2O_2$, is no longer observed.[22]

A more dramatic and immediate indication of the reducing agent/hole interaction is the observation that, with some reducing agents, the current is almost doubled at constant light intensity.[22] The addition of reducing agents in the solution obviously cannot increase the rate of production of photoproduced holes in the bulk material. Thus, there is only one possible source for the enhancement: electron injection. For example, with $HCOO^-$ (formate ion), the chemical reaction is assumed to be

$$p + HCOO^- \longrightarrow HCOO\cdot \qquad (6a)$$

$$HCOO\cdot \longrightarrow e + H^+ + CO_2. \qquad (6b)$$

The $HCOO\cdot$ radical, produced by the single hole capture (or, if decarboxylation occurs in eq. (6a), it could be the $H\cdot$ radical) is unstable and spontaneously injects an electron $e$ into the ZnO conduction band. For each hole reaching the surface, an electron is injected, leading to doubling of the current.

In terms of surface-state energy levels, it is assumed that, in order to account for electron injection, the energy level associated with the radical must be close to or more probably above the edge of the conduction band. In addition, the energy level of the formate ion, $HCOO^-$, is presumably somewhere in the band gap, since this ion does not spontaneously inject electrons, but will capture holes.

"Current doubling," as in the example represented chemically by eq. (6), is observed with several two-equivalent chemical species—species in which the intermediate valence state is unstable, so that it always tends to change valence by two electrons. Typical current-doubling species are $HCOO^-$, $CH_3OH$, $C_2H_5OH$, and $As^{3+}$. Chemical substances that are normally oxidized by one-equivalent, to a stable valence state, show no current doubling. Examples of such one-equivalent species are $Fe(CN)_6^{4-}$, $I^-$, and $Br^-$.[23–25]

Taking advantage of the current-doubling phenomenon, we can use a titration method to determine the relative hole-capture cross-section or "hole reactivity" of various species. The relative rate of hole capture is measured as follows: a solution of a current-doubling species is prepared with a concentration sufficient to attain essentially complete current doubling. To this solution is titrated a solution of a one-equivalent ion. As the concentration of the one-equivalent ion increases, the percentage of the holes captured by the one-equivalent species increases and the total current decreases. Eventually, the current reaches the value associated with the hole current alone. If the one-equivalent species has a low hole reactivity, a high concentration will be required to lower the current and vice versa. By comparing the current/concentration curves for various one-equivalent species, their relative hole reactivity can be evaluated. Similarly, using a single one-equivalent

| Ion | $k\sigma/k_{I^-}\sigma_{I^-}$ |
|---|---|
| $I^-$ | 1.00 (by definition) |
| $Br^-$ | 0.013 |
| $Cl^-$ | 0 |
| $CH_3OH$ | 0.09 |
| $C_2H_5OH$ | 0.18 |
| $(CH_3)_2CHOH$ | 0.18 |
| $SO_3^=$ | 0.23 |

species, we may compare the relative hole reactivity of a series of two-equivalent ions. Gomes, Freund, and Morrison[24] made such measurements and obtained the indicated values for $k\sigma/k_{I^-}\sigma_{I^-}$, the hole reactivity relative to iodide, where $\sigma$ is the hole-capture cross-section and $k$ is a constant relating the amount of ion adsorbed to the amount of ion in solution.

Such data provide information about the hole capture cross-section of foreign surface species but no information about the energy level except that it presumably lies within the band gap—a span of 3 volts or so in ZnO.

Another type of experiment utilizing the current-doubling phenomenon can elucidate interaction between surface species. It is found, when studying the current-doubling effect with formate (eq. (6)) that, if oxygen is bubbled through the solution, the current decreases again.[22] This behavior is interpreted according to

$$HCOOH + p \longrightarrow H^+ + HCOO\cdot$$

$$HCOO\cdot + O_2 \longrightarrow H^+ + CO_2 + O_2^-. \tag{6c}$$

Thus, the electron associated with the radical is captured by the oxygen (reduces the oxygen) and is not injected into the conduction band. Thus, we have a direct interaction of the oxygen with the intermediate radical, and the effect permits such an interaction to be investigated.

The only system that has been thoroughly investigated by this means is the $Cu(NH_3)^{2+}$ system studied by Gomes, Freund, and Morrison[24] using a series of alcohols and formate as the two-equivalent reducing agents. They found that the fraction of radicals that decomposed by reducing the $Cu^{2+}$ was proportional to the concentration of $Cu^{2+}$ (in solution) and independent of the concentration or type of the reducing agent. That is,

$$J_e/J_p = 1 - \gamma[Cu^{2+}] \tag{7}$$

with $J_p$ the photoproduced hole current, $J_e$ the resulting electron injection current and $\gamma$ the observed proportionality constant, found to be approximately $5 \times 10^4$ liters/mole, independent of reducing agent used.

This proportionality suggests that there must be a simple law governing the interaction between surface species; the observation that the behavior does not depend on the reducing agent used suggests that the intermediate radical is the same in all cases—perhaps H·.

One possible interpretation leading to eq. (7) is that the lifetime $\tau$ of the radical is very short, much too short for ion motion to occur. Then only if a molecule of $Cu(NH_3)_4^{2+}$ is within a given radius $\lambda$ of the radical, when the radical is formed (by hole capture), can the molecule be reduced; otherwise, the electron is injected in time $\tau$ into the conduction band. If such a model were verified, the value of $\lambda$ would be of interest in molecule/molecule electron exchange studies, and the value of $\tau$ would be of great interest as a parameter describing electron exchange between surface states and energy bands.

These studies of hole capture on ZnO and on germanium demonstrate how the phenomenon of current multiplication or current doubling can be utilized to provide information about the electronic and chemical behavior of surface states. The advantage of ZnO over germanium for these studies is that current multiplication (associated with the host lattice) does not occur. This permits simple studies of hole capture by foreign surface states, providing a measurement of hole-capture cross-section and a knowledge of the chemical behavior of the surface states following the hole-capture process.

In order to study hole-capture cross-sections for surface states on the germanium surface by the current-doubling method, a one-equivalent reducing agent would have to be present in concentration sufficient essentially to preclude current multiplication. Only then could current doubling by addition of a two-equivalent species be observed and the concentration required may be very high. Such experiments on germanium have not yet been performed. On the other hand, hole capture by germanium can be observed, as was done with ferrocyanide ions by Beck and Gerischer[10] via a measurement (with a *p*-type sample) of the current increase due to hole capture. Such data could be analyzed by the methods to be discussed in Section 4A. However, the unavoidable electrolytic etching of the sample may complicate the system too much. Thus, because of the competing current multiplication process, there seems to have been little work done on hole capture by foreign species at the germanium surface.

Current doubling is not restricted to the ZnO surface. It has been observed by Gerischer and Haberkorn on CdS, as reported in the discussion following the paper of Morrison and Freund,[23] and reported by Gerischer.[26] The analogous effect on *p*-type material has been observed by Gerischer[26] and by Memming.[27] Gerischer reports capture of photoelectrons by $Br_2$ on GaAs leads to hole injection, and Memming reports the same effect with $S_2O_8^=$ and $H_2O_2$ on GaP.

From a theoretical point of view, the study of holes as oxidizing agents opens up a field of great interest. First, we have that holes are electronic in nature and, therefore, can be generated and controlled by electronic means. Second, we have that holes are one-equivalent in nature, so hole capture by a two-equivalent species will always yield a radical of some form; the chemistry of such unstable radicals is a technology in its own right, and this means of radical generation and detection using solids should be of interest. Third, we can assume that holes from different semiconductors are of differing oxidative power (although the rules governing what will be oxidized are not known). The holes from ZnO oxidize many common reducing agents, including iodide and ferrocyanide, for example, but fail to oxidize water or the chloride ion. Holes in germanium have only been shown to oxidize germanium itself and to oxidize ferrocyanide.[10] Presumably holes in other semiconductors will have varying ability to oxidize species, and studies of hole reactivity may lead to a redox potential series for semiconductors analogous to that for ions in solution.

## 4. ELECTRON CAPTURE AND ELECTRON INJECTION

Electron capture by species on germanium was described in Section 3B together with hole injection. In this section we will discuss electron capture on ZnO, and electron injection into germanium and ZnO.

### A. Electron Capture from Zinc Oxide, Measurement of Capture Cross-section

Electron capture by oxidizing agents at the ZnO surface was examined quantitatively by Freund and Morrison[7] and by Morrison.[4] In these studies, one-equivalent oxidizing agents were used, so presumably the current was due to the reaction

$$e + \mathrm{X} \xrightarrow{k} \mathrm{X}^- \tag{8}$$

where X represents the solvated ion.

One purpose of these studies was to determine whether the rate of eq. (8) followed a first-order law, as would be expected on the simplest basis:

$$J = qk[\mathrm{X}]n_s = q\bar{c}\sigma[\mathrm{X}]n_s \tag{9}$$

that is, first order in available oxidizing species [X] and first order in the electron density at the surface, $n_s$. The first expression in eq. (9) is a first-order representation in terms of chemical kinetics, where $k$ is the rate constant, and the second expression in eq. (9) is a first-order representation in terms of solid-state parameters, where $\sigma$ is the capture cross-section for electrons, and $\bar{c}$ the mean electron velocity in the conduction band. In both cases, $q$ is the electronic charge and $J$ the current density.

The evaluation of $n_s$ is made using the capacitance measurement. Because of the surface barrier (Fig. 1), $n_s$ is given by

$$n_s = N_\mathrm{D} \exp(-qV_s/kT) \tag{10}$$

that is, $n_s$ is less than the donor density $N_\mathrm{D}$ by the Boltzmann factor associated with the surface barrier $V_s$. Now $V_s$ is related to the capacitance $C$ by eq. (3), so $n_s$ can be calculated quantitatively. Experimentally, equipment as in Fig. 2 is used, with the sample in the dark, and the dependence of the current density $J$ on $V_s$ or $n_s$ is observed by varying $V_m$. Then the parameter $\sigma[\mathrm{X}]$, which we can call the electron reactivity of the species X, can be evaluated from eq. (9).

It was found that $J$ is proportional to $n_s$ and to the concentration of oxidizing agent $C_x$ for many one-equivalent species. It was concluded that the theory of eq. (9) is supported and that the parameter $[X]$ is proportional to $C_x$.

A comparison of the values of the electron reactivity $\sigma[X]$ for various species is of particular interest, if the values are compared at a constant molarity $C_x$ of oxidizing agent in solution. If there is no specific adsorption, so that $[X]$ is the same for all species if $C_x$ is constant, then at constant molarity of oxidizing agent the variation in $\sigma[X]$ reflects only a variation in the electron-capture cross-section. A general model describing the electron-capture cross-section (electron reactivity) of various ions under various conditions was the overall objective of the study by Morrison.[4] An attempt was made to test whether the electron-capture cross-section by ions could be described by a model similar to that used for electron capture by levels in solids, where the dominant parameter is the electron energy level relative to the conduction band edge. In electrochemical work, if there is no specific adsorption, the dominant parameter would thus be the energy level on the ion in solution relative to the conduction band edge. Of course, in order to relate cross-section to energy level, it is first necessary to estimate the energy level of the ion.

Qualitatively, as was suggested by Gerischer and Beck[15], there should be a relationship between the redox potential and the energy level of a species. A strong reducing agent must be represented on the band model by a high occupied electronic energy level, as it tends to give up electrons. A strong oxidizing agent must be represented by a low unoccupied energy level as it tends to capture electrons. Morrison[4] derived a quantitative relationship for one-equivalent redox couples. To a first approximation it was shown:

$$E_x = E_x^0 + \text{const} \tag{11}$$

where $E_x$ is the energy level on the solvated species, and $E_x^0$ the standard redox potential of the couple $X/X^-$ (American convention, Latimer[19]). It was shown that more detailed consideration of the energy of hydration, hydrolysis, and other chemical transformations could be applied to improve the approximation (11). From eq. (11), variations in $E_x^0$ will reflect variations in $E_{CS} - E_x$, the surface-state energy, if $V_H$ is constant. $V_H$ was shown to be constant, independent of the oxidizing agent (an "insulator-type" of Helmholz layer), by measurements of $C$ and $V_m$ (Section 2B). From these arguments it was concluded there should be a simple relationship between $\sigma$ and $E^0$, if the capture cross-section depends primarily on the energy level of the species.

Figure 5 shows the relationship observed[4] between the experimentally determined electron reactivity, $\sigma[X]$ with $C_x = 10^{-2}$M (reflecting the capture cross-section $\sigma$), and the standard redox potential $E_x^0$ of various one-equivalent species (reflecting the energy level $E_x$). The results available seem consistent with a simple qualitative model. As discussed in the reference (and, in part, in

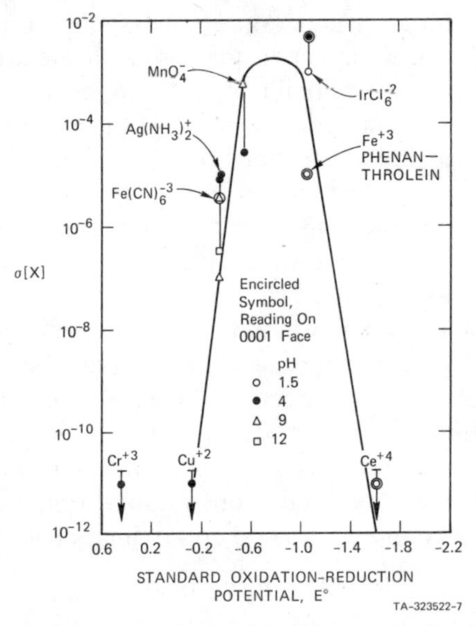


FIG. 5. The variation in electron reactivity (electron capture cross-section) as a function of the redox potential (the energy level) for various one-equivalent oxidizing agents. The value of pH is indicated. Normally the measurements were on the (000$\bar{1}$) face, and exceptions are indicated. From Morrison.[4]

Section 4B below), there is independent evidence that the effective redox potential for the ZnO conduction band edge $E_{CS}$ is between 0 and $-0.2$ eV. With the conduction band edge at this value, the shape of Fig. 5 becomes understandable. Oxidizing agents with an energy level (redox potential) greater than $E_{CS}$ ($E_x^0 > -0.1$ V) cannot be reduced by ZnO conduction band electrons, because the electron transfer would be to a higher energy level, so $\sigma[X]$ is low. Thus, for example, with $E_x^0 = 0.4$, $E_x - E_{CS} = 0.5$, $Cr^{+3}$ has an energy level well above $E_{CS}$, so it has a very low capture cross-section. As the energy level drops below $E_{CS}$ ($E_x^0 < -0.1$ V), electron capture becomes favorable. If the energy level drops too far below $E_{CS}$ ($E_x^0 < -1$ V), there is some indication that electron capture becomes unfavorable again. Such behavior with low $E^0$ is consistent with electron capture by surface states, for often the capture cross-section decreases if the dissipation of large amounts of energy is required.

These studies of electron capture, both from the valence band of germanium (Section 3B) and, in the present case, from the conduction band of ZnO, illustrate early attempts to utilize semiconductor electrochemistry to determine the energy levels of surface species. Although the information obtained is very qualitative in these particular studies, the fact that information can be obtained with such simple experimental methods suggests that with refinement in theory and experiment, the approach will be rewarding.

## B. Electron Injection into Germanium and Zinc Oxide

For redox couples on ZnO it is found that, if the oxidizing agent of the couple can capture electrons, the reducing agent is not found to inject.[7,21]

It is found that only very strong two-equivalent, or fairly strong one-equivalent, reducing agents will inject electrons into ZnO. Electron injection by surface species has been observed by Freund, Gomes, and Morrison[37] with $V^{2+}$, $Cr^{2+}$, $Sn^{2+}$, and ascorbic acid, and by Freund[28] with $Eu^{2+}$, ethylenediamine cobalt(II), and Ti(III). On the other hand, ions such as $Fe(CN)_6^{-4}$ and $Fe^{2+}$ do not inject.

This behavior is consistent with the model discussed above (eq. (11)) relating the redox potential $E_x^0$ (for one-equivalent ions in solution) to the energy level, and estimating that the ZnO conduction band edge corresponds to a redox potential of about $-0.1$ V. The injecting one-equivalent species have $E^0$ (energy level) more positive than this; the noninjecting one-equivalent species, $E^0$ more negative.

On germanium, Pleskov and Kabanov[29] found injection by $C_2O_4^{2-}$, $I_3^-$, and $V^{2+}$. Beck and Gerischer[10] found injection from $V^{2+}$, but not from $Cr^{2+}$. These latter two ions are both one-equivalent reducing agents at about the same redox potential, about 0.3 V (actually chromous is about 0.1 V higher), so both should inject if either does. The difference between the two may arise, either because of kinetic effects, or because the energy levels lie in the neighborhood of the germanium conduction band edge (the band edge thus corresponding to $E^0 = +0.3$ or so volts), and so the approximation of eq. (11) (neglecting hydrolysis, for example) is too coarse. There is slight evidence favoring the latter explanation, that the $E_{CS}$ for germanium may be at $E^0 \sim 0.3$ V. This is the observation of Beck and Gerischer[10] that holes can be freely exchanged either way between the iron cyanide couple and germanium. For such rapid exchange, one might assume $E_{VS}$ is near the $E^0$ of iron cyanide ($-0.35$ eV). With the 0.7-V band gap of germanium, these redox potentials for $E_{CS}$ and $E_{VS}$ are consistent. However, the evidence for this model is sparse (particularly as $V_H$ on germanium is variable, as will be discussed later), so the hypothesis needs direct testing.

Efimov and Erusalimchik[30] studied electron injection into germanium by oxalate $C_2O_4^=$ and iodide $I^-$. They found that during the oxidation of these two-equivalent species, the germanium dissolution rate increased. This may imply, with eq. (5), that hole injection by the reaction intermediate follows the electron injection. They also found that during electron injection by $I^-$, iodine was not formed. Turner[18] suggests that the $I^-$ injects an electron to the conduction band, then the neutral iodine captures an electron from the valence band, the molecule returning to the $I^-$ form. This appears unlikely, as it amounts to a large net gain in the energy of the system (creation of an electron–hole pair). It appears more likely that a net chemical change, perhaps involving the germanium, is required in the primary reaction to supply the energy of the hole–electron pair. Many possible reaction schemes could account for the result, for example:

$$I^- \longrightarrow I + e \qquad \text{(electron injection)}$$

$$I \longrightarrow I^+ + I^- \qquad \text{(disproportionation)}$$

$$6OH^- + 2I^+ + Ge \longrightarrow (GeO_3)^2 + 2I^- + 3H_2O$$

where the last reaction is still a complex, multistep process which may or may not involve free holes. If holes are produced, the energy in this model comes from the oxidation and solution of the germanium, which is included as a primary reaction. Another possibility, not involving the germanium in the primary process, might be oxidation of the iodine to an oxidation state greater than +1, with hole injection following.

Such complex reactions involving two-equivalent species are of general interest, and the electrochemical methods we have discussed, if chosen well, can be applied to probe the detailed reaction steps of such corrosion reactions.

A discussion of electron injection from solution cannot be complete without some reference to the recent studies, particularly by the groups of Hauffe[(31,32)] and Gerischer[(33–35)] on photoassisted injection by dyes in solution. These studies are important from a fundamental view, for they can provide information about both the photoexcitation process in dyes and charge transfer between energy levels in solution and in the solid. As found by Tributsch and Gerischer, for example, a single photoexcited dye molecule in solution can either cause electron injection (rhodamine B at the CdS or ZnO surface) or hole injection (rhodamine B at the GaP or $Cu_2O$ surface). Such studies are also important from a technological point of view, as they may help in the understanding of dye sensitization in photography and electrophotography.

We will not discuss dye injection in detail here, as the immediate problems to be resolved in such research appear to be related to the complicated photochemistry of the dyes and, until these problems are cleared up, the interpretation of the carrier injection mechanisms, of interest here, is not straightforward.

## 5. SIMULTANEOUS ELECTRON AND HOLE CAPTURE, RECOMBINATION

If the rate of electron reduction of an oxidizing agent (electron capture by an unoccupied surface state) equals the rate of hole oxidation of a reducing agent (hole capture by an occupied surface state), then we have two possibilities, catalytic recombination or simple noncatalytic hole–electron recombination. In this sense, we identify catalysis with a net chemical change in the system (the disappearance of a hole–electron pair is not a "chemical" change, but a return to the electronic ground state). If the reducing and oxidizing agents form a redox couple,

$$p + Fe^{2+} \longrightarrow Fe^{3+} \tag{12a}$$

$$e + Fe^{3+} \longrightarrow Fe^{2+} \tag{12b}$$

then there is no net chemical change and we have simple hole–electron recombination. If the reducing and oxidizing agents do not form a redox couple:

$$2e + H_2O_2 \longrightarrow 2OH^- \tag{13a}$$

$$2p + H_2O_2 \longrightarrow 2H^+ + O_2 \tag{13b}$$

or

$$2p + HCOOH \longrightarrow CO_2 + 2H^+ \tag{14a}$$

$$2e + O_2 \longrightarrow 2O^- \tag{14b}$$

then there is chemical change and the recombination is a catalytic process.

From a physicist's point of view, the simple electron–hole recombination represents alternate electron and hole capture at the same surface state (unoccupied if the species is $Fe^{3+}$, occupied if the species is $Fe^{2+}$). Catalysis represents electron and hole capture at different surface states, usually followed by removal of the products, a situation not considered in solid-state analysis to date.

Research on recombination on ZnO has been restricted to catalytic recombination. Research on germanium has to date been concerned with simple noncatalytic electron–hole recombination on surface states associated with copper and with free germanium surface bonds.

### A. Recombination on Zinc Oxide, Photocatalysis

It has been known for years that if ZnO in an aqueous solution is illuminated with light of energy $h\nu > 3$ eV (hole generation), it can catalyze the oxidation of many organic species, if oxygen and the organic are present in the solution. A list of references to nonelectrochemical work is given by Morrison and Freund.[22]

Morrison and Freund used a single crystal of ZnO with the electrochemical measurements, described in the preceding sections, to study the photocatalytic oxidation of formic acid:

$$HCOOH + O_2 \longrightarrow CO_2 + H_2O_2. \qquad (15)$$

It was assumed that oxygen, formic acid, and hydrogen peroxide were involved in the redox steps, and three types of measurements were made, anodic (hole reactions), cathodic (electron reactions), and catalytic (both together). The reaction of holes with formic acid (eq. (6a)) was observed by current doubling (Section 3C) and the observation that current doubling was quenched by oxygen or peroxide was interpreted in terms of eq. (6c). From these measurements it was concluded that eq. (14a) could *not* proceed – under no circumstances did a molecule of formic acid capture two holes.

The second type of measurement used was the method for exploring electron capture (Section 4A). It was found that both oxygen and hydrogen peroxide had a reasonable electron reactivity, with or without formic acid present.

The third type of measurements, that during catalytic recombination when hole capture and electron capture proceed at the same rate, was made as follows. Holes were produced at a known rate by illumination, but the external circuit was opened, so no net current flow was possible. Then the electron flow to the surface $J_n$ must equal the known hole flow $J_p$. Although the circuit was open, it was still possible to measure the capacity, so the value of $\sigma[X]$, the electron reactivity of the oxidizing agents, could be determined as in eq. (9) and the subsequent discussion.

It was found that when the ZnO is illuminated in the presence of formic acid, the value of $\sigma[X]$ for either oxygen or peroxide is anomalously low. The value of $\sigma[X]$ for oxygen or peroxide is high if holes are produced, but no formic acid is present. The value of $\sigma[X]$ for oxygen or peroxide is high if formic acid is present but no holes. With both holes and formic acid present, $\sigma[X]$ is low. These facts suggest that it is the product of holes plus formic acid, viz., the formyloxy radical HCOO· (eq. (6a)), which is the cause of the decreased electron reactivity of oxygen or peroxide. It was suggested that an irreducible complex was formed, between the radical and oxygen in some valence state, which prevented further reduction of the oxygen. The overall reaction scheme suggested was

$$O_2 + 2e \longrightarrow O_2^{=}$$

$$p + HCOOH \longrightarrow HCOO\cdot + H^+$$

$$HCOO\cdot + O_2^{=} \longrightarrow [\text{complex}]^{=}$$

$$H^+ + [\text{complex}]^{=} + p \longrightarrow CO_2 + H_2O_2$$

where hydrolysis is neglected in describing the steps in the scheme.

There are several details of this scheme that may have an alternate interpretation, and further research is necessary. But the overall research plan is valuable as an illustration of the use of semiconductor electrochemical techniques to clarify chemical reactions, even nonelectrochemical reactions, which can occur at a semiconductor/electrolyte interface.

Hauffe[36] examined other reactions occurring during hole–electron recombination at the ZnO surface. He examined the decarboxylation of methoxy acetic acid and determined that band-gap illumination (the presence of holes) was necessary. In this case, however, current doubling was not observed. Hauffe suggested that the free radical resulting from hole capture and decarboxylation is not further oxidized, as would be the case if electron injection were observed, but rather becomes reduced (captures an electron) to form dimethyl ether.

Hydroquinone oxidation was also studied by Hauffe. Here, in contrast to the reactions discussed above, the presence of holes (illumination) lowers the rate of the reaction. Hauffe suggested that in this case the active intermediate is $O_2^-$ (or the hydrolyzed form $O_2H$) formed by reduction of neutral oxygen. Holes coming to the surface neutralize the oxygen ions and, as the steady state density of $O_2^-$ (or $O_2H$) is lowered by the presence of holes, the oxidation of hydroquinone is lowered. This interpretation seems reasonable and implies a low hole-capture cross-section for hydroquinone.

Freund, Gomes, and Morrison[37] and Freund and Gomes[38] have discussed the general concepts of photocatalysis at the solid/electrolyte interface and the conditions under which catalytic recombination seems unavoidable.

### B. Recombination on Germanium, Noncatalytic

In semiconductor physics, the noncatalytic recombination of holes and electrons at a surface is described in terms of the surface recombination velocity *s*. This is a parameter with the units of velocity which represents the rate of flow of excess carriers to the surface to recombine. The rate of recombination (pairs/cm$^2$/sec) is assumed proportional to the excess density at the surface (pairs/cm$^3$), and the proportionality constant is *s* (cm/sec).

The measurement of surface recombination velocity *s* on a germanium surface, when measured as a function of the surface barrier,

$V_s$, yields information on the energy of the surface states involved. A plot of $s$ vs. $V_s$ yields a bell-shaped curve showing a maximum at some value of $V_s$, which is favorable for both hole and electron capture by the surface states. If $V_s$ is higher than the optimum value, electrons cannot reach the surface easily; if $V_s$ is lower than the optimum value, holes cannot reach the surface easily. Thus, there is an optimum $V_s$ and a maximum in the $s/V_s$ curve.

In electrochemical measurements, $V_s$ can be varied at will by changing the voltage $V_m$ applied to the cell. The value of $s$ is normally measured by standard nonelectrochemical techniques.

### (*i*) *Free Valences on Germanium*

The chemical processes giving rise to recombination centers, when peroxide is present in the solution at the germanium/electrolyte interface, are of substantial interest. The effects observed provide an excellent illustration of the complexities associated with surface species interacting with holes and electrons in the semiconductor and how these interactions may be quantitatively analyzed.

Initially, we will discuss the formation of adsorbed OH or H layers at the germanium surface, and the means to substitute the one layer for the other. It is during the transformation from the one form of adsorbed (Helmholz) layer to the other that the free valencies Ge· arise, which act as recombination centers. Following the discussion of adsorbed layers, we will discuss recombination on Ge· centers and the influence of $H_2O_2$ in controlling the concentration [Ge·].

The surface properties of the germanium electrode differ depending on whether a cathodic or an anodic current has been flowing across the interface. This is concluded from the fact that the minimum capacitance occurs at a different potential for the two cases (eq. (4)). There is a difference of about 0.4 V in $V_H$, the Helmholz potential, for the two cases. Gerischer, Maurer, and Mindt[39] conclude that, after a cathodic (germanium negative) treatment, hydrogen atoms H· are bonded to the germanium surface atoms (formation of chemisorbed hydrogen), whereas hydroxyl groups OH· are bonded to the surface after an anodic treatment. Thus,

$$\text{Ge–OH} + e + \text{H}^+ \rightleftharpoons \text{Ge}\cdot + \text{H}_2\text{O} \tag{16a}$$

$$\text{Ge}\cdot + \text{H}^+ \rightleftharpoons \text{Ge–H} + p \tag{16b}$$

are the surface reactions of interest. A cathodic current drives both reactions to the right, until the hydroxyl surface layer is replaced by the hydrogen layer; conversely a continuous anodic current forms the hydroxyl groups. It is of interest that such a chemical surface layer determines the Helmholz potential in a manner similar to an insulating material, but that the dominant layer can be switched by current flow.

With hydrogen peroxide present in the solution, Gerischer and Mindt[40] found that electrons can be captured by the peroxide, but only while the hydroxyl layer is present. When the hydroxyl layer is converted to the hydrogen layer, the electron capture ceases. This is shown in Fig. 6.

Figure 6 illustrates measurements of the current as a function of the potential while the potential is swept from anodic (hydroxyl) to cathodic (hydrogen) conditions and back with hydrogen peroxide in the system. The arrows on the curve show the direction in which

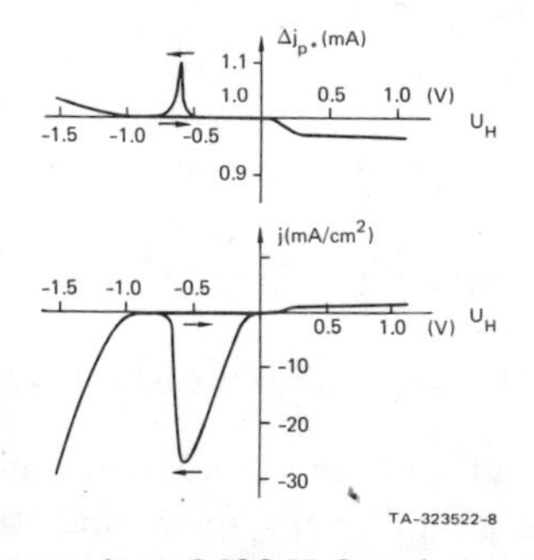


FIG. 6. Electron capture by a 0.2M $H_2O_2$ solution (lower curve) as the potential is swept between positive and negative as indicated by the arrows. Hole injection ($\Delta j_p$) during the same cycle is shown in the top curve. From Gerischer and Mindt.[40]

the applied potential is varied. The bottom curve shows the rate of electron capture; the top curve shows hole injection (by the transistor method described in Section 2B) during the same cycle. It is observed that, as the potential is swept negative, reduction of hydrogen peroxide occurs, i.e. there is a large negative current in the bottom curve until about $-0.5$ V, and then a pulse of holes is injected (top curve), changing the hydroxyl layer to the hydrogen layer as in eq. (16b). When these Ge–H bonds form, reduction stops, and the current (bottom curve) becomes very small.

Gerischer and Mindt[40] interpret these results as follows. They conclude that reaction (16c) occurs as the voltage is swept negative toward $-0.5$ V. The surface changes between the Ge· condition

$$Ge\cdot + H_2O_2 + e \longrightarrow Ge\text{–}OH + OH^- \qquad (16c)$$

(caused by eq. (16a)) and the Ge–OH condition (caused by eq. (16c)). The net effect of these two reaction steps is the reduction of $H_2O_2$ to $H_2O$ by electrons, leading to cathodic current. However, as the germanium is swept more negative, the Ge–H bonds are formed (eq. (16b)), and the reduction of peroxide by (16a) and (16c) can no longer continue. The cathodic current thus becomes very small. Eventually, at very negative (cathodic) potential, reduction begins, presumably of protons, even with the hydride layer present. (Memming and Neumann[41] discuss detailed mechanisms for hydrogen evolution from the Ge–H layer.)

If there is a high concentration of peroxide in the solution, however, the low values of cathodic current $J_e$ at $V \cong -0.8$ V do not occur. In this case, the reduction of peroxide (a high cathodic current) occurs at all negative potentials. This behavior suggests that a further reaction, not involving charge carriers,

$$Ge\text{–}H + H_2O_2 \longrightarrow Ge\text{–}OH + H_2O \qquad (16d)$$

maintains the hydroxyl surface layer, even under cathodic conditions, when there is a high peroxide concentration. If eq. (16d) can maintain the hydroxyl layer, reduction of peroxide proceeds rapidly.

Memming and Neumann[42] have examined the surface recombination velocity $s$, while the above processes are proceeding, using one of the standard solid-state techniques—the short-circuit photocurrent of a diode located adjacent to the germanium/electrolyte interface. They concluded that the Ge· form of the surface provided recombination centers; that is, these free valencies can alternately capture holes and electrons. One of the arguments was based on the fact that the appearance of recombination centers qualitatively coincided with the transient appearance of Ge· during the transformation between the two types of surface layer in eqs. (16a) and (16b) with no peroxide present. Using the known theories of surface recombination velocity $s$ as a function of surface barrier, they estimated the energy level of the recombination center to be 0.09 or 0.25 eV above the midgap position.[43]

Of particular interest was a study when hydrogen peroxide was present, permitting more quantitative control over the Ge·, in accordance with eqs. (16), and permitting steady-state values of Ge· in accordance with these equations.

The experimental method and results can be summarized as follows. The sample was held at −0.6 V for a time long enough to establish steady state in the four competing reactions. Then the potential was quickly swept anodic, and $s$ (measured by the diode photocurrent) was measured as a function of the potential. The usual bell-shaped curve of $s$ vs. surface barrier $V_s$ was obtained. The value of $s$ at the maximum (corresponding to the greatest capture probability for holes and electrons simultaneously) depended upon the peroxide concentration, as shown in Fig. 7.

Now $s$ depends on the surface barrier $V_s$ and is proportional to the density of recombination centers. The maximum in $s$ should occur at the same value of $V_s$ (assuming only one surface-state energy). Thus, the maximum value of $s$ should be proportional to the density of recombination centers. So in Fig. 7 we are effectively plotting the density of recombination centers vs. the peroxide concentration.

The interpretation of such a curve follows from eqs. (16a) through (16d). Steady-state analysis of the equations during a net cathodic current flow yields an expression of the form

$$[\mathrm{Ge\cdot}] = (A + B[\mathrm{H_2O_2}] + C/[\mathrm{H_2O_2}])^{-1} \qquad (17)$$

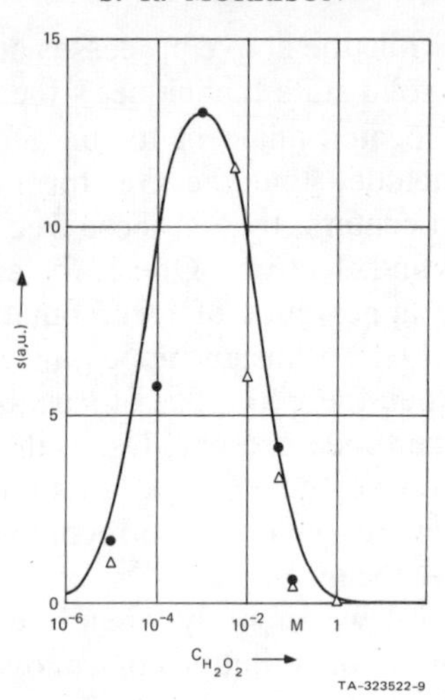


FIG. 7. The surface recombination maximum $s_{max}$ (arbitrary units) vs. $H_2O_2$ concentration. The solid line shows a theoretical curve. From Memming and Neumann.[42]

where the square brackets indicate concentrations, and where the parameters $A$, $B$, and $C$ are functions of the rate constants of eqs. (16). Qualitatively we see that if $[H_2O_2]$ is very high, [Ge·] decreases due to eq. (16c). If $[H_2O_2]$ is very low, [Ge·] decreases, because eq. (16b) is not compensated by eq. (16d). Thus, an overall dependence of [Ge·] and $s$ arises as in eq. (17) and Fig. 7. The solid line of Fig. 7 arises by fitting of the data to an expression like (17).

More recently, Plitt, Schulte and Seliger[44] have measured the effect of Ge· surface states on the "surface-state capacitance," a type of measurement which will be discussed in Section (ii) below. This work provides further verification of the general models proposed above.

It is of interest that with peroxide present, electron exchange in both directions apparently becomes rapid enough to determine the surface double-layer equilibrium, so that $V_H$ at zero net current is no longer determined by a history-dependent, insulator-type adsorption. Brouwer[45] reported this anomaly in 1966. However, the driving force for this electron exchange is presumably not a simple noncatalytic redox reaction. Equations (16a) to (16d) represent a rapid reduction of peroxide, which presumably is balanced (at zero net current) by the oxidation of germanium according to eq. (5). Brouwer[46] reports the dissolution of his germanium electrode.

This series of measurements of the various forms and characteristics of germanium surface layers provides an excellent example of the power of electrochemical tools to determine the details of free carrier/chemical reactant interactions at a semiconductor surface.

(*ii*) *Copper on Germanium*

Research on the gas/germanium interface[47,48] indicated that copper ions on the surface cause surface states that are active in surface recombination and in determining the surface barrier. These copper ions arise during the pretreatment of the germanium surface, i.e. during immersion of the germanium in water as a final rinse in the etching process. Thus, a study of the effect of copper in an aqueous solution by electrochemical techniques was indicated.

Boddy and Brattain[49] made a study of the effect of copper ions in solution by measuring the surface recombination velocity and the "surface-state capacitance." The surface recombination velocity was monitored by following the photoconductivity. The usual bell-shaped curve of $s$ vs. $V_s$ was observed when cupric ions were present in the solution. The energy level of the recombination center was calculated to be 1.2 or 1.8 kT above the midgap position.[43]

The study of surface states by measurements of capacitance is somewhat more complex. We will first discuss the "surface-state capacitance" measurement as a method of detecting surface states, and then discuss the results of such measurements with copper present.

In a capacitance measurement, the voltage is varied, $\Delta V$, and the corresponding charge stored, $\Delta Q$, is measured. Then

$$C = \Delta Q/\Delta V.$$

We will focus on the charge introduced through the semiconductor and stored at the interface. An equal charge (of opposite sign) must, of course, move through the solution to be stored at the solution side of the interface. From the mathematical model of the space-charge region, we can calculate the charge stored in the conduction and valence bands of the semiconductor, $\Delta Q_{sc}$, and thus calculate $C_{sc}$, the space-charge capacity. The capacitance measured at a semiconductor/electrolyte interface is usually associated only with charge storage in this space-charge region. If, however, surface states with special characteristics are present, extra charge $\Delta Q_{ss}$ can be stored there. Then

$$\Delta Q = \Delta Q_{sc} + \Delta Q_{ss} = C\Delta V$$

or

$$C = C_{sc} + C_{ss}$$

where $C_{ss}$ is the "surface-state capacitance." Thus, the presence of surface states can give rise to an increase in the measured capacitance. By detailed analysis of the increase in measured capacitance as a function of surface barrier $V_s$, the density and energy of the surface states can be determined.

The detection and measurement of surface states by their influence on capacity is fairly straightforward at the gas/solid interface, but presents a serious problem at the liquid/solid interface. The problem is associated with the requirement that the charge $\Delta Q_{ss}$ be "stored." When the voltage $\Delta V$ is removed, the charge stored from the semiconductor side must return to the semiconductor to contribute to $C_{ss}$. At the liquid/solid interface, unfortunately, two alternatives are available. The charge can be transferred to ions in solution (reduction or oxidation of these ions), or the species associated with the surface state can go into solution. The probability of the charge being carried into solution must be much lower than the probability of its return to the semiconductor, in order to measure the surface states by capacitance measurements. Only surface states with this special property, a low "reactivity" with the solution, can be observed.

Copper on germanium is one of the few systems in which surface states have been detectable by capacitance measurements at a liquid/solid interface. Figure 8 shows the results of capacitance measurements, with and without copper ions in solution, as a function of the surface barrier.[49] The extra capacitance is clearly observed. It is of interest that the added capacitance is relatively independent of the copper ion concentration.

Boddy and Brattain analyzed results of measurements of surface-state capacitance and $s$, and determined the energy and density of the surface states. They then compared the calculated surface-state energy $\nu$ and density $N_t$ of the two states they detected, for copper in solution, to values associated with copper at the gas/solid interface.[50] Table 2 is reproduced from their paper, to compare the results.

Gobrecht, Schaldach, Hein, Blaser and Wagemann[51] have questioned such straightforward interpretation of the excess capacity

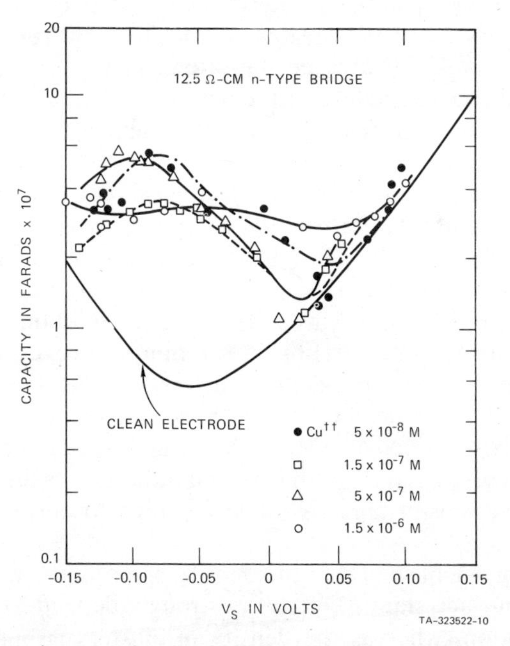


FIG. 8. Capacity vs. surface barrier $V_s$ for various concentrations of cupric ion. The expected minimum in $C$ vs. $V_s$ for a "clean electrode" is discussed in Section 2(ii). From Boddy and Brattain.[49]

TABLE 2. COMPARISON OF COPPER SURFACE STATES AT THE GE/ELECTROLYTE AND THE GE/GAS INTERFACES

| | | Surface state in electrolyte | Surface state at gas/solid interface |
|---|---|---|---|
| State I: | $\nu$ | 1.9 kT | 1.0 kT |
| Below midgap | $N_t$ | $10^{11}$ cm$^{-2}$ | 0.1 to 1 × $10^{11}$ cm$^{-2}$ |
| State II: | $\nu$ | −1.2 kT | −3.0 to 6.0 kT |
| Above midgap | $N_t$ | 2 × $10^{10}$ cm$^{-2}$ | 1 to 4 × $10^{11}$ cm$^{-2}$ |
| | $\sigma_p$ | 6 × $10^{-14}$ cm$^2$ | 6 to 15 × $10^{-15}$ cm$^2$ |

$\nu$ is the energy below the midgap, $N_t$ is the density of states, and $\sigma_p$ is the hole-capture cross-section.

and the surface recombination in terms of surface-state analysis. They observed odd behavior with the copper additive. For example, the excess capacity in the capacity/$V_s$ curve depends upon whether $V_s$ is being swept from negative to positive or from positive to negative. $C_{ss}$ also depends on the rate of change of $V_s$. Effects of this type may arise because of the behavior of the Helmholz layers (see the discussion of eqs. (16a) and (16b) above). It is clear a good analysis of such effects must be made before quantitative interpretation of surface-state energies and densities can be reliable.

The actual chemical form of the copper that gives rise to the surface states is not known. From the discussion of Memming[(52)] the following conclusions can be drawn. The deposition of copper is a monolayer (~ $10^{15}$ cm$^{-2}$) in a $10^{-5}$ molar solution, and varies linearly with the concentration in solution. However, most of the copper is in the form of metallic clusters, and the density of these clusters is of the order of $10^{11}$ cm$^{-2}$, the same order as the measured surface-state density. Memming[(52)] and Boddy and Brattain[(49)] suggest that dislocations are the precipitation centers for the clusters (actually the density appears much too high to be related to dislocations).

Memming established that the density of surface states (as determined by surface-state capacity) is strongly dependent on the pH of the solution, whereas the density of clusters is not. (For acid solutions, the surface-state density goes to zero.) He concluded, therefore, that the metallic clusters cannot be the surface states.

Thus, the identity of the molecule causing the active surface states due to copper on germanium, and perhaps even the energy level, remains unresolved. From a strictly physical point of view, however, it is hard to imagine how a cluster of $10^4$ copper atoms, which would be expected to exhibit fairly broad metallic energy bands, can avoid acting as a recombination center with some values of the surface barrier. The capacitance measurement in surface-state investigations at the liquid/solid interface may be the problem and the interpretation of this measurement should perhaps be examined more closely.

## 6. CONCLUDING REMARKS

From the studies reported above, it is apparent that a great deal has been learned by semiconductor electrochemistry methods about the capture and injection of holes and electrons by chemical species at the surface. It seems fair to expect that continued application of the techniques to a diversity of systems will result in continued improvement of this understanding. In the following, we discuss the value of such studies and the places where we may look for the most significant overall developments leading to an understanding of carrier/surface species interaction.

### A. The Application of the Models to the Solid/Gas System

Although the behavior at the solid/liquid interface is of great interest, the knowledge accumulated becomes of further interest if one can relate the behavior to that expected at the solid/gas interface.

Phenomenologically, in most cases the same behavior can be expected. For example, "current doubling" should often occur upon electron exchange with a two-equivalent species adsorbed at a gas/solid interface. There is no direct evidence that this phenomenon exists at the gas/solid interface, but there is nothing in the theory[53] to restrict it to the liquid/solid.

A more difficult comparison is the expected surface-state energy level (relative to the conduction band) for the "same" ion or molecule at a solid/gas interface and at a solid/liquid interface. There are many factors that can perturb the expected energy level, and these will differ for the two types of interface. Some of the factors that we will consider in evaluating the possible differences in the energy level are the Helmholz double layer, hydration, hydrolysis, and the difference in adsorption sites. The first three factors are associated with species in solution, the fourth more often with species at the solid/gas interface.

If the ion of interest is not adsorbed, a common situation in electrochemical studies, then it will be physically located beyond the Helmholz double-layer (the layer being only the order of an angstrom thick), and its energy with respect to the conduction band edge will be shifted by the Helmholz double-layer potential an amount up to a few tenths of an electron volt. Such a shift can, in principle, be determined, because independent measurements of the double-layer voltage can be made. In particular, conditions can be chosen such that the Helmholz potential is zero. On ZnO the Helmholz voltage is a function of pH and Parks[5] reports that it is zero at pH $\sim 9$.

Hydration refers to the electrical polarization of the liquid. In the present context, it effectively means that the whole system, semiconductor and ion, is in a medium (water) of dielectric constant about 80, rather than a medium (gas) of dielectric constant unity. The effect of this, when we compare two electronic energy levels (the energy of the electron in the conduction band vs. the energy of the electron on the ion), depends upon the radius of the electron wave function. We suggest that an analysis here will be similar to the analysis of the energy level of an impurity in a solid, where it has been shown that the energy of an electron in a dielectric medium is perturbed greatly if the orbital is large.[54] In cases where the electron orbitals of the solid and of the ion do not penetrate into the aqueous "dielectric" (highly ionic solids and surface species), there may be little effect, and the surface-state energy will not be greatly affected by hydration.

Hydrolysis (bonding of the ion to protons or hydroxyl groups), or other similar chemical reactions, can occur in solution, but not generally at the gas/solid interface. Obviously, any strong chemical reaction will alter the energy levels.[4] Local bonding during chemisorption of the ion or molecule, which is more probable at the solid/gas than at the solid/liquid interface, is another strong reaction which will alter the energy levels.

Undoubtedly, there are few cases to be expected where the energy level for a molecule should be identical at the solid/gas and at the solid/liquid interface. A series of studies in our group at SRI are in progress examining ferrocyanide and ferricyanide ions at the ZnO surface (both solid/gas and solid/electrolyte). From the above discussion the energy levels for this surface state might be similar, at pH 9. The lack of strong chemical bonding (hydrolysis or adsorption) may be expected, because of the shell of cyanide ligand groups surrounding the iron.

A good theoretical analysis (as opposed to the simple discussion above) of the various chemical mechanisms leading to differences between the solid/gas and solid/liquid systems would be very valuable.

## B. Objectives for Future Studies

### (*i*) *Experimental: Development of Phenomenological Models*

The studies discussed emphasize the need for a phenomenological understanding both of the driving force and the rate limitations of carrier/molecule interactions. First, we need a simple model that will predict which reactions will be favorable. For example, it would be desirable, as was attempted in some of the studies, to develop an "electromotive scale" for oxidation by holes and reduction by electrons from various semiconductors, similar to the redox scale for oxidation/reduction by chemical reactants. Such a scale would lead to a greatly improved understanding of hole and electron reactivity, and of the relationship between the band structure of a solid and the energy levels of surface species. Second, a simple model is needed to establish the rules (in terms of an energy level or equivalent model) governing the *rate* of the oxidation or reduction processes by holes and electrons. The question is when these rate laws will follow more closely the rate laws for carrier capture as in semiconductor physics, or the rate laws for electron exchange processes in chemistry. In the work discussed on reduction by electrons from ZnO, a qualitative agreement with electron capture rate laws from solid-state physics was claimed. However, analogous rate laws do not often apply in electron exchange between molecules; viz. strictly chemical systems, and the processes we are discussing presumably should show some characteristics of each of the phases involved.

#### (*ii*) *Theoretical: Development of a Theory of Semiconductor Electrochemistry*

The phenomenological approach has been adequate to describe the early experiments. However, improvement of the theoretical models for semiconductor electrochemistry appears necessary for more quantitative understanding. The classical electrochemical approach, describing the interactions in terms of the electrochemical potential of ions in solution, has not been adequate to describe the semiconductor/electrolyte interface, because it does not fit well when combined with the band model of semiconductors. The advantages of the sharp picture available from the theories of semiconductor surface physics are lost if, at the other side of the interface, one must use generalized formulations arising principally from thermodynamic arguments. Also, the band model deals with energy levels for electrons, and it is not easily combined in a single diagram with a classical electrochemical model, which deals with the free energy of systems and with "reaction coordinates."

Preliminary attempts to formulate electrochemical transitions in terms of electronic energy levels have been made by Gerischer and by Morrison and were referred to in Section 2B. Serious problems arise in such extensions of the band picture, for the band model is not convenient to illustrate the chemical changes (hydration, hydrolysis, dimerization, desorption, etc.) which can and do follow electron capture or loss by foreign species at the surface. If it could be developed in a simple form, a "band model" that will satisfactorily describe chemical transformations at the same time as it describes electronic transitions would be invaluable in all aspects of surface physics and chemistry.

The author is grateful to Dr. K. Sancier for a critical reading of the manuscript.

## REFERENCES

1. W. H. Brattain and C. G. B. Garrett, *B.S.T.J.* **34**, 129 (1955).
2. J. F. Dewald, *Proceedings of 2nd Conference on Semiconductor Surfaces*, Ed. J. Zemel (Pergamon, New York, 1960).
3. P. J. Boddy, *J. Electroanal. Chem.* **10**, 199 (1965).
4. S. R. Morrison, *Surface Science* **15**, 363 (1969).
5. G. A. Parks, *Chemical Reviews* **65**, 177 (1965).
6. F. Lohmann, *Ber. Bunsengesell.* **70**, 428 (1966).
7. T. Freund and S. R. Morrison, *Surface Science* **9**, 119 (1968).
8. K. Bohnenkamp and H. Engel, *Z. Elektrochem.* **61**, 1184 (1957).
9. H. Gerischer and W. Vielstich, *Z. Phys. Chem.* **3**, 16 (1955).
10. F. Beck and H. Gerischer, *Z. Elektrochem.* **63**, 943 (1959).
11. H. Gerischer, *Advances in Electrochemistry and Electrochemical Engineering*, Vol. 1, Ed. P. Delahay (Interscience, New York, 1961).
12. J. F. Dewald, *B.S.T.J.* **39**, 615 (1960).
13. V. A. Myamlin and Y. V. Pleskov, *Electrochemistry of Semiconductors* (Plenum Press, New York, 1967).
14. Y. V. Pleskov, *Dokl. Akad. Nauk SSSR* **132**, 1360 (1960).
15. H. Gerischer and F. Beck, *Z. Phys. Chem.* **13**, 389 (1957).
16. Y. V. Pleskov, *Dokl. Akad. Nauk SSSR* **126**, 111 (1959).
17. Y. V. Pleskov, *Proceedings Int. Conf. on Semiconductor Physics, Prague* (Academic Press, New York, 1961). In Russian.
18. D. R. Turner, *The Electrochemistry of Semiconductors*, Ed. P. J. Holms (Academic Press, London, 1962).
19. W. M. Latimer, *Oxidation Potentials* (Prentice-Hall, New York, 1952).
20. W. W. Harvey, *J. Phys. Chem.* **65**, 1641 (1961).
21. K. Hauffe and J. Range, *Ber. Bunsengesell.* **71**, 690 (1967).
22. S. R. Morrison and T. Freund, *J. Chem. Phys.* **47**, 1543 (1967).
23. S. R. Morrison and T. Freund, *Electrochim. Acta* **13**, 1343 (1968).
24. W. Gomes, T. Freund and S. R. Morrison, *J. Electrochem. Soc.* **115**, 818 (1968).
25. W. Gomes, T. Freund and S. R. Morrison, *Surface Science* **13**, 201 (1968).
26. H. Gerischer, *Surface Science* **18**, 97 (1969).
27. R. Memming, *J. Electrochem Soc.* **116**, 785 (1969).
28. T. Freund, *J. Phys. Chem.* **73**, 468 (1969).

29. Y. V. PLESKOV and B. N. KABANOV, *Dokl. Akad. Nauk SSSR* **123**, 884 (1958).
30. E. A. EFIMOV and I. G. ERUSALIMCHIK, *Dokl. Akad. Nauk SSSR*, **128**, 124 (1959).
31. K. HAUFFE and J. RANGE, *Z. Naturforsch.* **23b**, 736 (1968).
32. K. HAUFFE, H. PUSCH and J. RANGE, *Z. Phys. Chem.* **64**, 122 (1969).
33. H. GERISCHER and H. TRIBUTSCH, *Ber. Bunsengesell.* **72**, 437 (1968).
34. H. TRIBUTSCH and H. GERISCHER, *Ber. Bunsengesell.* **73**, 251 (1969).
35. H. TRIBUTSCH and H. GERISCHER, *Ber. Bunsengesell.* **73**, 850 (1969).
36. K. HAUFFE, *Rev. Pure and Appl. Chem.* **18**, 79 (1968).
37. T. FREUND, W. GOMES and S. R. MORRISON, *Proceedings of IVth International Congress on Catalysis*, Moscow, 1968.
38. T. FREUND and W. GOMES, *Catalysis Reviews* **3**, 1 (1969).
39. H. GERISCHER, A. MAURER and W. MINDT, *Surface Science* **4**, 431 (1966).
40. H. GERISCHER and W. MINDT, *Surface Science* **4**, 440 (1966).
41. R. MEMMING and G. NEUMANN, *J. Electroanal. Chem.* **21**, 295 (1969).
42. R. MEMMING and G. NEUMANN, *Surface Science* **10**, 1 (1968).
43. The two values represent two possible algebraic solutions.
44. U. PLITT, H. D. SCHULTE and H. SELIGER, *Phys. Letters* **30A**, 266 (1969).
45. G. BROUWER, *Phys. Letters* **21**, 399 (1966).
46. G. BROUWER, *J. Electrochem. Soc.* **114**, 743 (1967).
47. S. R. MORRISON, *Semiconductor Surfaces*, Ed. J. N. Zemel (Pergamon Press, Oxford, 1960).
48. D. R. FRANKL, *J. Electrochem. Soc.* **109**, 238 (1962).
49. P. J. BODDY and W. H. BRATTAIN, *J. Electrochem. Soc.* **109**, 812 (1962).
50. A. MANY, *J. Phys. Chem. Solids* **8**, 87 (1959).
51. H. GOBRECHT, M. SCHALDACH, F. HEIN, R. BLASER and H. G. WAGEMANN, *Ber. Bunsengesell.* **70**, 646 (1966).
52. R. MEMMING, *Surface Science* **2**, 436 (1964).
53. S. R. MORRISON, *Surface Science* **10**, 459 (1968).
54. N. B. HANNAY, *Semiconductors*, p. 21 (Reinhold, New York, 1960).

# A BIBLIOGRAPHY OF LOW ENERGY ELECTRON DIFFRACTION AND AUGER ELECTRON SPECTROSCOPY

T. W. HAAS, G. J. DOOLEY, III and J. T. GRANT

*Aerospace Research Laboratories (ARC)*
*Building 450, Wright-Patterson Airforce Base*
*Ohio 45433, U.S.A.*

and

A. G. JACKSON and M. P. HOOKER

*Systems Research Laboratories, Inc.*
*Dayton, Ohio 45440, U.S.A.*

## CONTENTS

# INTRODUCTION

The growing and widespread use of low energy electron diffraction (LEED) and Auger electron spectroscopy (AES) in surface studies is readily apparent from this bibliography. The numbers of papers in these subjects have increased rapidly in the last five years, so that a bibliography of these subjects should be quite useful. It is this need that we hope to satisfy with this work. In this regard we should state the criteria used in selecting papers for entry in this list. In the first place, we did not include government or company reports, abstracts (e.g. *Bull. Am. Phys. Soc.*), or other unpublished works. Secondly, we have limited ourselves to papers which involve either LEED or AES primarily, or those in which a major part of the work is to compare results from these techniques with another technique, or finally, a few papers which are obviously related background material to one of the subjects listed. Thus, the only high energy electron diffraction (HEED) papers included are those in which a significant proportion of the paper is devoted to a comparison of HEED and LEED structural data; some references on electron energy analyzers as related to AES are included; but, epitaxy studies involving electron microscopy or HEED, and not strongly related to LEED, were omitted.

The subject and material indices were included to facilitate locations of papers on various topics. Obviously, the number of subject headings could be enlarged enormously. However, when getting to very narrow subject listings, a great deal of subjectivity enters into the codification, and we felt it best not to extend this part too much. Materials are listed in more or less alphabetical order and, where possible, were listed according to the crystal face exposed. Polycrystal materials, spherical crystals, materials of indefinite composition, etc., are listed together. Adsorbates, such as $O_2$, CO, etc., are so listed, so that, if one is interested in oxidation studies of W(110), then a comparison of numbers common to the W(110) and $O_2$ listings will give these papers.

In any listing of this sort, there are bound to be a number of errors. We would appreciate being advised of these, and also express now our appreciation to those people who were kind enough to point out errors in our earlier version of this bibliography, which appeared as the Aerospace Research Laboratory report ARL 69-0003.

# REFERENCES BY YEAR


## 1927

27.001 H. BETHE. Scattering of Electrons by Crystals. *Naturwissenschaften* **15**, 786–8 (1927).

27.002 C. J. DAVISSON and L. H. GERMER. Diffraction of Electrons by a Crystal of Ni. *Phys. Rev.* **30**, 705–40 (1927).

27.003 C. J. DAVISSON and L. H. GERMER. Scattering of Electrons by a Crystal of Ni. *Nature* **119**, 558–60 (1927).

27.004 C. ECKART. The Reflection of Electrons from Crystals. *Proc. Nat. Acad. Sci.* **13**, 460–2 (1927).

27.005 H. E. FARNSWORTH. A Method of Obtaining an Intense Beam of Low Velocity Electrons. *J. Opt. Soc. Amer. & Rev. Sci. Instrum.* **15**, 290–4 (1927).

## 1928

28.001 H. BETHE. Scattering of Electrons by Crystals. *Naturwissenschaften* **16**, 333–4 (1928).

28.002 H. BETHE. Theorie der Beugung von Elektronen an Kristallen. *Ann. Phys.* **87**, 55–129 (1928).

28.003 C. J. DAVISSON. Are Electrons Waves? *J. Franklin Inst.* **205**, 597–623 (1928).

28.004 C. J. DAVISSON and L. H. GERMER. Reflection of Electrons by a Crystal of Ni. *Proc. Nat. Acad. Sci.* **14**, 317–22 (1928).

28.005 C. J. DAVISSON and L. H. CERMER. Reflection and Refraction of Electrons by a Crystal of Ni. *Proc. Nat. Acad. Sci.* **14**, 619–27 (1928).

28.006 G. P. THOMSON. Waves of an Electron. *Engineering* **126**, 79–80 (1928).

## 1929

29.001 C. J. DAVISSON and L. H. GERMER. Test for Polarization of Electron Waves by Reflection. *Phys. Rev.* **33**, 760–72 (1929).

29.002 H. E. FARNSWORTH. Electron Emission and Diffraction by a Cu Crystal. *Phys. Rev.* **34**, 679–96 (1929).

29.003 H. E. FARNSWORTH. Diffraction of Electrons by a Cu Crystal. *Nature* **123**, 941–2 (1929).

29.004 L. H. GERMER. An Application of Electron Diffraction to the Study of Gas Adsorption. *Z. Phys.* **54**, 408–21 (1929).

29.005 S. B. HENDRICKS. Electron Diffraction by a Cu Crystal. *Phys. Rev.* **34**, 1287–8 (1929).

**1930**

30.001 H. E. FARNSWORTH. Certain Effects Accompanying Electron Diffraction. *Phys. Rev.* **35**, 1131–3 (1930).

30.002 H. E. FARNSWORTH. The Inner Potential of a Cu Crystal. *Phys. Rev.* **36**, 1799 (1930).

30.003 P. M. MORSE. The Quantum Mechanics of Electrons in Crystals. *Phys. Rev.* **35**, 1310–24 (1930).

30.004 E. RUPP. Electron Diffraction by Adsorbed Gas Layers on Metals. *Ann. Physik* **5**, 453–74 (1930).

**1931**

31.001 M. VON LAUE. The Diffraction of an Electron-wave at a Single Layer of Atoms. *Phys. Rev.* **37**, 53–9 (1931).

**1932**

32.001 H. E. FARNSWORTH. Diffraction of Low-speed Electrons by Single Crystals of Cu and Ag. *Phys. Rev.* **40**, 684–712 (1932).

**1933**

33.001 H. E. FARNSWORTH. Fine Structure of Electron Diffraction Beams from a Au Crystal and from a Ag Film on a Au Crystal. *Phys. Rev.* **43**, 900–6 (1933).

33.002 H. E. FARNSWORTH. Concerning W. T. Sproull's Article on "Diffraction of Low Speed Electrons by a W Single Crystal". *Phys. Rev.* **44**, 417 (1933).

33.003 V. E. LASHKAREV, E. W. BÄRENGARTEN and G. A. KUZMIN. Diffraction of Slow Electrons by Single Crystals of Graphite. *Z. Phys.* **85**, 631–46 (1933).

33.004 W. T. SPROULL. Diffraction of Low Speed Electrons by a W Single Crystal. *Phys. Rev.* **43**, 516–26 (1933).

33.005 W. T. SPROULL. A New Type of Apparatus for Experiments in Secondary Electron Diffraction. *Rev. Sci. Instrum.* **4**, 193–6 (1933).

33.006 G. P. THOMSON. Comments on W. T. Sproull's Article "Diffraction of Low Speed Electrons by a W Single Crystal". *Phys. Rev.* **44**, 417–18 (1933).

**1934**

34.001 W. EHRENBERG. A New Method of Investigating the Diffraction of Slow Electrons by Crystals. *Phil. Mag.* **18**, 878–901 (1934).

34.002 V. E. LASHKAREV and G. A. KUZMIN. The Influence of Temperature on the Diffraction of Slow Electrons by a Graphite Crystal. *Physik. Z. Sowjetunion* **6**, 211–23 (1934).

34.003 V. E. LASHKAREV and G. A. KUZMIN. Effect of Temperature on Diffraction of Slow Electrons and its Application. *Nature* **134**, 62 (1934).

**1935**

35.001 V. J. KASSATOCHKIN. Reflection of Slow Electrons on Sublimed W. *Acta Physicochim. URSS* **2**, 317–36 (1935).

35.002 V. E. LASHKAREV. Inner Potentials of Crystals and the Electron Diffraction. *Trans. Faraday Soc.* **31**, 1081–95 (1935).

**1936**

36.001 H. E. FARNSWORTH. Investigation of Certain Effects Accompanying the Diffraction of Low Speed Electrons. *Phys. Rev.* **49**, 598–605 (1936).

36.002 H. E. FARNSWORTH. Penetration of Low Speed Diffracted Electrons. *Phys. Rev.* **49**, 605–9 (1936).

36.003 S. G. KALASHNIKOV and I. A. YAKOVLEV. Diffraction of Slow Electrons by Zn Single Crystals. *Physik. Z. Sowjetunion* **9**, 13–26 (1936).

**1938**

38.001 J. C. TURNBULL and H. E. FARNSWORTH. Inelastic Scattering of Slow Electrons from a Ag Single Crystal. *Phys. Rev.* **54**, 509–14 (1938).

38.002 YA. FRENKEL and S. RYZHANOV. Theory of the Diffraction of Electrons from Crystals. *J. Exptl. Theoret. Phys. (U.S.S.R.)* **8**, 1255–66 (1938).

**1939**

39.001 S. G. KALASHNIKOV. Apparatus for the Study of the Diffraction of Slow Electrons at Different Temperatures. *J. Exptl. Theoret. Phys. (U.S.S.R.)* **9**, 1405–7 (1939).

39.002 S. G. KALASHNIKOV and O. I. ZAMSHA. Influence of the Temperature on the Diffraction of Slow Electrons from a Ag Mono-crystal. *J. Exptl. Theoret. Phys. (U.S.S.R.)* **9**, 1408–14 (1939).

39.003 G. P. THOMSON and W. COCHRANE. *Theory and Practice of Electron Diffraction* (Macmillan, London, 1939).

**1941**

41.001 S. G. KALASHNIKOV. Diffraction of Slow Electrons as a Surface Effect. *J. Exptl. Theoret. Phys. (U.S.S.R.)* **11**, 385–402 (1941).

**1949**

49.001 P. P. REICHERTZ and H. E. FARNSWORTH. Inelastic Scattering of Low Speed Electrons from a Cu Single Crystal. *Phys. Rev.* **75**, 1902–8 (1949).

**1950**

50.001 H. E. FARNSWORTH. A Simple Contamination Free Electron Gun. *Rev. Sci. Instrum.* **21**, 102 (1950).

**1951**

51.001 V. F. G. TULL. Some Experimental Measurements of the Inner Potentials of Various Crystals. *Proc. Roy. Soc.* **A206**, 219–32 (1951).

51.002 V. F. G. TULL. The Calculation of the Inner Potential of a Crystal. *Proc. Roy. Soc.* **A206**, 232–41 (1951).

**1952**

52.001 E. N. CLARKE and H. E. FARNSWORTH. Observations on the Photoelectric Work Functions and Low Speed Electron Diffraction from Thin Films of Ag on the (100) Face of a Ag Single Crystal. *Phys. Rev.* **85**, 484–5 (1952).

**1953**

53.001 J. J. LANDER. Auger Peaks in the Energy Spectra of Secondary Electrons from Various Materials. *Phys. Rev.* **91**, 1382–7 (1953).

53.002 Z. G. PINSKER. *Electron Diffraction* (Butterworths Scientific Publications, London, 1953).

**1954**

54.001 R. E. SCHLIER and H. E. FARNSWORTH. LEED Investigation of Chemisorbed Gases on (100) Faces of Cu and Ni Single Crystals. *J. Appl. Phys.* **25**, 1333–6 (1954).

**1955**

55.001 H. E. FARNSWORTH, R. E. SCHLIER, T. H. GEORGE and R. M. BURGER. Ion Bombardment-cleaning of Ge and Ti as Determined by LEED. *J. Appl. Phys.* **26**, 252–3 (1955).

**1956**

56.001 G. A. HARROWER. Auger Electron Emission in the Energy Spectra of Secondary Electrons from Mo and W. *Phys. Rev.* **102**, 340–7 (1956).

**1957**

57.001 J. A. DILLON and H. E. FARNSWORTH. Work Function Studies of Ge Crystals Cleaned by Ion-bombardment. *J. Appl. Phys.* **28**, 174–84 (1957).

57.002 R. E. SCHLIER and H. E. FARNSWORTH. LEED Study of Oxygen Adsorption and Oxide Formation on a (100) Crystal Face of Ni Cleaned Under High Vacuum Conditions. *Adv. in Catalysis* **9**, 434–40 (1957); see comments pp. 493–6.

57.003 R. E. SCHLIER and H. E. FARNSWORTH. LEED Studies of Cleaned and Gas-covered Ge(100) Surfaces. *Semiconductor Surface Physics*, ed. R. H. Kingston (Univ. Pa. Press, Phila. 1957), pp. 3–22.

**1958**

58.001 H. E. FARNSWORTH, R. E. SCHLIER, T. H. GEORGE and R. M. BURGER. Application of the Ion Bombardment

Cleaning Method to Ti, Ge, Si and Ni as Determined by LEED. *J. Appl. Phys.* **29**, 1150–61 (1958).

58.002 H. E. FARNSWORTH, R. E. SCHLIER and J. TUUL. Tests with LEED for Adsorbed Hydrogen and Nickel Hydroxide Formed on the (100) Surface of a Ni Crystal at 25°C. *J. Phys. Chem. Solids* **9**, 57–9 (1958).

58.003 H. A. FOWLER and H. E. FARNSWORTH. Reflection of Very Slow Electrons. *Phys. Rev.* **111**, 103–12 (1958).

58.004 R. E. SCHLIER and H. E. FARNSWORTH. Structure of Epitaxed Cu Deposited by Evaporation onto a Clean (00.1) Face of a Ti Crystal. *J. Phys. Chem. Solids* **6**, 271–6 (1958).

## 1959

59.001 J. A. DILLON, R. E. SCHLIER and H. E. FARNSWORTH. Some Surface Properties of SiC Crystals. *J. Appl. Phys.* **30**, 675–9 (1959).

59.002 H. E. FARNSWORTH. Clean Surfaces. *Surface Chemistry of Metals and Semiconductors*, ed. H. C. Gatos (J. Wiley, N.Y., 1959), pp. 21–37.

59.003 H. E. FARNSWORTH and H. H. MADDEN. Chemisorption of Oxygen by Ni Films Deposited by Evaporation onto a (100) Face of a Ni Crystal. *Structure and Properties of Thin Films*, ed. C. A. Neugebauer, J. B. Newkirk and D. A. Vermilyea (J. Wiley, N.Y., 1959), pp. 517–24.

59.004 H. E. FARNSWORTH, R. E. SCHLIER and J. A. DILLON. Surface Structure and Work Function Determinations for Si Crystals. *J. Phys. Chem. Solids* **8**, 116–18 (1959).

59.005 H. E. FARNSWORTH and J. TUUL. Influence of Lattice Defect Density on the Chemisorption and Oxide Formation on a Clean (100) Crystal Face of Ni as Determined by LEED. *J. Phys. Chem. Solids* **9**, 48–56 (1959).

59.006 T. H. GEORGE, H. E. FARNSWORTH and R. E. SCHLIER. Some Measurements of Adsorption of Nitrogen and Oxygen on a (0001) Ti Surface Using LEED. *J. Chem. Phys.* **31**, 89–90 (1959).

59.007 G. J. OGILVIE. The Surface Structure of Ag Crystals

After Argon-ion Bombardment. *J. Phys. Chem. Solids* **10**, 222–8 (1959).

59.008 R. E. Schlier and H. E. Farnsworth. Structure and Adsorption Characteristics of Clean Surfaces of Ge and Si. *J. Chem. Phys.* **30**, 917–26 (1959).

**1960**

60.001 H. E. Farnsworth. Comments on the Paper by Scheibner, Germer and Hartman "Apparatus for Direct Observation of LEED Patterns". *Rev. Sci. Instrum.* **31**, 795 (1960).

60.002 D. C. Gazis, R. Herman and R. F. Wallis. Surface Elastic Waves in Cubic Crystals. *Phys. Rev.* **119**, 533–44 (1960).

60.003 L. H. Germer and C. D. Hartman. Structures of Monolayers of Adsorbed Gases. *J. Phys. Chem. Solids* **14**, 75–6 (1960).

60.004 L. H. Germer and C. D. Hartman. Improved LEED Apparatus. *Rev. Sci. Instrum.* **31**, 784 (1960).

60.005 L. H. Germer and C. D. Hartman. Oxygen on Ni. *J. Appl. Phys.* **31**, 2085–95 (1960).

60.006 L. H. Germer, E. J. Scheibner and C. D. Hartman. Study of Adsorbed Gas Films by Electron Diffraction. *Phil. Mag.* **5**, 222–36 (1960).

60.007 D. Haneman. Comparison of Structures of Surfaces Prepared in High Vacuum by Cleaving and by Ion Bombardment and Annealing. *Phys. Rev.* **119**, 563–6 (1960).

60.008 D. Haneman. Adsorption and Bonding Properties of Cleavage Surfaces of $Bi_2Te_3$. *Phys. Rev.* **119**, 567–9 (1960).

60.009 D. Haneman. Structure and Adsorption Characteristics of (111) and $(11\bar{1})$ Surfaces of InSb Cleaned by Ion Bombardment and Annealing. *J. Phys. Chem. Solids* **14**, 162–8 (1960).

60.010 E. J. Scheibner, L. H. Germer and C. D. Hartman. Apparatus for Direct Observation of LEED Patterns. *Rev. Sci. Instrum.* **31**, 112–14 (1960).

**1961**

61.001 E. BAUER. Interpretation of LEED Patterns of Adsorbed Gases. *Phys. Rev.* **123**, 1206–8 (1961).

61.002 H. E. FARNSWORTH and H. H. MADDEN. The Mechanism of Oxygen Chemisorption on Ni. *Adv. in Chem. Series* **33**, 114–21 (1961).

61.003 H. E. FARNSWORTH and H. H. MADDEN. Mechanism of Chemisorption, Place Exchange and Oxidation on a (100) Ni Surface. *J. Appl. Phys.* **32**, 1933–7 (1961).

61.004 L. H. GERMER and A. U. MACRAE. New Low-energy Diffraction Technique Having Possible Application to Catalysis. *R. A. Welch Found. Res. Bull.* No. 11, 5–26 (1961).

61.005 L. H. GERMER, A. U. MACRAE and C. D. HARTMAN. (110) Ni Surface. *J. Appl. Phys.* **32**, 2432–9 (1961).

61.006 D. A. GORODETSKII and A. M. KORNEV. Apparatus for Visual Observation of the Diffraction of Slow Electrons. *Ukr. Fiz. Zh.* **6**, 422–4 (1961).

61.007 D. A. GORODETSKII and A. M. KORNEV. Diffraction of Slow Electrons from W Surfaces Coated with Thin Adsorbed Films of Ba and BaO. *Sov. Phys. Solid State* **3**, 997–1003 (1961).

61.008 D. HANEMAN. Surface Structures and Properties of Diamond Structure Semiconductors. *Phys. Rev.* **121**, 1093–1100 (1961).

61.009 D. HANEMAN. Structure and Adsorption Characteristics of Clean (111) and $(11\bar{1})$ Surfaces of GaSb. *Proceedings of the International Conference on Semiconductor Physics*, Prague, 1960 (Czech. Acad. Sci., Prague, 1961), pp. 540–3.

61.010 J. P. HOBSON and I. H. KHAN. Specular Reflection in the Diffraction of Slow Electrons Near Normal Incidence. *Phys. Rev.* **123**, 1241–2 (1961).

61.011 H. H. MADDEN and H. E. FARNSWORTH. Tests for Chemisorption of Nitrogen on a Clean (100) Ni Surface. *J. Chem. Phys.* **34**, 1186 (1961).

61.012 J. A. SIMPSON. Design of Retarding Field Energy Analyzers. *Rev. Sci. Instrum.* **32**, 1283–93 (1961).

**1962**

62.001 A. A. Babad-Zakhryopin, N. S. Gorbunov and V. I. Izvekov. Experimental Techniques of LEED. *Sov. Phys. USP* **5**, 711–22 (1962).

62.002 H. E. Farnsworth. Surface Migrations and Place Exchange. *1962 Transactions of the 9th National Vacuum Symposium* (Macmillan, N.Y., 1962), pp. 68–73.

62.003 H. E. Farnsworth, J. B. Marsh and J. Toots. Surface Structures of Atomically Clean Diamond and Cleaved (111) Ge. *Proceedings of the International Conference on Semiconductors*, Exeter (Institute of Physics and Physical Society, London, 1962), pp. 836–41.

62.004 D. C. Gazis and R. F. Wallis. Surface Vibrational Modes in Crystal Lattices with Complex Interatomic Interactions. *J. Math. Phys.* **3**, 190–9 (1962).

62.005 L. H. Germer. A New Electron Diffraction Technique, Potentially Applicable to Research in Catalysis. *Adv. Catalysis* **13**, 191–201 (1962).

62.006 L. H. Germer and A. U. MacRae. Adsorption of Hydrogen on a (110) Ni Surface. *J. Chem. Phys.* **37**, 1382–6 (1962).

62.007 L. H. Germer and A. U. MacRae. Nitrogen on Ni Surfaces. *J. Chem. Phys.* **36**, 1555–6 (1962).

62.008 L. H. Germer and A. U. MacRae. Oxygen Nickel Structures on the (110) Face of Clean Ni. *J. Appl. Phys.* **33**, 2923–32 (1962).

62.009 L. H. Germer and A. U. MacRae. Surface Reconstruction Caused by Adsorption. *Proc. Nat. Acad. Sci.* **48**, 997–1000 (1962).

62.010 L. H. Germer and A. U. MacRae. Low-energy Diffraction Studies of Adsorbed Gases. *J. Phys. Soc. Japan*, Suppl. B, **17**, 286–8 (1962).

62.011 M. Green and R. Seiwatz. Model for the (100) Surfaces of Si and Ge. *J. Chem. Phys.* **37**, 458–9 (1962).

62.012 D. Haneman. Free Bonds in Semiconductors. *Proceedings of the International Conference on Semiconductors*,

Exeter (Institute of Physics and Physical Society, London, 1962), pp. 842–7.

62.013 J. J. LANDER and J. MORRISON. LEED Study of Si Surface Structures. *J. Chem. Phys.* **37**, 729–46 (1962).

62.014 J. J. LANDER and J. MORRISON. Low Voltage Electron Diffraction Study of the Oxidation and Reduction of Si. *J. Appl. Phys.* **33**, 2089–92 (1962).

62.015 J. J. LANDER, J. MORRISON and F. UNTERWALD. Improved Design and Method of Operation of LEED Equipment. *Rev. Sci. Instrum.* **33**, 782–3 (1962).

62.016 A. U. MACRAE and L. H. GERMER. Thermal Vibrations of Surface Atoms. *Phys. Rev. Letters* **8**, 489–90 (1962).

62.017 J. MORRISON. Ultrahigh Vacuum System for LEED Studies. *1961 Transactions of the 8th National Vacuum Symposium* (Macmillan, N.Y., 1962), Vol. 2, pp. 1183–5.

62.018 C. W. TUCKER. Gas Atom Scattering in LEED. *Appl. Phys. Letters* **1**, 34–5 (1962).

## 1963

63.001 F. CALOGERO. Novel Approach to Elementary Scattering Theory. *Nuovo Cimento* **27**, 261–302 (1963).

63.002 J. A. DILLON. The Nature of a Clean Surface. *Ann. N.Y. Acad. Sci.* **101**, 634–46 (1963).

63.003 H. E. FARNSWORTH. LEED from a Cleaved Ge Surface. *Ann. N.Y. Acad. Sci.* **101**, 658–66 (1963).

63.004 H. E. FARNSWORTH. The Formation of $Ni_3O$ in Single Crystals. *Appl. Phys. Letters* **2**, 199–200 (1963).

63.005 L. H. GERMER. Slow Electron Diffraction. *Ann. N.Y. Acad. Sci.* **101**, 599–604 (1963).

63.006 L. H. GERMER, R. M. STERN and A. U. MACRAE. Beginning of the Oxidation of Metal Surfaces. *Metal Surfaces: Structure, Energetics, and Kinetics* (Amer. Soc. Met., Metals Park, Ohio, 1963), pp. 287–303.

63.007 S. GOLDSZTAUB and B. LANG. Appareil Metallique

Démontable pour la Diffraction des Electrons de Faible Énergie. *C.R. Acad. Sci.* **257**, 1908–11 (1963).

63.008 I. H. KHAN, J. P. HOBSON and R. A. ARMSTRONG. Reflection and Diffraction of Slow Electrons from Single Crystals of W. *Phys. Rev.* **129**, 1513–23 (1963).

63.009 J. J. LANDER and J. MORRISON. Structures of Clean Surfaces of Ge and Si. I. *J. Appl. Phys.* **34**, 1403–10 (1963).

63.010 J. J. LANDER and J. MORRISON. LEED Study of the Surface Reactions of Ge with Oxygen and with Iodine. II. *J. Appl. Phys.* **34**, 1411–15 (1963).

63.011 J. J. LANDER, G. W. GOBELI and J. MORRISON. Structural Properties of Cleaved Si and Ge Surfaces. *J. Appl. Phys.* **34**, 2298–306 (1963).

63.012 J. J. LANDER and J. MORRISON. Scattering Factors and Other Properties of LEED. *J. Appl. Phys.* **34**, 3517–35 (1963).

63.013 J. J. LANDER and J. MORRISON. LEED Study of Si Surface Structures. *Ann. N.Y. Acad. Sci.* **101**, 605–26 (1963).

63.014 A. U. MACRAE. The Epitaxial Growth of NiO on a (111) Ni Surface. *Appl. Phys. Letters* **2**, 88–90 (1963).

63.015 A. U. MACRAE. LEED. *Science* **139**, 379–88 (1963).

63.016 A. U. MACRAE and L. H. GERMER. The Interatomic Spacings at the Surface of a Clean Ni Crystal. *Ann. N.Y. Acad. Sci.* **101**, 627–33 (1963).

63.017 A. A. MARADUDIN and P. A. FLINN. Anharmonic Contributions to the Debye–Waller Factor. *Phys. Rev.* **129**, 2529–47 (1963).

63.018 J. MORRISON. The Detection of Monolayer Adsorption on Si and Ge by LEED. *1963 Transactions of the 10th National Vacuum Symposium* (Macmillan, N.Y., 1963), pp. 440–3.

63.019 R. L. PARK and H. E. FARNSWORTH. The Interaction of Oxygen with a Clean (111) Ni Surface. *Appl. Phys. Letters* **3**, 167–8 (1963).

63.020 C. W. TUCKER. Multiple Scattering in LEED from Oxygen Covered Pt. *Appl. Phys. Letters* **3**, 98–100 (1963).

**1964**

64.001 J. E. BOGGIO and H. E. FARNSWORTH. LEED and Photoelectric Study of (110) Ta as a Function of Ion Bombardment and Heat Treatment. *Surface Sci.* **1**, 399–406 (1964).

64.002 J. CELY. Der Debye-Waller-Faktor für die Diffraktion Langsammer Elektronen. *Phys. Stat. Solidi* **4**, 521–6 (1964).

64.003 J. M. CHARLOT and R. DEGEILH. Étude De La Face (110) D'un Monocristal De Cuivre Par Diffraction Des Électrons De Basse Énergie. *C.R. Acad. Sci.* **259**, 2977–9 (1964).

64.004 P. J. ESTRUP and J. MORRISON. Studies of Monolayers of Pb and Sn on Si(111) Surfaces. *Surface Sci.* **2**, 465–72 (1964).

64.005 H. E. FARNSWORTH. The Clean Single-crystal-surface Approach to Surface Reactions. *Adv. Catalysis* **15**, 31–63 (1964).

64.006 G. GAFNER. Calculations on the Wavelength Dependance of the Intensity of Back Reflected Electrons from Ni(111). *Surface Sci.* **2**, 534–43 (1964).

64.007 L. H. GERMER. LEED. *Physics Today* **17** (7), 19–23 (1964).

64.008 S. GOLDSZTAUB and B. LANG. Diffraction des Électrons de Faible Énergie Par Les Cristaux Isolants. *C.R. Acad. Sci.* **258**, 117–18 (1964).

64.009 D. A. GORODETSKII, A. M. KORNEV and YU. P. MELNIK. The Structure of Ba Films Adsorbed on the (110) Face of a W Single Crystal. *Bull. Acad. Sci. U.S.S.R. Phys. Series* **28**, 1241–4 (1964).

64.010 N. R. HANSEN and D. HANEMAN. Interpretation of LEED Data to Predict Surface Atom Arrangements. *Surface Sci.* **2**, 566–74 (1964).

64.011 C. A. HAQUE and H. E. FARNSWORTH. Cleaning Methods, Gas Adsorption and Corrosion Characteristics of a (100) Cr Single Crystal Using LEED. *Surface Sci.* **1**, 378–86 (1964).

64.012 J. J. LANDER. Chemisorption and Ordered Surface Structures. *Surface Sci.* **1**, 125–64 (1964).

64.013 J. J. LANDER and J. MORRISON. Surface Reactions of Si with Al and In. *Surface Sci.* **2**, 553–65 (1964).

64.014 J. J. LANDER and J. MORRISON. LEED Study of Graphite. *J. Appl. Phys.* **35**, 3593–8 (1964).

64.015 A. U. MACRAE. Adsorption of Oxygen on the (111), (100) and (110) Surfaces of Clean Ni. *Surface Sci.* **1**, 319–48 (1964).

64.016 A. U. MACRAE. Surface Atom Vibrations. *Surface Sci.* **2**, 522–6 (1964).

64.017 A. U. MACRAE and G. W. GOBELI. LEED Study of the Cleaved (110) Surfaces of InSb, InAs, GaAs, GaSb. *J. Appl. Phys.* **35**, 1629–38 (1964).

64.018 A. A. MARADUDIN and J. MELNGAILIS. Some Dynamical Properties of Surface Atoms. *Phys. Rev.* **133**, 1188–93 (1964).

64.019 E. G. MCRAE and C. W. CALDWELL. LEED Study of LiF(100) Surface. *Surface Sci.* **2**, 509–15 (1964).

64.020 J. B. MARSH and H. E. FARNSWORTH. LEED Studies of (100) and (111) Surfaces of Semiconducting Diamond. *Surface Sci.* **1**, 3–21 (1964).

64.021 K. MÜLLER. Beugungsexperimente Mit langsamen Elecktronen an Silber Aufdampfschichten. *Z. Naturforsch.* **19a**, 1234–5 (1964).

64.022 P. W. PALMBERG, D. W. C. JOHNSON and H. J. BOLL. Construction of Hemispherical Grids. *Rev. Sci. Instrum.* **35**, 244–5 (1964).

64.023 R. L. PARK and H. E. FARNSWORTH. CO Adsorption and Interaction with Oxygen on (110) Ni. *J. Chem. Phys.* **40**, 2354–7 (1964).

64.024 R. L. PARK and H. E. FARNSWORTH. Pulsed-beam LEED System for Rapid Precision Measurements. *Rev. Sci. Instrum.* **35**, 1592–5 (1964).

64.025 R. L. PARK and H. E. FARNSWORTH. Interaction of Oxygen with (110) Ni. *J. Appl. Phys.* **35**, 2220–6 (1964).

64.026 R. L. PARK and H. E. FARNSWORTH. The Structures of

Clean Ni Crystal Surfaces. *Surface Sci.* **2**, 527–33 (1964).

64.027 R. Seiwatz. Possible Structures for Clean, Annealed Surfaces of Ge and Si. *Surface Sci.* **2**, 473–83 (1964).

64.028 R. M. Stern. W(110) Surface Characteristics in LEED. *Appl. Phys. Letters* **5**, 218–20 (1964).

64.029 N. J. Taylor. A LEED Study of the Structural Effect of Oxygen on the (111) Face of a W Crystal. *Surface Sci.* **2**, 544–52 (1964).

64.030 C. W. Tucker. LEED Studies of Gas Adsorption on Pt (100), (110), and (111) Surfaces. *J. Appl. Phys.* **35**, 1897–909 (1964).

64.031 C. W. Tucker. LEED of CO Adsorption on the (100) Face of Pt. *Surface Sci.* **2**, 516–21 (1964).

64.032 E. A. Wood. Vocabulary of Surface Crystallography. *J. Appl. Phys.* **35**, 1306–12 (1964).

64.033 E. A. Wood. The 80 Diperiodic Groups in Three Dimensions. *Bell System Tech. J.* **43**, 541–59 (1964).

**1965**

65.001 J. Aldag and R. M. Stern. Surface Thermal Diffuse Scattering from W. *Phys. Rev. Letters* **14**, 857–60 (1965).

65.002 J. Anderson and W. E. Danforth. LEED Study of the Adsorption of Oxygen on a (100) W Surface. *J. Franklin Inst.* **279**, 160–8 (1965).

65.003 J. Anderson, P. J. Estrup and W. E. Danforth. Structure of a Th Monolayer on the W(100) Surface. *Appl. Phys. Letters* **7**, 122–3 (1965).

65.004 E. Bauer. The Study of the Degree of Order in Adsorption by Electron Diffraction. Symposium: "Adsorption et Croissance Cristalline", Nancy, 6–12 June, 1965, pp. 20–48 (Editions du Centre National de la Recherche Scientifique).

65.005 J. E. Boggio and H. E. Farnsworth. LEED Study of the Formation of TaO(111) on Ta(110). *Surface Sci.* **3**, 62–70 (1965).

65.006 C. BURGGRAF and S. GOLDSZTAUB. Bragg Diffraction of Electrons in the 300 to 2300 eV Range by Ionic Crystals. *C.R. Acad. Sci.* **260**, 1115–18 (1965).

65.007 C. BURGGRAF, S. GOLDSZTAUB and B. LANG. Diffraction of Low-energy Electrons. "International Conference on Electron Diffraction and the Nature of Defects in Crystals", Melbourne, Australia, 1965, paper IL-3.

65.008 C. W. CALDWELL. Double Grid Repeller System to Improve Electron Resolution in LEED Equipment. *Rev. Sci. Instrum.* **36**, 1500–1 (1965).

65.009 B. C. CLARK, R. HERMAN and R. F. WALLIS. Theoretical Mean Square Displacements for Surface Atoms in Face Centered Cubic Lattices with Applications to Ni. *Phys. Rev.* **139A**, 860–7 (1965).

65.010 B. C. DICKER. Universal Motion Specimen Manipulator for use with an Ultra High Vacuum System. *J. Sci. Instrum.* **42**, 887–8 (1965).

65.011 H. E. FARNSWORTH. Considerations on the Interpretation of LEED Data. "International Symposium on Basic Problems in Thin Film Physics" (Clausthel-Zellerfeld and Göttingen, Germany, 1965), pp. 6–9.

65.012 H. E. FARNSWORTH and C. A. HAQUE. LEED Observations for Calibrated Ni Films on (111) Cu, "International Conference on Electron Diffraction and the Nature of Defects in Crystals", Melbourne, Australia, 1965, paper IL-1.

65.013 L. G. FEINSTEIN and D. P. SHOEMAKER. LEED Study of Vacuum Cleaved (110) CdTe. *Surface Sci.* **3**, 294–7 (1965).

65.014 L. H. GERMER. The Structure of Crystal Surfaces. *Scientific American* **212** (3), 32–41 (1965).

65.015 L. H. GERMER and J. W. MAY. Structure of Crystal Surfaces. Adsorption of Oxygen and Carbon Monoxide on a (110) W Surface. "International Conference on Electron Diffraction and the Nature of Defects in Crystals", Melbourne, Australia, 1965, paper IK-1.

65.016 S. HAGSTROM, H. B. LYON and G. A. SOMORJAI. LEED

Study of Pt Single Crystal Surfaces. "International Conference on Electron Diffraction and the Nature of Defects in Crystals", Melbourne, Australia, 1965, paper IL-2.

65.017 S. HAGSTROM, H. B. LYON and G. A. SOMORJAI. Surface Structures on the Clean Pt (100) Surface. *Phys. Rev. Letters* **15**, 491–3 (1965).

65.018 R. L. JACOBSON and G. K. WEHNER. Study of Ion-bombardment Damage on a Ge (111) Surface by LEED. *J. Appl. Phys.* **36**, 2674–81 (1965).

65.019 F. JONA. Preparation of Atomically Clean Surfaces of Si and Ge by Heating in Vacuum. *Appl. Phys. Letters* **6**, 205–8 (1965).

65.020 F. JONA. Observations of "Clean" Surfaces of Si, Ge and GaAs by LEED. *IBM J. of Res. Develop.* **9**, 375–87 (1965).

65.021 J. J. LANDER. LEED and Surface Structural Chemistry. *Progress in Solid State Chemistry*, ed. H. Reiss (Pergamon Press, N.Y., 1965), Vol. 2, pp. 26–116.

65.022 J. J. LANDER. LEED Studies of Surface Structure. "International Symposium on Basic Problems in Thin Film Physics" (Clausthel-Zellerfeld and Göttingen, Germany, 1965), pp. 1–5.

65.023 J. J. LANDER and J. MORRISON. Surface Reactions of Si(111) with Al and In. *J. Appl. Phys.* **36**, 1706–13 (1965).

65.024 R. N. LEE and H. E. FARNSWORTH. LEED Studies of Adsorption on Clean (100) Cu Surfaces. *Surface Sci.* **3**, 461–79 (1965).

65.025 A. MANY, Y. GOLDSTEIN and N. B. GROVER. *Semiconductor Surfaces* (North-Holland, Amsterdam, 1965), Chapter 3.

65.026 J. W. MAY. Electron Diffraction and Surface Chemistry. *Industr. Engng. Chem.* **57** (7), 18–39 (1965).

65.027 D. MENZEL. The Importance of Electron Adsorbate Interactions in LEED Experiments. *Surface Sci.* **3**, 424–6 (1965).

65.028 E. MENZEL and O. SCHOTT. Diffraction of Slow Electrons on Spherical Cu Crystals with Untouched Surfaces. *Z. Naturforsch.* **20a**, 1221–3 (1965).

65.029 D. F. MITCHELL, G. W. SIMMONS and K. R. LAWLESS. Comparison of LEED and Low-angle, HEED Techniques for the Observation of Chemisorption. *Appl. Phys. Letters* **7**, 173–5 (1965).

65.030 S. MIYAKE and K. HAYAKAWA. On Nature of LEED by Crystal. "International Conference on Electron Diffraction and the Nature of Defects in Crystals", Melbourne, Australia, 1965, paper IK-3.

65.031 K. MÜLLER. Orientierte Sauerstoffadsorption an Silber Aufdampfschichten. *Z. Naturforsch.* **20a**, 153–4 (1965).

65.032 R. L. PARK and H. E. FARNSWORTH. Work Function Changes Resulting from the Interaction of Oxygen with Clean Ni Surfaces. *Surface Sci.* **3**, 287–9 (1965).

65.033 R. L. PARK and H. E. FARNSWORTH. Adsorption and Oxidation of Carbon Monoxide on (100) Ni. *J. Chem. Phys.* **43**, 2351–4 (1965).

65.034 A. J. PIGNOCCO and G. E. PELLISSIER. LEED Study of Oxygen Adsorption and Oxide Formation on a (001) Fe Surface. *J. Electrochem. Soc.* **112**, 1188–94 (1965).

65.035 P. B. SEWELL, E. G. BREWER and M. COHEN. Reflection Electron Diffraction Studies of Gas Adsorption Phenomena. "International Conference on Electron Diffraction and the Nature of Defects in Crystals", Melbourne, Australia, 1965, paper IL-4.

65.036 P. B. SEWELL and M. COHEN. The Observation of Gas Adsorption Phenomena by Reflection HEED. *Appl. Phys. Letters* **7**, 32–5 (1965).

65.037 R. J. SZOSTAK and K. MOLIÈRE. Diffraction Experiments with Cold Electrons on Fe. "International Symposium on Basic Problems in Thin Film Physics" (Clausthel-Zellerfeld and Göttingen, Germany, 1965), pp. 10–17.

65.038 C. J. TODD. Spherically Shaped Grids for a LEED System. *J. Sci. Instrum.* **42**, 755 (1965).

65.039 C. W. TUCKER. The Importance of Electron Adsorbate Interactions in LEED Experiments. *Surface Sci.* **3**, 427–8 (1965).

**1966**

66.001 R. A. Armstrong. The Reflection of Slow Electrons by Hydrogen-covered Single Crystals of W. *Canad. J. Phys.* **44**, 1753–64 (1966).

66.002 J. Bailleul-Langlais. Use of LEED for the Study of Semiconductor Surfaces. *Vide* **21** (special issue), 147–52 (1966).

66.003 E. Bauer. Comments on "The Uncertainty Regarding Reconstructed Surfaces" by L. H. Germer. *Surface Sci.* **5**, 152–4 (1966).

66.004 E. Bauer, A. K. Green and K. M. Kunz. Epitaxial Growth of Single Crystal Metal Films Free of Impurities. *Appl. Phys. Letters* **8**, 248–9 (1966).

66.005 L. de Bersuder. Nouvel Appareil pour l'étude de la Diffraction Électronique á Faible et Moyenne Énergie. *C.R. Acad. Sci.* **262B**, 1056–8 (1966).

66.006 C. Burggraf and S. Goldsztaub. Electron Diffraction in the Range of 800–2300 eV at Grazing Incidence. *C.R. Acad. Sci.* **262B**, 1118–20 (1966).

66.007 W. P. Ellis. Surface Configurations of the (111) Plane of $UO_2$. *Fundamentals of Gas-surface Interactions*, eds. H. Saltsburg, J. N. Smith and M. Rogers (Academic Press, N.Y., 1966), pp. 89–101.

66.008 J. Escard, S. Goldsztaub and G. David. Examen d'échantillons de Graphite Pyrolytique, par Diffraction des Électrons Leuts. *C.R. Acad. Sci.* **262B**, 966–7 (1966).

66.009 P. J. Estrup and J. Anderson. Chemisorption of Hydrogen on W(100). *J. Chem. Phys.* **45**, 2254–60 (1966).

66.010 P. J. Estrup, J. Anderson and W. E. Danforth. LEED Studies of Th Adsorption on W. *Surface Sci.* **4**, 286–98 (1966).

66.011 D. G. Fedak and N. A. Gjostein. Structure and Stability of the (100) Surface of Au. *Phys. Rev. Letters* **16**, 171–2 (1966).

66.012 K. Fujiwara, K. Hayakawa and S. Miyake. A New Slow Electron Diffraction Apparatus. *Japan. J. Appl. Phys.* **5**, 295–8 (1966).

66.013 L. H. Germer. The Uncertainty Regarding Reconstructed Surfaces. *Surface Sci.* **5**, 147–51 (1966).

66.014 L. H. Germer. Present and Proposed Uses of LEED in Studying Surfaces. *Fundamental Phenomena in the Materials Sciences*, eds. L. J. Bonis, P. L. de Bruyn, and J. J. Duga (Plenum Press, N.Y., 1966), pp. 23–39.

66.015 L. H. Germer and C. C. Chang. Secondary Scattering of Low Energy Electrons by Rows of Atoms. *Surface Sci.* **4**, 498–501 (1966).

66.016 L. H. Germer, S. Goldsztaub, J. Escard, G. David and J. P. Deville. Étude de l'intensité des Électrons Leuts Diffractés par le Pyrographite. *C.R. Acad. Sci.* **262B**, 1059–62 (1966).

66.017 L. H. Germer and J. W. May. Diffraction Study of Oxygen Adsorption on a (110) W Face. *Surface Sci.* **4**, 452–70 (1966).

66.018 G. W. Gobeli, J. J. Lander and J. Morrison. LEED Study of the Adsorption of Cs on the (111) Surface of Si. *J. Appl. Phys.* **37**, 203–6 (1966).

66.019 S. Goldsztaub, G. David, J. P. Deville and B. Lang. Low-energy Electron Study of a Muscovite Crystal Cleaved in Ultra-high Vacuum. *C.R. Acad. Sci.* **262B**, 1718–20 (1966).

66.020 D. A. Gorodetskii and Yu. P. Melnik. Structure of BaO Films on the (110) Face of a W Single Crystal. *Sov. Phys. Solid State* **7**, 2248–54 (1966).

66.021 T. W. Haas. A Study of the Nb(110) Surface Using LEED Techniques. *Surface Sci.* **5**, 345–58 (1966).

66.022 T. W. Haas and A. G. Jackson. LEED Study of Mo (110) Surfaces. *J. Chem. Phys.* **44**, 2921–5 (1966).

66.023 S. Hagstrom. LEED. *Svensk. Kem. Tidskr.* **78**, 246–59 (1966).

66.024 C. A. Haque and H. E. Farnsworth. LEED Observations of Calibrated Epitaxial Ni Films on (111) Cu. *Surface Sci.* **4**, 195–200 (1966).

66.025 J. C. Helmer. Intensity of Electron Beams. *Amer. J. Phys.* **34**, 222–7 (1966).

66.026 K. Hirabayashi and Y. Takeishi. Diffraction of Low-energy Electrons at Crystal Surfaces. *Surface Sci.* **4**, 150–60 (1966).

66.027 D. C. JOHNSON and A. U. MACRAE. Kikuchi Bands in LEED. *J. Appl. Phys.* **37**, 1945–6 (1966).

66.028 D. C. JOHNSON and A. U. MACRAE. LEED Study of the Cleaved Surfaces of PbS, PbSe, and PbTe. *J. Appl. Phys.* **37**, 2298–303 (1966).

66.029 F. JONA. Study of the Early Stages of the Epitaxy of Si on Si. *Appl. Phys. Letters* **9**, 235–7 (1966).

66.030 E. R. JONES, J. T. MCKINNEY and M. B. WEBB. Surface Lattice Dynamics of Ag. I. Low-energy Electron Debye–Waller Factor. *Phys. Rev.* **151**, 476–83 (1966).

66.031 P. KOFSTAD. *High Temperature Oxidation of Metals* (J. Wiley, N.Y., 1966), Chapter II.

66.032 J. J. LANDER. Properties of Ordered Physisorbed Phases Observed with LEED. *Fundamentals of Gas-surface Interactions*, eds. H. Saltsburg, J. N. Smith, and M. Rogers (Academic Press, N.Y., 1966), pp. 25–41.

66.033 J. J. LANDER and J. MORRISON. Ordered Physisorbed Layers on Graphite. *Surface Sci.* **4**, 103–7 (1966).

66.034 J. J. LANDER and J. MORRISON. LEED Study of the (111) Diamond Surface. *Surface Sci.* **4**, 241–6 (1966).

66.035 H. B. LYON, A. M. MATTERA and G. A. SOMORJAI. LEED Study of F.C.C. Metal Surfaces. *Fundamentals of Gas-surface Interactions*, eds. H. Saltsburg, J. N. Smith, and M. Rogers (Academic Press, N.Y., 1966), pp. 102–15.

66.036 H. B. LYON and G. A. SOMORJAI. Surface Debye Temperatures of the (100), (111) and (110) Faces of Pt. *J. Chem. Phys.* **44**, 3707–11 (1966).

66.037 A. U. MACRAE. LEED Study of the Polar (111) Surfaces of GaAs and GaSb. *Surface Sci.* **4**, 247–64 (1966).

66.038 A. U. MACRAE. LEED. NATO Advanced Study Inst. *Use of Thin Films in Physical Investigation*, ed. J. C. Anderson (Academic Press, N.Y., 1966), pp. 149–61.

66.039 A. U. MACRAE and G. W. GOBELI. LEED Studies. *Semiconductors and Semimetals*, Vol. 2, *Physics of III–V Compounds*, ed. R. K. Willardson and A. C. Beer (Academic Press, N.Y., 1966), pp. 149–61.

66.040 I. Marklund and S. Andersson. LEED Study of NaCl (100) Surface. *Surface Sci.* **5**, 197–202 (1966).

66.041 J. W. May and L. H. Germer. Adsorption of Carbon Monoxide on a W(110) Surface. *J. Chem. Phys.* **44**, 2895–902 (1966).

66.042 J. W. May, L. H. Germer and C. C. Chang. Coadsorption of Oxygen and Carbon Monoxide upon a (110) W Surface. *J. Chem. Phys.* **45**, 2383–9 (1966).

66.043 R. E. McCurry. Moiré Patterns with Display Type LEED Apparatus. *J. Appl. Phys.* **37**, 473–9 (1966).

66.044 E. G. McRae. Wave-interference Mechanism for Fractional Order Peaks in LEED Intensity Curves. *Fundamentals of Gas-surface Interactions*, eds. H. Saltsburg, J. N. Smith, and M. Rogers (Academic Press, N.Y., 1966), pp. 116–31.

66.045 E. G. McRae. Multiple-scattering Treatment of LEED Intensities. *J. Chem. Phys.* **45**, 3258–76 (1966).

66.046 S. Miyake and K. Hayakawa. Nature of Slow Electron Diffraction by Crystals. *J. Phys. Soc. Japan* **21**, 363–78 (1966).

66.047 S. Miyake and K. Hayakawa. Kikuchi Patterns in Slow Electron Diffraction. *J. Phys. Soc. Japan* **21**, 1454 (1966).

66.048 J. Morrison. Low Temperature Specimen Holder for Use in LEED Systems. *Rev. Sci. Instrum.* **37**, 1263–4 (1966).

66.049 J. Morrison and J. J. Lander. Ordered Physisorption of Xe on Graphite. *Surface Sci.* **5**, 163–5 (1966).

66.050 K. Müller. Oberflächenstrukturen von Silber-Aufdampfschichten. Eine Untersuchung mit Interferenzen Langsamer Electronen. *Z. Phys.* **195**, 105–24 (1966).

66.051 K. Müller and H Viefhaus. Orientierung von Goldaufdampfschichten. *Z. Naturforsch.* **21a**, 1726 (1966).

66.052 Y. H. Ohtsuki and S. Yanagawa. Dynamical Theory of Diffraction. I. Electron Diffraction. *J. Phys. Soc. Japan* **21**, 326–35 (1966).

66.053 R. L. Park. LEED Beam Broadening as a Result of Sputtering Damage. *J. Appl. Phys.* **37**, 295–8 (1966).

66.054 J. L. ROBINS, R. L. GERLACH and T. N. RHODIN. Kikuchi Effects from Low-energy Electrons in Ni. *Appl. Phys. Letters* **8**, 12–14 (1966).

66.055 A. M. RUSSELL. Structure of an Adsorbed Layer on a Ti(0001) Surface. *Appl. Phys. Letters* **8**, 177–8 (1966).

66.056 G. SHINODA, S. KOTO, S. FUJIMOTO and H. KOBAYASHI. Apparatus for Alternative Observation of LEED and Photoelectric Emission. *Rev. Sci. Instrum.* **37**, 533–4 (1966).

66.057 Y. TAKEISHI, I. SASAKI and Y. IOKI. An Ordered Structure on a Tantalum-adsorbed Si(111) Surface. *Surface Sci.* **4**, 317–19 (1966).

66.058 H. TAUB and R. M. STERN. Nearest-neighbor Electron Scattering in Silicon. *Appl. Phys. Letters* **9**, 261–3 (1966).

66.059 N. J. TAYLOR. A LEED Study of the Epitaxial Growth of Cu on the (110) Surface of W. *Surface Sci.* **4**, 161–94 (1966).

66.060 C. W. TUCKER. Interaction of Boron with W Single Crystal Substrates. *Surface Sci.* **5**, 179–86 (1966).

66.061 C. W. TUCKER. Stability of Coincidence Lattices for Chemisorbed Structures. *J. Appl. Phys.* **37**, 528–31 (1966).

66.062 C. W. TUCKER. Chemisorbed Coincidence Lattices on Rh. *J. Appl. Phys.* **37**, 3013–9 (1966).

66.063 C. W. TUCKER. Chemisorbed Oxygen Structures on the Rh (110) Surface. *J. Appl. Phys.* **37**, 4147–55 (1966); Corrections, *ibid.* **38**, 2696–7 (1967).

66.064 R. F. WALLIS and A. A. MARADUDIN. Theory of Surface Effects on the Thermal Diffuse Scattering of X-rays and Electrons from Crystal Lattices. *Phys. Rev.* **148**, 962–7 (1966).

66.065 G. K. ZYRYANOV. Diffraction of Slow Electrons by NaCl. *Vestn. Leningr. Univ.* **21** (No. 4, Issue 1), 34–40 (1966).

66.066 G. K. ZYRYANOV. Interpretation of the Pattern of Slow Electron Diffraction by NaCl. *Vestn. Leningr. Univ.* **21** (No. 16, Issue 3), 70–4 (1966).

66.067 G. K. ZYRYANOV. Diffraction of Slow Electrons by PbS. *Vestn. Leningr. Univ.* **21** (No. 22, Issue 4), 80–4 (1966).

**1967**

67.001 R. O. ADAMS. Cleaning a (0001) Be Surface. *Mater. Res. Bull.* **2**, 649–72 (1967).

67.002 G. ALLIE, R. RIWAN and J. J. TRILLAT. LEED by Growth Faces of Cd. *C.R. Acad. Sci.* **265B**, 1452–5 (1967).

67.003 J. ANDERSON and P. J. ESTRUP. Adsorption of Carbon Monoxide on a W(100) Surface. *J. Chem. Phys.* **46**, 563–7 (1967).

67.004 N. T. BATKIN and R. J. MADDIX. A Study of the Gaseous Etching of Ge by Oxygen. *Surface Sci.* **7**, 109–14 (1967).

67.005 E. BAUER. Multiple Scattering Versus Superstructures in LEED. *Surface Sci.* **7**, 351–64 (1967).

67.006 W. BERNDT. Beugung Langsamer Elektronen an Kugelförmigen Oxydierten Kupfer-Einkristallen. *Z. Naturforsch.* **22a**, 1655–71 (1967).

67.007 L. DE BERSUDER. Observations on the Secondary Emission of the (001) Face of Al Related to the Diffraction of Electrons of Low and Medium Energy. *C.R. Acad. Sci.* **265B**, 885–8 (1967).

67.008 H. P. BONZEL and N. A. GJOSTEIN. Use of a Laser Diffraction Pattern to Study Surface Self-diffusion of Metals. *Appl. Phys. Letters* **10**, 258–60 (1967).

67.009 D. S. BOUDREAUX and V. HEINE. Band Structure Treatment of LEED Intensities. *Surface Sci.* **8**, 426–44 (1967).

67.010 J. J. BURTON and G. JURA. Phase Transformations on Solid Surfaces. *Phys. Rev. Letters* **18**, 740–3 (1967).

67.011 J. J. BURTON and G. JURA. Surface Distortion in F.C.C. Solids. *J. Phys. Chem.* **71**, 1937–9 (1967).

67.012 C. C. CHANG and L. H. GERMER. Oxidation of the (112) Face of W. *Surface Sci.* **8**, 115–29 (1967).

67.013 J. M. CHARIG. LEED Observations of $\alpha$-Alumina. *Appl. Phys. Letters* **10**, 139–40 (1967).

67.014 A. CHUTJIAN. Coherence Area in LEED. *Phys. Letters* **24A**, 615–16 (1967).

67.015 C. COROTTE, P. DUCROS and D. LAFEUILLE. Study of the (111) Face of Ag by LEED. Importance of Incident

Beam Direction on Secondary Emission, Particularly on Kikuchi Scattering. *C.R. Acad. Sci.* **265B**, 1040–3 (1967).

67.016 G. DAVID, J. ESCARD and S. GOLDSZTAUB. LEED by Molybdenite. *C.R. Acad. Sci.* **265B**, 1449–51 (1967).

67.017 T. A. DELCHAR and F. C. TOMPKINS. Chemisorption of Oxygen and Hydrogen on Ni Films. Reconstitution of the Adsorbate–Adsorbent Interface. *Surface Sci.* **8**, 165–72 (1967).

67.018 J. P. DEVILLE, J. P. EBERHART and S. GOLDSZTAUB. Diffraction of Low-energy Electrons by a Phlogopite Mica Crystal. *C.R. Acad. Sci.* **264B**, 124–5 (1967).

67.019 J. P. DEVILLE, J. P. EBERHART and S. GOLDSZTAUB. LEED by a Muscovite Mica Crystal. *C.R. Acad. Sci.* **264B**, 289–91 (1967).

67.020 J. L. DOMANGE and J. OUDAR. Structure of The Single Layer of Adsorbed Sulfur on the (100) Face of Cu. *C.R. Acad. Sci.* **264C**, 35–7, 951–3 (1967).

67.021 V. F. DVORYANKIN and A. YU. MITYAGIN. Some Problems in the Geometry of Diffraction Patterns from Slow Electrons. *Kristallografiya* **12**, 966–70 (1967).

67.022 V. F. DVORYANKIN and A. YU. MITYAGIN. Diffraction of Slow Electrons: A Method of Studying the Atomic Structure of Surfaces. *Kristallografiya* **12**, 1112–34 (1967).

67.023 V. F. DVORYANKIN, A. YU. MITYAGIN and S. A. AKHUNDOV. A Criterion for the Applicability of the Born Approximation to Electron-diffraction Structural Studies. *Kristallografiya* **12**, 579–83 (1967).

67.024 W. ECKSTEIN. Measurement of the Spin Polarization of Slow Electrons after Elastic Scattering from Solid Hg. *Z. Physik.* **203**, 59–65 (1967).

67.025 G. ERTL. Untersuchung Von Oberflächenreaktionen Mittels Beugung Langsamer Elektronen (LEED). I. Wechselwirkung von $O_2$ und $N_2O$ mit (110)-, (111)- und (100)-Kupfer-Oberflachen. *Surface Sci.* **6**, 208–32 (1967).

67.026 G. ERTL. Untersuchung Von Oberflächenreaktionen an Kupfer Mittels Beugung Langsamer Electronen (LEED). II. *Surface Sci.* **7**, 309–31 (1967).

67.027 P. J. ESTRUP and J. ANDERSON. LEED Studies of the Adsorption Systems W(100) + $N_2$ and W(100) + $N_2$ + CO. *J. Chem. Phys.* **46**, 567–70 (1967).

67.028 P. J. ESTRUP and J. ANDERSON. Problem of Adsorbate Arrangement in the W(100) $(1 \times 4) - (N + CO)$ Structure. *J. Chem. Phys.* **47**, 1881–2 (1967).

67.029 P. J. ESTRUP and J. ANDERSON. Th Layers on W(100); Structure and LEED Intensities. *Surface Sci.* **7**, 255–8 (1967).

67.030 P. J. ESTRUP and J. ANDERSON. Characterization of Chemisorption by LEED. *Surface Sci.* **8**, 101–14 (1967).

67.031 H. E. FARNSWORTH. Atomically Clean Solid Surfaces—Preparation and Evaluation. *The Solid–Gas Interface*, ed. E. A. Flood (M. Dekker, N.Y., 1967), Chapter 13.

67.032 H. E. FARNSWORTH and K. HAYEK. Investigation of Surface Bombardment Damage by LEED. *Surface Sci.* **8**, 35–56 (1967).

67.033 H. E. FARNSWORTH and K. HAYEK. Surface Structures Due to Interaction of Oxygen with Mo Crystals. *Supplemento Al Nuovo Cimento* **V**, 452–8 (1967).

67.034 D. G. FEDAK and N. A. GJOSTEIN. A LEED Study of the (100), (110) and (111) Surfaces of Au. *Acta Metallurgica* **15**, 827–40 (1967).

67.035 D. G. FEDAK and N. A. GJOSTEIN. On the Anomalous Surface Structures of Au. *Surface Sci.* **8**, 77–97 (1967).

67.036 T E. FEUCHTWANG. Dynamics of a Semi-infinite Crystal Lattice in a Quasiharmonic Approx. I. *Phys. Rev.* **155**, 715–30 (1967).

67.037 T. E. FEUCHTWANG. Dynamics of a Semi-infinite Crystal Lattice in a Quasiharmonic Approx. II. *Phys. Rev.* **155**, 731–44 (1967).

67.038 L. FIERMANS and J. VENNIK. Influence of the LEED Beam on the Structure of the $V_2O_5$ (010) Surface. *Phys. Letters* **25A**, 687–8 (1967).

67.039 R. L. GERLACH and T. N. RHODIN. Surface Asymmetry Mechanism for Fractional Order Peaks Observed in LEED Beams from Metals. *Surface Sci.* **8**, 1–13 (1967).

67.040 L. H. GERMER, J. W. MAY and R. J. SZOSTAK. Thermally Ordered Oxygen on a Ni Surface. *Surface Sci.* **7**, 430–47 (1967).

67.041 D. A. GORODETSKII, YU. P. MELNIK and A. A. YASKO. Structure of Li Films on a (110) W Surface. *Ukr. Fiz. Zh.* **12**, 644–9 (1967).

67.042 D. A. GORODETSKII, YU. P. MELNIK and A. A. YASKO. Structure of (110) W Surface Covered with a Film of Adsorbed Oxygen. *Ukr. Fiz. Zh.* **12**, 967–73 (1967).

67.043 T. W. HAAS, A. G. JACKSON and M. P. HOOKER. Adsorption on Nb(110), Ta(110), V(110) Surfaces. *J. Chem. Phys.* **46**, 3025–33 (1967).

67.044 T. A. HEPPELL. A Combined Low Energy and Reflection High Energy Electron Diffraction Apparatus. *J. Sci. Instrum.* **44**, 686–8 (1967).

67.045 K. HIRABAYASHI. On Edge Dislocations in a Possible Relation to Structure of a Crystal Surface. *J. Phys. Soc. Japan* **22**, 590–7 (1967).

67.046 F. HOFMANN and H. P. SMITH. Dynamical Theory of LEED. *Phys. Rev. Letters* **19**, 1472–4 (1967).

67.047 D. L. HUBER. Thermal Diffuse Scattering of Low-energy Electrons. *Phys. Rev.* **153**, 772–9 (1967).

67.048 A. G. JACKSON and M. P. HOOKER. A LEED Study of Carbon Monoxide and Carbon Dioxide Adsorption on Mo(110). *Surface Sci.* **6**, 297–308 (1967).

67.049 A. G. JACKSON, M. P. HOOKER and T. W. HAAS. LEED Study of the Growth of Al Films on the Ta(110) Surface. *J. Appl. Phys.* **38**, 4998–5004 (1967).

67.050 F. JONA. LEED Study of the Epitaxy of Si on Si. *Proceedings of the 13th Sagamore Army Materials Research Conference*, eds. J. J. Burke, N. L. Reed, and V. Weiss (Univ. Press, Syracuse, N.Y., 1967), pp. 399–434.

67.051 F. JONA. Preparation and Properties of Clean Surfaces of Al. *J. Phys. Chem. Solids* **28**, 2155–60 (1967).

67.052 F. JONA. LEED Study of Surfaces of Sb and Bi. *Surface Sci.* **8**, 57–76 (1967).

67.053 F. JONA. A Note on the Sensitivity of LEED to Surface Perfection. *Surface Sci.* **8**, 478–84 (1967).

67.054 K. KAMBE. Theory of LEED. *Z. Naturforsch.* **22a**, 322–30 (1967).

67.055 K. KAMBE. Theory of Electron Diffraction by Crystals. *Z. Naturforsch.* **22a**, 422–31 (1967).

67.056 H. KRUPP. Surface Physics and Chemical Engineering. *Chem.-Ing.-Tech.* **39**, 1226–30 (1967).

67.057 C. E. KUYATT and J. A. SIMPSON. Electron Monochromator Design. *Rev. Sci. Instrum.* **38**, 103–11 (1967).

67.058 J. J. LANDER and J. MORRISON. A LEED Investigation of Physisorption. *Surface Sci.* **6**, 1–32 (1967).

67.059 R. LOTH. Measurement of the Spin Polarization of 900 eV Electrons After Elastic Scattering From Heavy Metal Films as a Function of Target Temperature. *Z. Physik.* **203**, 66–70 (1967).

67.060 H. B. LYON and G. A. SOMORJAI. LEED Study of the Clean (100), (111), and (110) Faces of Pt. *J. Chem. Phys.* **46**, 2539–50 (1967).

67.061 A. U. MACRAE. Techniques for Studying Clean Surfaces. *Proceedings of the 13th Sagamore Army Materials Research Conference*, eds. J. J. Burke, N. L. Reed and V. Weiss (Univ. Press, Syracuse, N.Y., 1967), pp. 29–52.

67.062 A. U. MACRAE. Surface Measurements Using LEED. *Measurement Techniques for Thin Films*, ed. B. and N. Schwartz (Electrochem. Soc., N.Y., 1967), pp. 38–48.

67.063 A. M. MATTERA, R. M. GOODMAN and G. A. SOMORJAI. LEED Study of the (100) Face of Ag, Au and Pd. *Surface Sci.* **7**, 26–40 (1967).

67.064 J. T. MCKINNEY, E. R. JONES and M. B. WEBB. Surface Lattice Dynamics of Ag. II. Low-energy Electron Thermal Diffuse Scattering. *Phys. Rev.* **160**, 523–30 (1967).

67.065 E. G. MCRAE. Self-consistent Multiple-scattering Approach to the Interpretation of LEED. *Surface Sci.* **8**, 14–34 (1967).

67.066 E. G. McRae and C. W. Caldwell. Observation of Multiple Scattering Resonance Effects in LEED Studies of LiF, NaF and Graphite. *Surface Sci.* **7**, 41–67 (1967).

67.067 E. Menzel and O. Schott. LEED with Spherical Shaped Cu Crystals. *Surface Sci.* **8**, 217–22 (1967).

67.068 D. L. Mills. Magnetic Scattering of Low-energy Electrons from the Surface of a Ferromagnetic Crystal. *J. Phys. Chem. Solids* **28**, 2245–55 (1967).

67.069 J. Morrison. LEED. *Industrial Research*, Feb. 1967, pp. 84–7.

67.070 H. E. Neustadter and R. J. Bacigalupi. Dependence of Adsorption Properties on Surface Structure for B.C.C. Substrates. *Surface Sci.* **6**, 246–60 (1967).

67.071 M. Onchi. Studies of the Gas Adsorption and Reaction on the Surface of Ni Single Crystal by Means of LEED. *Shokubai (Tokyo)* **9**, 179–200 (1967).

67.072 P. W. Palmberg. Low Temperature Cleavage Manipulator for LEED Apparatus. *Rev. Sci. Instrum.* **38**, 834–5 (1967).

67.073 P. W. Palmberg. Secondary Emission Studies on Ge and Na-covered Ge. *J. Appl. Phys.* **38**, 2137–47 (1967).

67.074 P. W. Palmberg and W. T. Peria. LEED Studies on Ge and Na-covered Ge. *Surface Sci.* **6**, 57–97 (1967).

67.075 P. W. Palmberg and T. N. Rhodin. Surface Structure of Clean Au(100) and Ag(100) Surfaces. *Phys. Rev.* **161**, 586–8 (1967).

67.076 P. W. Palmberg, T. N. Rhodin and C. J. Todd. LEED Studies of Epitaxial Growth of Ag and Au in Ultrahigh Vacuum. *Appl. Phys. Letters* **10**, 122–4 (1967).

67.077 P. W. Palmberg, T. N. Rhodin and C. J. Todd. Epitaxial Growth of Au and Ag on MgO Cleaved in Ultrahigh Vacuum. *Appl. Phys. Letters* **11**, 33–5 (1967).

67.078 A. J. Pignocco and G. E. Pellissier. LEED Studies of Oxygen Adsorption and Oxide Formation on an (011) Fe Surface. *Surface Sci.* **7**, 261–78 (1967).

67.079 H. Raether. Surface Plasma Oscillations as a Tool for Surface Examinations. *Surface Sci.* **8**, 233–46 (1967).

67.080 J. C. RICHARD and P. SAGET. Slow Electron Diffraction Diagrams Obtained on Alkali Metal Halide Single Crystals by Means of a Simple Experimental Device. *Vide* **22**, 99–104 (1967).

67.081 E. J. SCHEIBNER and L. N. THARP. Inelastic Scattering of Low-energy Electrons from Surfaces. *Surface Sci.* **8**, 247–65 (1967).

67.082 O. SCHOTT. Diffraction of Slow Electrons on Spherical Cu Monocrystals. *Z. Angew Phys.* **22**, 63–8 (1967).

67.083 P. B. SEWELL and M. COHEN. Reflection HEED and X-ray Emission Analysis of Surfaces and Their Reaction Products. *Appl. Phys. Letters* **11**, 298–9 (1967).

67.084 K. SIEGBAHN, C. NORDLING, A. FAHLMAN, R. NORDBERG, K. HAMRIN, J. HEDMAN, G. JOHANSSON, T. BERGMARK, S. KARLSSON, I. LINDGREN and B. LINDBERG. *ESCA-Atomic, Molecular and Solid State Structure Studied by Means of Electron Spectroscopy* (Almqvist and Wiksells Boktrycheri AB, Uppsala, 1967).

67.085 B. M. SIEGEL and J. F. MENADUE. Quantitative Reflection Electron Diffraction in an Ultra High Vacuum Camera. *Surface Sci.* **8**, 206–22 (1967).

67.086 G. W. SIMMONS, D. F. MITCHELL and K. R. LAWLESS. LEED and HEED Studies of the Interaction of Oxygen with Single Crystal Surfaces of Cu. *Surface Sci.* **8**, 130–64 (1967).

67.087 G. A. SOMORJAI. Note to the paper by D. G. Fedak and N. A. Gjostein, "On the Anomalous Surface Structures of Au". *Surface Sci.* **8**, 98–100 (1967).

67.088 K. SPIEGEL. Untersuchungen zum Schichtwachstum von Silber auf der Silizium(111)-Oberfläche Durch Beugung Langsamer Electronen. *Surface Sci.* **7**, 125–56 (1967).

67.089 R. SZOSTAK and K. MOLIÈRE. Beugung Langsamer Elektronen an Oktaederflächen von Magnetit. *Z. Naturforsch.* **22a**, 1615–20 (1967).

67.090 Y. TAKEISHI, I. SASAKI and K. HIRABAYASHI. A LEED Study of the Epitaxial Si Layers on a Ge(111) Surface. *Appl. Phys. Letters* **11**, 330–2 (1967).

67.091 L. N. THARP and E. J. SCHEIBNER. Energy Spectra of

Inelastically Scattered Electrons and LEED Studies of W. *J. Appl. Phys.* **38**, 3320–30 (1967).

67.092 R. N. THOMAS and M. H. FRANCOMBE. A LEED Study of the Homoepitaxial Growth of Thick Si Films. *Appl. Phys. Letters* **11**, 108–10 (1967).

67.093 R N. THOMAS and M. H. FRANCOMBE. A New Mechanism for Stacking Fault Generation in Epitaxial Growth of Si in Ultra-high Vacuum. *Appl. Phys. Letters* **11**, 134–6 (1967).

67.094 L. TREPTE, C. MENZEL-KOPP and E. MENZEL. Surface Structures on Spherical Cu Crystals After Adsorption of Oxygen. *Surface Sci.* **8**, 223–32 (1967).

67.095 C. W. TUCKER. LEED from Faceted Surfaces. *J. Appl. Phys.* **38**, 1988–9 (1967).

67.096 C. W. TUCKER. Corrections to "Chemisorbed Oxygen Structures on the Rh(110) Surface". *J. Appl. Phys.* **38**, 2696–7 (1967).

67.097 C. W. TUCKER. Oxygen Faceting on Rh(210) and (100) Surfaces. *Acta Metallurgica* **15**, 1465–74 (1967).

67.098 C. W. TUCKER. Suggestions Regarding Certain Oxygen Structures Formed on the W(110) Surface. *Surface Sci.* **6**, 124–6 (1967).

67.099 C. W. TUCKER. Comment on LEED Studies of the Adsorption Systems W(100)+$N_2$ and W(100)+$N_2$+CO. *J. Chem. Phys.* **47**, 1880–1 (1967).

67.100 A. J. VAN BOMMEL and F. MEYER. LEED Measurement of $H_2S$ and $H_2Se$ Adsorption on Ge(111). *Surface Sci.* **6**, 391–4 (1967).

67.101 A. J. VAN BOMMEL and F. MEYER. A LEED Study of the $PH_3$ Adsorption on the Si(111) Surface. *Surface Sci.* **8**, 381–98 (1967).

67.102 A. J. VAN BOMMEL and F. MEYER. LEED Study of a Ni Induced Surface Structure on Si(111). *Surface Sci.* **8**, 467–12 (1967).

67.103 R. E. DE WAMES and L. A. VREDEVOE. Electron-magnon Scattering and Polarization of the Scattered Beam. *Phys. Rev. Letters* **18**, 853–5 (1967).

67.104 R. E. WEBER and W. T. PERIA. Use of LEED Apparatus

for the Detection and Identification of Surface Contaminants. *J. Appl. Phys.* **38**, 4355–8 (1967).

67.105 H. YAMAZAKI, H. OEDA, K. KANAYA and K. TANAKA. Space Charge Effect on the Minimum Spot Size of a Low-energy Electron Probe. *Bull. Electrotech. Lab. (Japan)* **31**, 134–47 (1967).

67.106 J. ZELL, G. BOUCHET, A. J. MASCALL and B. MUTAFTSCHIEV. Slow Electron Diffraction Study of the Growth Surfaces of Cd. *C.R. Acad. Sci.* **265B**, 1452–5 (1967).

## 1968

68.001 H. C. ABBINK, R. M. BROUDY and G. P. MCCARTHY. Surface Processes in the Growth of Si on (111) Si in Ultrahigh Vacuum. *J. Appl. Phys.* **39**, 4673–81 (1968).

68.002 D. ABERDAM, G. BOUCHET and P. DUCROS. Slow Electron Diffraction Observation of Linear Disorder on the (001) Face of Barium Titanate Single Crystals Heated in an Ultravacuum. *C.R. Acad. Sci.* **266B**, 1334–6 (1968).

68.003 D. ABERDAM, C. COROTTE, P. DUCROS and D. LAFEUILLE. LEED Study of the Rearrangement of the (111) Face of Ag During Oxidation. *Kristallogr.* **8**, 271–2 (1968).

68.004 G. ALLIE. Examination of the (100) Face of NiO Crystals with Slow Electron Diffraction. The Effect of Temperature. *C.R. Acad. Sci.* **266B**, 1568–70 (1968).

68.005 G. F. AMELIO and E. J. SCHEIBNER. Auger Spectroscopy of Graphite Single Crystals with Low Energy Electrons. *Surface Sci.* **11**, 242–54 (1968).

68.006 J. ANDERSON and P. J. ESTRUP. Effect of Electrons on $NH_3$ Adsorbed on a W(100) Surface. *Surface Sci.* **9**, 463–7 (1968).

68.007 S. ANDERSSON, D. ANDERSSON and I. MARKLUND. Clean Te Surfaces Studied by LEED. *Surface Sci.* **12**, 284–98 (1968).

68.008 S. ANDERSSON, I. MARKLUND and J. MARTINSSON. Epitaxial Growth of Te on the Cu(111) Surface Studied by LEED. *Surface Sci.* **12**, 269–83 (1968).

68.009 G. Appelt. Fine Structure Measurements in the Energy Angular Distribution of Secondary Electrons from a (110) Face of Cu. *Phys. Stat. Solidi* **27**, 657–69 (1968).

68.010 L. Bakker. Comments on a paper by K. Müller: "Surface Structure of Evaporated Ag Films, on Investigation by Means of LEED". *Z. Phys.* **212**, 104–6 (1968).

68.011 R. F. Barnes, M. G. Lagally and M. B. Webb. Multi-phonon Scattering of Low-energy Electrons. *Phys. Rev.* **171**, 627–33 (1968).

68.012 G. C. Barton and J. Ferrante. Faceting from CO Adsorption on Mo(111). *J. Chem. Phys.* **48**, 4791–3 (1968).

68.013 R. Baudoing and R. M. Stern. Metastable Surface Structure of the W(110) Face. *Surface Sci.* **10**, 392–8 (1968).

68.014 R. Baudoing, R. M. Stern and H. Taub. Inner Sources in LEED: W(110). *Surface Sci.* **11**, 255–64 (1968).

68.015 E. Bauer. On the Nature of Annealed Semiconductor Surfaces. *Phys. Letters* **26A**, 530–1 (1968).

68.016 S. M. Bedair, F. Hofmann and H. P. Smith. LEED Studies of Oxygen Adsorption on the (100) Face of Al. *J. Appl. Phys.* **39**, 4026–8 (1968).

68.017 J. L. Beeby. The Diffraction of Low-energy Electrons by Crystals. *J. Phys.* C *(Proc. Phys. Soc.)* **1**, 82–7 (1968).

68.018 L. de Bersuder. Observation and Geometric Interpretation of Kikuchi Lines Associated with a Two-dimensional Lattice. *C.R. Acad. Sci.* **266B**, 1489–91 (1968).

68.019 H. P. Bonzel and N. A. Gjostein. Diffraction Theory of Sinusoidal Gratings and Application to *in situ* Surface Self-diffusion Measurements. *J. Appl. Phys.* **39**, 3480–91 (1968).

68.020 G. A. Boutry, H. Dormont, J. C. Richard and P. Saget. Study of K and Rb Films by LEED. *C.R. Acad. Sci.* **267B**, 255–8 (1968).

68.021 R. M. Broudy and H. C. Abbink. Si Surface Structure. *Appl. Phys. Letters* **13**, 212–13 (1968).

68.022 C. W. Caldwell and K. Müller. LEED Apparatus Design for Thin Film Use. *Rev. Sci. Instrum.* **39**, 860–3 (1968).

68.023 B. D. CAMPBELL and W. P. ELLIS. Laser Simulation of LEED Patterns. *Surface Sci.* **10**, 124–7 (1968).

68.024 B. D. CAMPBELL and H. E. FARNSWORTH. Studies of Structure and Oxygen Adsorption of (0001) CdS Surfaces by LEED. *Surface Sci.* **10**, 197–214 (1968).

68.025 C. C. CHANG. LEED Studies of the (0001) Face of $\alpha$-Alumina, *J. Appl. Phys.* **39**, 5570–3 (1968).

68.026 C. C. CHANG. LEED Studies, Adsorption of Carbon Monoxide on the W(112) Face. *J. Electrochem. Soc.* **115**, 354–8 (1968).

68.027 C. COROTTE, P. DUCROS and A. MASCALL. Changes in the Surface State of the (111) Face of Ag Observed by Slow-electron Diffraction. *C.R. Acad. Sci.* **267B**, 507–10 (1968).

68.028 A. E. CURZON and T E. GALLON. A LEED Study of the Effect of Argon-ion Bombardment on Ag. *J. Phys.* D *(Brit. J. Appl. Phys.)* **1**, 437–40 (1968).

68.029 G. DAVID, J. ESCARD and S. GOLDSZTAUB. Oxidation of Graphite Studied by Slow Electron Diffraction. *C.R. Acad. Sci.* **266C**, 1406–8 (1968).

68.030 J. P. DEVILLE. Spectroscopy of Auger Electrons. *Rev. Phys. Appl.* **3**, 351–5 (1968).

68.031 J. L. DORNANGE and J. OUDAR. Structure et Conditions de Formation de la Couche d'Adsorption du Soufre sur le Cuivre. *Surface Sci.* **11**, 124–42 (1968).

68.032 P. DUCROS. Commentaire sur "The Uncertainty Regarding Reconstructed Surfaces" by L. H. Germer and E. Bauer. *Surface Sci.* **10**, 118–23 (1968).

68.033 P. DUCROS. Possibility of Relating the Jahn–Teller Effect to Permanent Deformations Detected by the Diffraction of Slow Electrons on the Dense Faces of Si and Ge. *Surface Sci.* **10**, 295–8 (1968).

68.034 V. F. DVORYANKIN and A. YU. MITYAGIN. Elastic Scattering of Slow Electrons by Isolated Atoms Undergoing Isotropic Thermal Motion. *Kristallografiya* **13**, 5–9 (1968).

68.035 V. F. DVORYANKIN and A. YU. MITYAGIN. Scattering of

Slow Electrons by a Static Gaussian Atomic Potential. *Kristallografiya* **13**, 221–4 (1968).

68.036 W. P. ELLIS. LEED Studies of $UO_2$. *J. Chem. Phys.* **48**, 5695–701 (1968).

68.037 W. P. ELLIS and B. D. CAMPBELL. Laser Techniques in the Interpretation of LEED Patterns. *Trans. Am. Cryst. Assoc.* **4**, 97–108 (1968).

68.038 W. P. ELLIS and R. L. SCHWOEBEL. LEED From Surface Steps on $UO_2$ Single Crystals. *Surface Sci.* **11**, 82–98 (1968).

68.039 P. J. ESTRUP and J. ANDERSON. Adsorption and Decomposition of $NH_3$ on a Single-crystal W(100) Surface. *J. Chem. Phys.* **49**, 523–8 (1968).

68.040 H. E. FARNSWORTH. LEED. *Experimental Methods in Catalytic Research*, ed. R. B. Anderson (Academic Press, N.Y., 1968), pp. 265–85.

68.041 H. E. FARNSWORTH and J. BELLINA. The Application of LEED to the Study of Bombardment Damage at Crystal Surfaces. *Trans. Am. Cryst. Assoc.* **4**, 45–58 (1968).

68.042 D. G. FEDAK, T. E. FISCHER and W. D. ROBERTSON. Surface-structure Analysis by Optical Simulation of LEED Patterns. *J. Appl. Phys.* **39**, 5658–68 (1968).

68.043 L. FIERMANS and J. VENNIK. LEED Study of the $V_2O_5$ (010) Surface. *Surface Sci.* **9**, 187–97 (1968).

68.044 T. E. GALLON, I. G. HIGGINBOTHAM, M. PRUTTON and H. TOKUTAKA. Growth of Ag on KCl Observed by LEED and Auger Emission Spectroscopy. *Thin Solid Films* **2**, 369–73 (1968).

68.045 R. L. GERLACH and T. N. RHODIN. One-dimensionally Incoherent Surface Structures. *Surface Sci.* **10**, 446–58 (1968).

68.046 J. GERSTNER and P. H. CUTLER. Slow Electron Scattering from One-dimensional Crystal Models. *Surface Sci.* **9**, 198–216 (1968).

68.047 A. GERVAIS, R. M. STERN and M. MENES. Multiple Diffraction Origin of LEED Intensities. *Acta Cryst.* **A24**, 191–9 (1968).

68.048 R. M. Goodman, H. H. Farrell and G. A. Somorjai. Mean Displacement of Surface Atoms in Pd and Pb Single Crystals. *J. Chem. Phys.* **48**, 1046–51 (1968).

68.049 R. M. Goodman, H. H. Farrell and G. A. Somorjai. Properties of the Specular Low-energy Electron Beam Scattered by F.C.C. Metal Single-crystal surfaces. *J. Chem. Phys.* **49**, 692–700 (1968).

68.050 D. A. Gorodetskii and A. A. Yas'ko. Slow Electron Diffraction Study of Sc Films Adsorbed on the (110) Face of W. *Fiz. Tverd. Tela.* **10**, 2302–10 (1968).

68.051 D. A. Gorodetskii and A. A. Yas'ko. Structure of Lithium Oxygen Double Films on the (110) Face of W. *Ukr. Fiz. Zh. (Ukr. Ed.)* **13**, 557–66 (1968).

68.052 T. W. Haas. Low-energy Electron Scattering from Clean and Hydrogen Covered Nb(110) Surfaces. *J. Appl. Phys.* **39**, 5854–8 (1968).

68.053 H. Hafner, J. A. Simpson and C. E. Kuyatt. Comparison of the Spherical Deflector and the Cylindrical Mirror Analyzers. *Rev. Sci. Instrum.* **39**, 33–5 (1968).

68.054 L. A. Harris. Analysis of Materials by Electron-excited Auger Electrons. *J. Appl. Phys.* **39**, 1419–27 (1968).

68.055 L. A. Harris. Some Observations of Surface Segregation by Auger Electron Emission. *J. Appl. Phys.* **39**, 1428–31 (1968).

68.056 L. A. Harris. Carbon Evaporation from a Th Dispenser Cathode Observed by Auger Electron Emission. *J. Appl. Phys.* **39**, 4862 (1968).

68.057 K. Hayakawa. Diffuse Patterns in Slow Electron Diffraction from Cleaned Faces of Zincblende Crystals. *J. Phys. Soc. Japan* **25**, 1647–53 (1968).

68.058 K. Hayek, H. E. Farnsworth and R. L. Park. Interaction of Oxygen, Carbon Monoxide and Nitrogen with (001) and (110) Faces of Mo. *Surface Sci.* **10**, 429–45 (1968).

68.059 K. Hirabayashi. Pseudopotential Approach to Diffraction Intensity of Low-energy Electrons by Crystals. *J. Phys. Soc. Japan* **24**, 846–54 (1968).

68.060 K. Hirabayashi. Resonance Effects in Diffraction of

Low-energy Electrons by Crystals. *J. Phys. Soc. Japan* **25**, 856–61 (1968).

68.061 S. INO, S. OGAWA, M. UCHIYAMA and Z. ODA. LEED Study of Molybdenite Cleavage Surface. *Japan J. Appl. Phys.* **7**, 308–9 (1968).

68.062 N. ITOH. Degradation of Low Energy Electrons in Alkali Halides. *Phys. Stat. Solidi* **30**, 199–207 (1968).

68.063 A. G. JACKSON, M. P. HOOKER and T. W. HAAS. $O_2$ and CO Interaction with Al Films Grown on Ta(110). *Surface Sci.* **10**, 308–10 (1968).

68.064 F. JONA. LEED – Circa 1968. *Helvetica Physica Acta* **41**, 960–4 (1968).

68.065 F. JONA, R. F. LEVER and J. B. GUNN. Comments on "Multiple Scattering Versus Superstructures in LEED" by E. Bauer. *Surface Sci.* **9**, 468–70 (1968).

68.066 L. K. JORDAN and E. J. SCHEIBNER. Characteristic Energy Loss Spectra of Cu Crystals with Surfaces Described by LEED. *Surface Sci.* **10**, 373–91 (1968).

68.067 K. KAMBE. Theory of LEED. *Z. Naturforsch.* **23A**, 1208–94 (1968).

68.068 H. KOBAYASHI and S. KATO. Observations on the Photoelectric Work Function and LEED Pattern From the (100) Surface of an Fe Single Crystal. *Surface Sci.* **12**, 398–402 (1968).

68.069 M. G. LAGALLY and M. B. WEBB. Experimental Determination of the Effective Atomic Scattering Factor and Rigid Lattice Interference Function in LEED. *Phys. Rev. Letters* **21**, 1388–91 (1968).

68.070 B. LANG. Charging of Insulating Crystals Using LEED. *C.R. Acad. Sci.* **267B**, 1325–34 (1968).

68.071 R. N. LEE. Electron Gun for LEED Applications. *Rev. Sci. Instrum.* **39**, 1306–12 (1968).

68.072 R. N. LEE. Chemisorption of Oxygen on (100) PbS. *J. Phys. (Paris), Colloq.* **29**, 43–5 (1968).

68.073 H. K. LINTZ. Comment on "LEED Study of Ni Induced Surface Structure on Si (111)" by A. J. Van Bommel and F. Meyer. *Surface Sci.* **12**, 390 (1968).

68.074 A. A. LUCAS. Contribution of the Electrostatic Potential to the LEED Intensities at Alkali-halide Distorted Surfaces. *Surface Sci.* **11**, 19–24 (1968).

68.075 P. M. MARCUS and D. W. JEPSEN. Accurate Calculation of LEED Intensities by the Propagation Matrix Method. *Phys. Rev. Letters* **20**, 925–9 (1968),

68.076 I. MARKLUND, S. ANDERSSON and J. MARTINSSON. Scattering of Low-energy Electrons from a Cu(111) Surface. *Arkiv för Fysik.* **37**, 127–39 (1968).

68.077 J. W. MAY and L. H. GERMER. Incipient Oxidation of (110) Ni. *Surface Sci.* **11**, 443–64 (1968).

68.078 E. G. MCRAE. Electron Diffraction at Crystal Surfaces. I. Generalization of Darwin's Dynamical Theory. *Surface Sci.* **11**, 479–91 (1968).

68.079 E. G. MCRAE. Electron Diffraction at Crystal Surfaces. II. The Double-diffraction Picture. *Surface Sci.* **11**, 492–507 (1968).

68.080 E, G. MCRAE, P. J. JENNINGS and D. E. WINKEL. Computations of LEED for a Model Crystal. *Trans. Am. Cryst. Assoc.* **4**, 1–13 (1968).

68.081 A. J. MELMED, H. P. LAYER and J. KRUGER. Ellipsometry, LEED and Field Electron Microscopy Combined. *Surface Sci.* **9**, 476–83 (1968).

68.082 J. M. MORABITO and G. A. SOMORJAI. LEED: The Technique and Its Application to Metallurgical Science. *J. Metals* **20**, 17–24 (1968).

68.083 A. E. MORGAN and G. A. SOMORJAI. Adsorption Studies on Pt Single Crystal Surfaces. *Trans. Am. Cryst. Assoc.* **4**, 59–71 (1968).

68.084 A. E. MORGAN and G. A. SOMORJAI. LEED Studies of Gas Adsorption on the Pt(100) Single Crystal Surface. *Surface Sci.* **12**, 405–25 (1968).

68.085 K. MÜLLER and C. C. CHANG. LEED Observations of Electric Dipoles on Mica Surfaces. *Surface Sci.* **9**, 455–8 (1968).

68.086 R. H. MULLER. Ellipsometer for Use with LEED Chamber. *Rev. Sci. Instrum.* **39**, 1593–4 (1968).

68.087 A. G. NAUMOVETS and A. G. FEDORUS. Structure of

Platinum Films on a (110) Face of a W Crystal. *Fiz. Tverd. Tela* **10**, 2570–2 (1968).

68.088 Y. H. OHTSUKI. The Theory of the Slow Electron Diffraction. I. Strong Adsorption. *J. Phys. Soc. Japan* **24**, 1116–24 (1968).

68.089 Y. H. OHTSUKI. Theory of LEED. II. OPW Method. *J. Phys. Soc. Japan* **25**, 481–8 (1968).

68.090 M. ONCHI and H. E. FARNSWORTH. Adsorption and Oxidation of Carbon Monoxide on (100) Ni. II. Combined LEED and Mass Spectrometer Measurements. *Surface Sci.* **11**, 203–15 (1968).

68.091 M. ONCHI and H. E. FARNSWORTH. Adsorption of Carbon Monoxide on Clean (100) Ni. *Phys. Letters* **26A**, 349–50 (1968).

68.092 M. ONO and K. NAKAYAMA. Recent Studies of Solid Surfaces by Low-energy Electron Beams. *Oyo Butsure* **37**, 181–3 (1968).

68.093 J. OUDAR. LEED Study of Metallic Surfaces. *Rev. Phys. Appl.* **3**, 337–42 (1968).

68.094 P. W. PALMBERG. Structure Transformations on Cleaved and Annealed Ge(111) Surfaces. *Surface Sci.* **11**, 153–8 (1968).

68.095 P. W. PALMBERG. Atomic Arrangement of Au(100) and Related Metal Overlayer Surface Structures. *Trans. Am. Cryst. Assoc.* **4**, 89–96 (1968).

68.096 P. W. PALMBERG. Optimization of Auger Electron Spectroscopy in LEED Systems. *Appl. Phys. Letters* **13**, 183–5 (1968).

68.097 P. W. PALMBERG and T. N. RHODIN. Auger Electron Spectroscopy of F.C.C. Metal Surfaces. *J. Appl. Phys.* **39**, 2425–32 (1968).

68.098 P. W. PALMBERG and T. N. RHODIN. Atomic Arrangement of Au(100) and Related Metal Overlayer Surface Structures. I. *J. Chem. Phys.* **49**, 134–46 (1968).

68.099 P. W. PALMBERG and T. N. RHODIN. Analysis of LEED Patterns from Simple Overlayer Surface Structures. II. *J. Chem. Phys.* **49**, 147–55 (1968).

68.100 P. W. PALMBERG and T. N. RHODIN. Surface Dissociation of KCl by Low-energy Electron Bombardment. *J. Phys. Chem. Solids* **29**, 1917–24 (1968).

68.101 P. W. PALMBERG, C. J. TODD and T. N. RHODIN. Role of Surface Defects in the Epitaxial Growth of Some FCC Metals on KCl Cleaved in Ultrahigh Vacuum. *J. Appl. Phys.* **39**, 4650–62 (1968).

68.102 P. W. PALMBERG, R. E. DE WAMES and L. A. VREDEVOE. Direct Observation of Coherent Exchange Scattering by LEED from Antiferromagnetic Nickel Oxide. *Phys. Rev. Letters* **21**, 682–5 (1968).

68.103 R. L. PARK and H. H. MADDEN. Annealing Changes on the (100) Surface of Pd and Their Effect on CO Adsorption. *Surface Sci.* **11**, 188–202 (1968).

68.104 R. PENELLE. LEED and Its Applications to the Study of Chemisorption. *Metaux* **43**, 339–54 (1968).

68.105 M. PERDEREAU. Structural Study of the Adsorption of Sulfur on the (100) Face of Ni. *C.R. Acad. Sci.* **267B**, 1107–9 (1968).

68.106 M. R. PIGGOTT. An Electron Diffraction Anomaly Resulting from Secondary Electron Emission. *J. Appl. Phys.* **39**, 4438–43 (1968).

68.107 A. P. POGANY and P. S. TURNER. Reciprocity in Electron Diffraction and Microscopy. *Acta Cryst.* **A24**, 103–9 (1968).

68.108 J. H. POLLARD and W. E. DANFORTH. LEED Observations of Epitaxially Grown Th on (100) Ta. *J. Appl. Phys.* **39**, 4019–20 (1968).

68.109 R. RIWAN and G. ALLIE. Examination of the (100) Face of Nickel Oxide Crystals by Slow-electron Diffraction. Effect of Temperature. *C.R. Acad. Sci.* **266B**, 1568–70 (1968).

68.110 G. ROVIDA and E. ZANAZZI. LEED Study of Si (111) Surface Contamination by Carbon. *Ric. Sci.* **38**, 356–61 (1968).

68.111 P. B. SEWELL, D. F. MITCHELL and M. COHEN. HEED

and X-ray Analysis of Surfaces and Their Reaction Products. *Develop. Appl. Spectrosc.* **7A**, 61–79 (1968).

68.112 G. W. SIMMONS and K. R. LAWLESS. LEED Studies of the Interaction of Oxygen with Cu Single Crystal Surfaces. *Trans. Am. Cryst. Assoc.* **4**, 72–88 (1968).

68.113 G. A. SOMORJAI. Surface Chemistry. *Ann. Rev. Phys. Chem.* **19**, 251–72 (1968).

68.114 G. A. SOMORJAI. Small Angle X-ray Scattering and LEED Studies of Catalyst Surfaces. *Progress in Analytical Chemistry*, Vol. 1, *X-ray and Electron Methods of Analysis*, eds. H. van Olphen and W. Parrish (Plenum Press, N.Y., 1968), Chapter VI.

68.115 R. M. STERN. Three Dimensional Dynamical LEED. *Trans. Am. Cryst. Assoc.* **4**, 14–44 (1968).

68.116 R. M. STERN and H. TAUB. Origin of the Angular Dependence of Secondary Emission of Electrons from W. *Phys. Rev. Letters* **20**, 1340–3 (1968).

68.117 Y. TAKEISHI and K. HIRABAYASHI. LEED Study of the Super-structure of a Clean Si (111) Surface. *Lattice Defects in Semiconductors*, Vol. V, ed. R. R. Hasiguti (Tokyo Univ. Press, 1968), pp. 455–7.

68.118 A. TALONI and D. HANEMAN. Computer Calculations of Semiconductor Surface Structures. *Surface Sci.* **10**, 215–31 (1968).

68.119 J. THIRLWELL. Characteristic Energy Losses of Low-energy Electrons Reflected from Al and Cu. *J. Phys. C (Proc. Phys. Soc.)* **1**, 979–89 (1968).

68.120 G. P. THOMSON. The Early History of Electron Diffraction. *Contemp. Physics* **9**, 1–15 (1968).

68.121 H. TOKUTAKA and M. PRUTTON. LEED Observations of the (11) Beam from the (100) Surfaces of Some NaCl-type Crystals. *Surface Sci.* **11**, 216–26 (1968).

68.122 J. C. TRACY. Temperature Controller for LEED Experiments. *Rev. Sci. Instrum.* **39**, 1300–3 (1968).

68.123 A. J. VAN BOMMEL and F. MEYER. Reply to Comment of H. K. Lintz Concerning "LEED Study of a Ni Induced

Surface Structure on Si (111)". *Surface Sci.* **12**, 391–2 (1968).

68.124 L. A. VREDEVOE and R. E. DE WAMES. Magnetic Scattering of Electrons from Crystals and Polarization of the Scattered Beam. *Phys. Rev.* **176**, 684–7 (1968).

68.125 C. M. K. WATTS. LEED from Crystals Surfaces. I. A Matrix Formulation. *J. Phys.* C *(Proc. Phys. Soc.)* **1**, 1237–45 (1968).

## 1969

69.001 D. ABERDAM, G. BOUCHET, P. DUCROS, J. DAVAL and G. GRUNBERG. LEED Study of the (0001) Face of $YMnO_3$ in Order to Epitaxy CdSe on this Surface. *Surface Sci.* **14**, 121–40 (1969).

69.002 R. O. ADAMS. LEED and Adsorption Studies of Be. *The Structure and Chemistry of Solid Surfaces*, ed. G. A. Somorjai (John Wiley, N.Y., 1969), pp. 70–1 to 70–9.

69.003 R. E. ALLEN, G. P. ALLDREDGE and F. W. DE WETTE. Surface Modes of Vibration in Monatomic Crystals. *Phys. Rev. Letters* **23**, 1285–6 (1969).

69.004 G. ALLIE and A. GERVAIS. LEED Study of a Superstructure on the (100) Face of Ni (II) Oxide. *C.R. Acad. Sci.* **269B**, 1212–14 (1969).

69.005 S. ANDERSSON. LEED Intensities from the Clean Cu (001) Surface. *Surface Sci.* **18**, 325–40 (1969).

69.006 S. ANDERSSON, I. MARKLUND and D. ANDERSSON. Clean Te Surfaces Studied by LEED. *The Structure and Chemistry of Solid Surfaces*, ed. G. A. Somorjai (John Wiley, N.Y. 1969), pp. 72–1 to 72–17.

69.007 R. A. ARMSTRONG. The Interaction of Slow Electrons with CO Adsorbed on Ni(100). *The Structure and Chemistry of Solid Surfaces*, ed. G. A. Somorjai (John Wiley, N.Y., 1969), pp. 52–1 to 52–12.

69.008 E. BAUER. On the Interpretation of Complex LEED Patterns. *The Structure and Chemistry of Solid Surfaces*, ed. G. A. Somorjai (John Wiley, N.Y., 1969), pp. 23–1 to 23–25.

69.009 E. BAUER. LEED. *Tech. Metals Res.* **11** (Pt. 2), 559–639 (1969).

69.010 E. BAUER. Inelastic Scattering of Slow Electrons in Solids. *Z. Physik.* **224**, 19–44 (1969).

69.011 S. M. BEDAIR and H. P. SMITH. Atomically Clean Surfaces by Pulsed Laser Bombardment. *J. Appl. Phys.* **40**, 4776–82 (1969).

69.012 R. BEHRISCH, G. MÜHLBAUER and B. M. U. SCHERZER. A Goniometer for Ultra-high Vacuum. *J. Phys.* E *(J. Sci. Instrum.)* **2**, 381–2 (1969).

69.013 K. H. BEHRNDT. Film Formation and Structure-sensitive Properties. An Introduction. *J. Vac. Sci. Technol.* **6**, 439–42 (1969).

69.014 L. DE BERSUDER, C. COROTTE, P. DUCROS and D. LAFEUILLE. LEED Studies of the (001) Face of Al and the (111) Face of Ag. *The Structure and Chemistry of Solid Surfaces*, ed. G. A. Somorjai (John Wiley, N.Y., 1969), pp. 30–1 to 30–10.

69.015 H. BETHGE. Nucleation and Surface Conditions. *J. Vac. Sci. Technol.* **6**, 460–7 (1969).

69.016 M. BEVIS. The Geometry of Lattice Planes. *Acta Cryst.* **A25**, 370–2 (1969).

69.017 H. E. BISHOP and J. C. RIVIÈRE. Comments on "Angular Dependences in Electron-excited Auger Emission" by L. A. Harris. *Surface Sci.* **17**, 446–7 (1969).

69.018 H. E. BISHOP and J. C. RIVIÈRE. Auger Spectroscopy of Si. *Surface Sci.* **17**, 462–5 (1969).

69.019 H. E. BISHOP and J. C. RIVIÈRE. Segregation of Au to the Si(111) Surface Observed by Auger Emission Spectroscopy and by LEED. *J. Phys.* D *(Brit. J. Appl. Phys.)* **2**, 1635–42 (1969).

69.020 H. E. BISHOP and J. C. RIVIÈRE. LEED Examination of the PbO Surface. *The Structure and Chemistry of Solid Surfaces*, ed. G. A. Somorjai (John Wiley, N.Y., 1969), pp. 37–1 to 37–7.

69.021 H. P. BONZEL and N. A. GJOSTEIN. *In Situ* Measurements of Surface Self Diffusion of Metals. *Molecular Processes on Solid Surfaces*, eds. E. Drauglis, R. Gretz and R. Jaffee (McGraw-Hill, New York, 1969), pp. 533–68.

69.022 G. A. BOOTSMA and F. MEYER. Measurements of Adsorption on Semiconductors by Ellipsometry and Other Methods. *Surface Sci.* **18**, 123–9 (1969).

69.023 M. BOUDART and D. F. OLLIS. LEED Investigation of the Carburization of W Single Crystals above 2000°K. *The Structure and Chemistry of Solid Surfaces*, ed. G. A. Somorjai (John Wiley, N.Y., 1969), pp. 63–1 to 63–14.

69.024 D. S. BOUDREAUX and U. HOFFSTEIN. Some New Results from the Matching-band Structure Treatment of LEED Intensities. *The Structure and Chemistry of Solid Surfaces*, ed. G. A. Somorjai (John Wiley, N.Y., 1969), pp. 4–1 to 4–7.

69.025 D. BUCKLEY. Effect of Various Properties of FCC Metals on their Adhesion as Studied with LEED. *J. Adhesion* **1**, 264–81 (1969).

69.026 C. BURGGRAF and A. MOSSER. (100) Face of Pt Observed by LEED. *C.R. Acad. Sci.* **268AB**, 1167–8 (1969).

69.027 J. J. BURTON. Equilibrium Concentration of Impurities on the Surface of a Crystal. *Phys. Rev.* **177**, 1346–8 (1969).

69.028 J. J. BURTON and G. JURA. Adsorption and Diffusion on Three Structures of the (100) Surface of a FCC Crystal. *Surface Sci.* **13**, 414–24 (1969).

69.029 J. J. BURTON and G. JURA. Surface Phase Transformations: An Interpretation of LEED Results. *The Structure and Chemistry of Solid Surfaces*, ed. G. A. Somorjai (John Wiley, N.Y., 1969), pp. 21–1 to 21–7.

69.030 T. A. CALCOTT and A. U. MACRAE. Photoemission from Clean and Cs-covered Ni Surfaces. *Phys. Rev.* **178**, 966–78 (1969).

69.031 B. D. CAMPBELL, C. A. HAQUE and H. E. FARNSWORTH. LEED Studies of the Polar {0001} Surfaces of the II–VI Compounds CdS, CdSe, ZnO and ZnS. *The Structure and Chemistry of Solid Surfaces*, ed. G. A. Somorjai (John Wiley, N.Y., 1969), pp. 33–1 to 33–17.

69.032 G. CAPART. Band Structure Calculations of LEED at Crystal Surfaces. *Surface Sci.* **13**, 361–76 (1969).

69.033 J. J. CARROLL and A. J. MELMED. Ellipsometry–LEED Study of the Adsorption of Oxygen on (011) W. *Surface Sci.* **16**, 251–64 (1969).

69.034 C. C. CHANG. LEED Studies of Thin Film Si Overgrowths on α-Alumina. *The Structure and Chemistry of Solid Surfaces*, ed. G. A. Somorjai (John Wiley, N.Y., 1969), pp. 77–1 to 77–14.

69.035 J. M. CHARIG and D. K. SKINNER. Carbon Contamination of Si(111) Surfaces. *Surface Sci.* **15**, 277–85 (1969).

69.036 J. M. CHARIG and D. K. SKINNER. The (0001) Surface of α-Alumina-LEED Observations. *The Structure and Chemistry of Solid Surfaces*, ed. G. A. Somorjai (John Wiley, N.Y., 1969), pp. 34–1 to 34–20.

69.037 J. M. CHEN. LEED Intensity Calculations using Kambe's Theory. *The Structure and Chemistry of Solid Surfaces*, ed. G. A. Somorjai (John Wiley, N.Y., 1969), pp. 3–1 to 3–14.

69.038 Y. C. CHENG. On the Influence of Lattice Vibrations on the Chemisorption Energy. *Physica* **43**, 603–14 (1969).

69.039 A. Y. CHO. Epitaxy by Periodic Annealing. *Surface Sci.* **17**, 494–503 (1969).

69.040 J. M. COWLEY and P. M. WARBURTON. Intensities in Reflection Electron Diffraction Patterns. *The Structure and Chemistry of Solid Surfaces* ed. G. A. Somorjai (John Wiley, N.Y., 1969), pp. 6–1 to 6–11.

69.041 M. I. DATSIEV. Chemisorption of Oxygen at the (111) Face of Ge and Si Single Crystals Studied by Irradiation with Slow Electrons. *Zh. Tekh. Fiz.* **39**, 1284–92 (1969).

69.042 J. P. DEVILLE and S. GOLDSZTAUB. Study of the Cleavage Face of Muscovite by Auger Electron Spectroscopy. *C.R. Acad. Sci.* **268B**, 629–30 (1969).

69.043 P. J. DOBSON and C. R. SERNA. LEED and RHEED Studies of the Growth of Ni on a Ag(111) Surface. *J. Phys.* D *(Brit. J. Appl. Phys.)* **2**, 1779–80 (1969).

69.044 J. L. DOMANGE, J. OUDAR and J. BENARD. Growth Mechanism and Structure of Adsorption Layers. *Molec-*

*ular Processes on Solid Surfaces*, eds. E. Drauglis, R. Gretz and R. Jaffee. (McGraw-Hill, New York, 1969), pp. 353–66.

69.045 C. B. DUKE and C. W. TUCKER. Multiple-scattering Description of Intensity Profiles Observed in LEED from Solids. *Phys. Rev. Letters* **23**, 1163–6 (1969).

69.046 C. B. DUKE and C. W. TUCKER. Inelastic-collision Model of LEED from Solid Surfaces. *Surface Sci.* **15**, 231–56 (1969).

69.047 V. F. DVORYANKIN, A. YU. MITYAGIN and V. P. ORLOV. Geometry of the Diffraction Patterns of Slow Electrons Reflected by the (100) Surface of Crystals of the NaCl Lattice Type. *Kristallografiya* **14**, 902–4 (1969).

69.048 V. F. DVORYANKIN, A. YU. MITYAGIN and K. S. POGOREL'SKII. Nature of Secondary Bragg Peaks in LEED Patterns. *Fiz. Tverd. Tela* **11**, 2444–51 (1969).

69.049 T. EDMONDS and R. C. PITKETHLY. The Adsorption of CO and $CO_2$ at the (111) Face of Ni Observed by LEED. *Surface Sci.* **15**, 137–63 (1969).

69.050 T. EDMONDS and R. C. PITKETHLY. The Interaction Between CO and $O_2$ and Between $CO_2$ and $O_2$ at the (111) Face of Ni Observed by LEED. *Surface Sci.* **17**, 450–7 (1969).

69.051 G. ERTL. LEED Studies of Molecular Interactions on Metal Surfaces. *Molecular Processes on Solid Surfaces*, eds. E. Drauglis, R. Gretz and R. Jaffee (McGraw-Hill, New York, 1969), pp. 147–65.

69.052 G. ERTL and P. RAN. Chemisorption und Katalytische Reaktion von Sauerstoff und Kohlenmonoxid an Einer Pd (110) – Oberfläche. *Surface Sci.* **15**, 443–65 (1969).

69.053 P. J. ESTRUP. The Effect of Temperature on LEED Intensities from Adsorbed Structures. *The Structure and Chemistry of Solid Surfaces*, ed. G. A. Somorjai (John Wiley, N.Y., 1969), pp. 19–1 to 19–9.

69.054 H. E. FARNSWORTH, C. A. HAQUE, D. M. ZEHNER and G. BARTON. Influence of High-Temperature Heat Treatment on Surface Order and Faceting of some Mo, W, Ta and Re Crystal Surfaces. *Surface Sci.* **17**, 1–6 (1969).

69.055 H. E. FARNSWORTH and M. ONCHI. Some Applications of LEED to Surface Problems. *Molecular Processes on Solid Surfaces*, eds. E. Drauglis, R. Gretz and R. I. Jaffee (McGraw-Hill, New York, 1969), pp. 31–48.

69.056 H. E. FARNSWORTH and D. M. ZEHNER. LEED Structures Due to Interaction of CO and $O_2$ with (0001)Re. *Surface Sci.* **17**, 7–31 (1969).

69.057 H. H. FARRELL and G. A. SOMORJAI. Properties of the Nonspecular LEED Beams Scattered by the (100) Face of FCC Metal Single Crystals. *Phys. Rev.* **182**, 751–9 (1969).

69.058 D. G. FEDAK, J. V. FLORIO and W. D. ROBERTSON. The Interaction between Chlorine and the (100) Surface of Au. *The Structure and Chemistry of Solid Surfaces*, ed. G. A. Somorjai (John Wiley, N.Y., 1969), pp. 74–1 to 74–18.

69.059 L. G. FEINSTEIN and E. BLANC. Determination of Nickel–Molybdenum Surface Alloy Structures Using LEED. *Surface Sci.* **18**, 350–6 (1969).

69.060 L. G. FEINSTEIN and M. S. MACRAKIS. LEED Study of Steps and Facets on a Re Surface. *Surface Sci.* **18**, 277–92 (1969).

69.061 L. FIERMANS and J. VERNIK. Particular LEED Features on the $V_2O_5$(010) Surface and Their Relation to the LEED Beam Induced Transition $V_2O_5 \rightarrow V_{12}O_{26}$. *Surface Sci.* **18**, 317–24 (1969).

69.062 J. V. FLORIO and W. D. ROBERTSON. Chlorine Reactions on the Si(111) Surface. *Surface Sci.* **18**, 398–427 (1969).

69.063 M. H. FRANCOMBE. Pitfalls in the Interpretation of Structure-sensitive Properties. *J. Vac. Sci. Technol.* **6**, 448–54 (1969).

69.064 S. FRIEDMAN and R. M. STERN. LEED Rocking Curves. *Surface Sci.* **17**, 214–31 (1969).

69.065 Y. FUJINAGA, S. INO and S. OGAWA. A LEED Study

of a Si(111) Surface in Oxygen–Nitrogen Mixed Gas. *Japan. J. Appl. Phys.* **8**, 815–16 (1969).

69.066 J. W. GADZUK. Resonance Transmission in Electron Emission from Surfaces with Adsorbed Atoms. *Surface Sci.* **18**, 193–203 (1969).

69.067 G. GAFNER. Calculation of Back-reflected LEED Intensities Using a Plane-wave Multiple Scattering Mechanism. *The Structure and Chemistry of Solid Surfaces*, ed. G. A. Somorjai (John Wiley, N.Y., 1969), pp. 2–1 to 2–10.

69.068 T. E. GALLON. A Simple Model for the Dependence of Auger Intensities on Specimen Thickness. *Surface Sci.* **17**, 486–9 (1969).

69.069 T. E. GALLON, I. G. HIGGINBOTHAM, and M. PRUTTON. An Improved Apparatus for the Measurement of Auger Electron Spectra. *J. Phys.* E *(J. Sci. Instrum.)* **2**, 894–6 (1969).

69.070 R. L. GERLACH and T. N. RHODIN. Structure Analysis of Alkali Metal Adsorption on Single Crystal Ni Surfaces. *Surface Sci.* **17**, 32–68 (1969).

69.071 R. L. GERLACH and T. N. RHODIN. Alkali Atom Adsorption on Single Crystal Ni Surfaces, Surface Structure and Work Function. *The Structure and Chemistry of Solid Surfaces*, ed. G. A. Somorjai (John Wiley, N.Y., 1969), pp. 55–1 to 55–25.

69.072 S. GOLDSZTAUB. Relations éntre la Symétrie des Diagrammes de Diffraction des Électrons Lents et celle du Cristal. *Acta Cryst.* **A25**, 306–8 (1969).

69.073 D. A. GORODETSKII and YU. P. MEL'NIK. Structure of Barium and Barium Oxide Films on a W(100) Face. *Izv. Akad. Nauk SSSR, Ser. Fiz.* **33**, 462–6 (1969).

69.074 D. A. GORODETSKII and PHUNG-HO. Structure of Al, Mg, and Sn Films on the (110) W Face. *Ukr. Fiz. Zh.* **14**, 94–104 (1969).

69.075 D. A. GORODETSKII and A. A. YAS'KO. Structure of Pb Films on a (110) Face of W. *Fiz. Tverd. Tela* **11**, 790–2 (1969).

69.076 D. A. GORODETSKII and A. A. YAS'KO. Sc and Y Films

on the (100) Face of W. *Fiz. Tverd. Tela* **11**, 2513–19 (1969).

69.077 U. GRADMANN. Pseudomorphic Growth of Ni on Cu. *Surface Sci.* **13**, 498–501 (1969).

69.078 J. T. GRANT. A LEED Study of the Ir(100) Surface. *Surface Sci.* **18**, 228–38 (1969).

69.079 J. T. GRANT and T. W. HAAS. On the Nature of Si(111) Surfaces. *Appl. Phys. Letters* **15**, 140–1 (1969).

69.080 J. T. GRANT and T. W. HAAS. Some Studies of the Cr(100) and Cr(110) Surfaces. *Surface Sci.* **17**, 484–5 (1969).

69.081 J. T. GRANT and T. W. HAAS. The Structure of the Pt(100) Surface. *Surface Sci.* **18**, 457–61 (1969).

69.082 T. W. HAAS. LEED Study of the Ta(112) Surface. *The Structure and Chemistry of Solid Surfaces*, ed. G. A. Somorjai (John Wiley, N.Y., 1969), pp. 31–1 to 31–15.

69.083 T. W. HAAS and J. T. GRANT. Chemical Shifts in Auger Electron Spectroscopy from the Initial Oxidation of Ta(110). *Phys. Letters* **30A**, 272 (1969).

69.084 T. HANAWA and K. TAKEDA. LEED Study of the Epitaxial Growth of Au on PbS(100) Surface. *Appl. Phys. Letters* **15**, 360–2 (1969).

69.085 D. HANEMAN and D. L. HERON. Nature of Clean Cleaved Si Surfaces with Wave Function Overlap Calculations. *The Structure and Chemistry of Solid Surfaces*, ed. G. A. Somorjai (John Wiley, N.Y., 1969), pp. 24–1 to 24–11.

69.086 L. A. HARRIS. Angular Dependences in Electron-excited Auger Emission. *Surface Sci.* **15**, 77–93 (1969).

69.087 L. A. HARRIS. Reply to Comments of H. E. Bishop and J. C. Rivière on "Angular Dependences in Electron-excited Auger Emission". *Surface Sci.* **17**, 448–9 (1969).

69.088 K. HAYAKAWA. Studies of Crystal Surfaces by LEED. *Nippon Kessho Gakkaisho* **11**, 266–72 (1969).

69.089 R. HECKINGBOTTOM. LEED Patterns and Surface Perfection. *Surface Sci.* **17**, 394–401 (1969).

69.090 R. HECKINGBOTTOM. A LEED Investigation of the Nitridation of the Si(111) Surface. *The Structure and Chemistry of Solid Surfaces*, ed. G. A. Somorjai (John Wiley, N.Y., 1969), pp. 78–1 to 78–19.

69.091 V. HEINE. Electrons at Clean Surfaces. *The Structure and Chemistry of Solid Surfaces*, ed. G. A. Somorjai (John Wiley, N.Y., 1969), pp. 1–1 to 1–31.

69.092 R. C. HENDERSON and W. J. POLITO. Reflection HEED of Si Fractional Order Structures. *Surface Sci.* **14**, 473–7 (1969).

69.093 M. HENZLER. Correlation between Surface Structure and Surface States at the Clean Ge(111) Surface. *J. Appl. Phys.* **40**, 3758–65 (1969).

69.094 A. T. HOANG and G. K. ZYRYANOV. Variation in the Intensity of Diffraction Images of Slow Electrons in Relation to Crystal Temperature and Gas Adsorption on $(10\bar{1}0)$ CdS. *Vestn. Leningrad Univ., Fiz. Khim.* **2**, 73–6 (1969).

69.095 F. HOFMANN and H. P. SMITH. Calculation of LEED Intensities Using Dynamical Theory. *The Structure and Chemistry of Solid Surfaces*, ed. G. A. Somorjai (John Wiley, N.Y., 1969), pp. 5–1 to 5–18.

69.096 J. HÖLZL, H. MAYER and K. W. HOFFMAN. Charakteristische Energieverluste bei der Reflexion von Langsamer Elektronen an Natrium-Oberflächen. *Surface Sci.* **17**, 232–9 (1969).

69.097 J. E. HOUSTON and R. L. PARK. Auger Excitation by Internal Secondary Electrons. *Appl. Phys. Letters* **14**, 358–60 (1969).

69.098 M. ICHIKAWA and Y. OHTSUKI. The Correction of the Mean Inner Potential in Electron Diffraction. *J. Phys. Soc. Japan* **27**, 953–6 (1969).

69.099 A. G. JACKSON. Simplification in the Calculation of Kambe Structure Constants. *Surface Sci.* **17**, 482–3 (1969).

69.100 A. G. JACKSON and M. P. HOOKER. LEED Study of the Growth of Sn Films on Nb(110). *The Structure and Chemistry of Solid Surfaces*, ed. G. A. Somorjai (John Wiley, N.Y., 1969), pp. 73–1 to 73–13.

69.101 F. JONA and H. R. WENDT. Sample Holder for Diffraction and Other Studies in Ultrahigh Vacuum. *Rev. Sci. Instrum.* **40**, 1172–3 (1969).

69.102 R. O. JONES and J. A. STROZIER. Inelastic Effects in LEED. *Phys. Rev. Letters* **22**, 1186–8 (1969).

69.103 N. V, JOSHI. Recalculation of Plasma Energy in Cu. *Surface Sci.* **15**, 175–6 (1969).

69.104 C. W. JOWETT, P. J. DOBSON and B. J. HOPKINS. The Surface Potential of Water Vapor on Epitaxially Grown (110) Oriented Films of Pt. *Surface Sci.* **17**, 474–81 (1969).

69.105 B. A. JOYCE, J. H. NEAVE and B. E. WATTS. The Influence of Substrate Surface Conditions on the Nucleation and Growth of Epitaxial Si Films. *Surface Sci.* **15**, 1–13 (1969).

69.106 H. K. A. KAN and S. FEUERSTEIN. LEED Studies of the Interaction of $O_2$ with a Mo(100) Surface. *J. Chem. Phys.* **50**, 3618–23 (1969).

69.107 E. KASPER. Theory of Inelastic Scattering of Electrons in Crystals. *Z. Physik.* **222**, 225–42 (1969).

69.108 S. KATO, H. KOBAYASHI, T. KANAYA and K. MURATA. LEED Study of Ni Single Crystal Surfaces. *Tech. Rep. Osaka Univ.* **19**, 361–70 (1969).

69.109 H. KOBAYASHI and S. KATO. Observations on the Photoelectric Work Function and LEED Pattern from the (111) Surface of an Fe Single Crystal. *Surface Sci.* **18**, 341–9 (1969).

69.110 G. O. KRAUSE. On the Structure of Annealed Si Surfaces. *Phys. Stat. Solidi* **35**, K59–62 (1969).

69.111 G. O. KRAUSE. Characterization of Si Surfaces by RHEED. *Semiconductor Silicon*, ed. R. R. Haberecht and E. L. Kern (Electrochemical Society, N.Y., 1969), pp. 574–84.

69.112 M. G. LAGALLY and M. B. WEBB. Effects of Phonon Scattering and the Atomic Scattering Factor in LEED. *The Structure and Chemistry of Solid Surfaces*, ed. G. A. Somorjai (John Wiley, N.Y., 1969), pp. 20–1 to 20–17.

69.113 J. J. LANDER and J. MORRISON. Cesium Plasma Spectra in the System W(100)–Cs. *Surface Sci.* **14**, 465–72 (1969).

69.114 J. J. LANDER and J. MORRISON. Chemisorption of Iodine on W(100). *Surface Sci.* **17**, 469–73 (1969).

69.115 J. D. LEVINE. The Nature of Surface States on III–V and II–VI Semiconductors. *J. Vac. Sci. Technol.* **6**, 549–51 (1969).

69.116 A. U. MACRAE. The Location of Atoms at Surfaces. *Surface Sci.* **13**, 130–3 (1969).

69.117 A. U. MACRAE, K. MÜLLER, J. J. LANDER and J. MORRISON. An Electron Diffraction Study of Cs Adsorption on W. *Surface Sci.* **15**, 483–97 (1969).

69.118 A. U. MACRAE, K. MÜLLER, J. J. LANDER, J. MORRISON and J. C. PHILLIPS. Electronic and Lattice Structure of Cs Films Adsorbed on W. *Phys. Rev. Letters* **22**, 1048–51 (1969).

69.119 P. M. MARCUS, D. W. JEPSEN and F. JONA. Angular Dependence of the LEED Spectrum of Al(001). *Surface Sci.* **17**, 442–5 (1969).

69.120 P. M. MARCUS, F. JONA and D. W. JEPSEN. Energy Diagram Method for Bragg Reflections in LEED Spectra. *I.B.M.J. Research Develop.* **13**, 646–61 (1969).

69.121 J. W. MAY. Platinum Surface LEED Rings. *Surface Sci.* **17**, 267–70 (1969).

69.122 J. W. MAY. A Mechanism for Surface Reconstruction at Room Temperature. *Surface Sci.* **18**, 431–6 (1969).

69.123 J. W. MAY and L. H. GERMER. Hydrogen and Oxygen on a (110) Ni Surface. *The Structure and Chemistry of Solid Surfaces*, ed. G. A. Somorjai (John Wiley, N .Y., 1969), pp. 51–1 to 51–24.

69.124 J. W. MAY, R. J. SZOSTAK and L. H. GERMER. Thermal Breakup of $NH_3$ Adsorption on W(211). *Surface Sci.* **15**, 37–76 (1969).

69.125 A. J. MELMED. Single-specimen FEM-LEED studies: Carbon on W. *J. Appl. Phys.* **40**, 2330–4 (1969).

69.126 A. J. MELMED. Surface Characterization by Ellipsometry, LEED and Field-electron Microscopy. *Molecular Processes on Solid Surfaces*, eds. E. Drauglis, R. Gretz and R. Jaffee (McGraw-Hill, New York 1969), pp. 105–27.

69.127 J. J. McCarroll, T. Edmonds and R. C. Pitkethly. Interpretation of a Complex LEED Pattern: Carbonaceous and Sulfur-containing Structures on Ni(111). *Nature* **223**, 1260–2 (1969).

69.128 E. G. McRae. Electron Diffraction at Crystal Surfaces. *Molecular Processes on Solid Surfaces*, eds. E. Drauglis, R. Gretz and R. Jaffee (McGraw-Hill, New York, 1969), pp. 81–104.

69.129 E. G. McRae and P. J. Jennings. Surface-state Resonances in LEED. *Surface Sci.* **15**, 345–8 (1969).

69.130 E. G. McRae and P. J. Jennings. Model Computations of Inelastic Scattering of Low-energy Electrons by Crystals. *The Structure and Chemistry of Solid Surfaces*, ed. G. A. Somorjai (John Wiley, N.Y., 1969), pp. 7–1 to 7–18.

69.131 E. G. McRae and D. E. Winkel. Electron Diffraction at Crystal Surfaces. III. Effect of a Layer of Foreign Atoms on Secondary Peaks in Low Energy Diffraction Intensity Curves. *Surface Sci.* **14**, 407–14 (1969).

69.132 K. Molière. LEED Study of Surface Structure. *Cesk. Cas. Fys.* **19**, 181–9 (1969).

69.133 K. Molière and F. Portele. LEED Study of Oxygen Adsorption and First Stages of Oxide Epitaxy on (110) Surfaces of α-Iron. *The Structure and Chemistry of Solid Surfaces*, ed. G. A. Somorjai (John Wiley, N.Y., 1969), pp. 69–1 to 69–21.

69.134 J. M. Morabito, R. Steiger, R. Muller and G. A. Somorjai. LEED and Ellipsometry Studies of Physical Adsorption on a (110) Ag Surface at Low Temperatures. *The Structure and Chemistry of Solid Surfaces*, ed. G. A. Somorjai (John Wiley, N.Y., 1969), pp. 50–1 to 50–35.

69.135 J. M. Morabito, R. F. Steiger and G. A. Somorjai. Studies of the Mean Displacement of Surface Atoms in the (100) and (110) Faces of Ag Single Crystals at Low Temperatures. *Phys. Rev.* **179**, 638–44 (1969).

69.136 A. E. Morgan and G. A. Somorjai. LEED Studies of the Adsorption of Unsaturated Hydrocarbons and CO on the Pt(111) and (100) Single-crystal Surfaces. *J. Chem. Phys.* **51**, 3309–20 (1969).

69.137 J. MORRISON and J. J. LANDER. The Adsorption of Ionic Salts on a W (100) Surface. *Surface Sci.* **18**, 428–30 (1969).

69.138 J. MORRISON and J. J. LANDER. Application of the Triple Grid LEED System to Auger Spectrum Analyses. *J. Vac. Sci. Technol.* **6**, 338–40 (1969).

69.139 A. R. L. MOSS and B. H. BLOTT. The Epitaxial Growth of Copper on the (110) Surface of a W Single Crystal by LEED, Auger Electron and Work Function Techniques. *Surface Sci.* **17**, 240–61 (1969).

69.140 K. MÜLLER. LEED Observations of Electric Dipole Fields on Mica and ZnO. *The Structure and Chemistry of Solid Surfaces*, ed. G. A. Somorjai (John Wiley, N.Y., 1969), pp. 35–1 to 35–13.

69.141 K. MÜLLER and C. C. CHANG. Electric Dipoles on Clean Mica Surfaces. *Surface Sci.* **14**, 39–51.

69.142 R. H. MULLER, R. F. STEIGER, G. A. SOMORJAI and J. M. MORABITO. Gas Adsorption Studies by Ellipsometry in Combination with LEED and Mass Spectrometry. *Surface Sci.* **16**, 234–50 (1969).

69.143 K. OHTAKA, T. FUJIWARA and S. YANAGAWA. Born Approximation in Electron Diffraction. *Oyo Butsuri* **38**, 47–54 (1969).

69.144 M. ONCHI and H. E. FARNSWORTH. Interactions of $N_2O$, NO and $CO_2$ with (100) Ni Using Combined LEED and Mass Spectrometer Measurements. *Surface Sci.* **13**, 425–45 (1969).

69.145 Y. H. OTSUKI and T. URAGAMI. Temperature Dependence of Resonance Intensity Profile of LEED. *Phys. Letters* **28A**, 545 (1969).

69.146 J. OUDAR. Chemisorption of Gas on Metals Studied by LEED. *Vide* **24**, 45–8 (1969).

69.147 P. W. PALMBERG. Auger Electron Spectroscopy in LEED Systems. *The Structure and Chemistry of Solid Surfaces*, ed. G. A. Somorjai (John Wiley, N.Y., 1969), pp. 29–1 to 29–18.

69.148 P. W. PALMBERG, G. K. BOHN and J. C. TRACY. High Sensitivity Auger Electron Spectrometer. *Appl. Phys. Letters* **15**, 254–5 (1969).

69.149 P. W. PALMBERG and T. N. RHODIN. Surface Dissociation of KCl by Low-energy Electron Bombardment. *J. Phys. Chem. Solids* **29**, 1917–24 (1969).

69.150 P. W. PALMBERG, R. E. DE WAMES, L. A. VREDEVOE and T. WOLFRAM. Coherent Exchange Scattering of Low-energy Electrons by Antiferromagnetic Crystals. *J. Appl. Phys.* **40**, 1158–63 (1969).

69.151 R. L. PARK. LEED Studies of Surface Imperfections. *The Structure and Chemistry of Solid Surfaces*, ed. G. A. Somorjai (John Wiley, N.Y., 1969), pp. 28–1 to 28–17.

69.152 R. L. PARK and J. E. HOUSTON. The Effect of Registry Degeneracy on LEED Beam Profiles. *Surface Sci.* **18**, 213–27 (1969).

69.153 J. B. PENDRY. The Application of Pseudopotentials to LEED. I. Calculation of the Potential and "Inner Potential". *J. Phys.* C **2**, 1215–21 (1969).

69.154 J. B. PENDRY. The Application of Pseudopotentials to LEED. II. Calculation of the Reflected Intensities. *J. Phys.* C **2**, 2273–82 (1969).

69.155 J. B. PENDRY. The Application of Pseudopotentials to LEED. III. The Simplifying Effect of Inelastic Scattering. *J. Phys.* C **2**, 2283–9 (1969).

69.156 J. H. POLLARD and W. E. DANFORTH. A LEED Study of Th Overlayers on a Ta(100) Substrate and Evidence for Initial Pseudomorphism Followed by Th Clustering. *The Structure and Chemistry of Solid Surfaces*, ed. G. A. Somorjai (John Wiley, N.Y., 1969), pp. 71–1 to 71–15.

69.157 F. PORTELE. Sauerstoffadsorption und Orientierung dünner Oxidschichten auf (100)–Oberflächen von $\alpha$-Eisen, untersucht mit Hilfe der Beugung langsamer Elektronen. (Oxygen Adsorption and Orientation of Thin Oxide Layers on (100) Surfaces of $\alpha$-iron, Investigated with LEED.) *Z. Naturforsch.* **24a**, 1263–77 (1969).

69.158 R. J. REID and H. MYKURA. LEED Patterns from Faceted Metal Surfaces. *J. Phys.* D **2**, 145–6 (1969).

69.159 G. E. RHEAD. Ring Structures in LEED Diagrams and Surface Contamination by a Graphite Powder. *C.R. Acad. Sci.* **268C**, 1817–20 (1969).

69.160 G. RHEAD and J. PERDEREAU. Slow Electron-diffraction Study of Neighboring Surfaces. Diagrams for Surfaces with Single Steps. *C.R. Acad. Sci.* **269C**, 1183–5 (1969).

69.161 G. RHEAD and J. PERDEREAU. Slow-electron Diffraction Study of Vicinal Surfaces. Interpretation of the Diagrams. *C.R. Acad. Sci.* **269C**, 1261–4 (1969).

69.162 G. RHEAD and J. PERDEREAU. Slow Electron Diffraction Study of Vicinal Surfaces. Diagrams for Surfaces having Complex Steps. *C.R. Acad. Sci.* **269C**, 1425–8 (1969).

69.163 T. N. RHODIN, P. W. PALMBERG and E. W. PLUMMER. Some Atomistic Considerations of Surface Binding on Metals. *The Structure and Chemistry of Solid Surfaces*, ed. G. A. Somorjai (John Wiley, N.Y., 1969), pp. 22–1 to 22–28.

69.164 T. N. RHODIN, P. W. PALMBERG and C. J. TODD. Surface Point Defects and Epitaxial Growth on Alkali Halides. *Molecular Processes on Solid Surfaces*, eds. E. Drauglis, R. Gretz and R. Jaffee (McGraw-Hill, New York, 1969), pp. 499–530.

69.165 J. W. T. RIDGWAY and D. HANEMAN. Silicon (111) 7 × 7 Structure. *Appl. Phys. Letters* **14**, 265–6 (1969).

69.166 J. W. T. RIDGWAY and D. HANEMAN. Correlation of LEED Surface Structures and Surface Tear Marks on Cleaved Si Surfaces. *Surface Sci.* **18**, 441–5 (1969).

69.167 G. ROVIDA, E. ZANAZZI and E. FERRONI. Measurements of Oxygen Adsorption on Si(111) Surfaces by LEED. *Surface Sci.* **14**, 93–102 (1969).

69.168 G. SCHÖN. LEED Studies of Faceted Ag Foils. *Surface Sci.* **18**, 437–40 (1969).

69.169 M. P. SEAH. Comments on the Paper Titled "Fine Structure Measurements in the Energy Angular Distribution of Secondary Electrons from a (110) Face of Copper" by G. Appelt. *Phys. Stat. Solidi* **31**, K123–25 (1969).

69.170 M. P. SEAH. Slow Electron Scattering from Metals. I. The Emission of True Secondary Electrons. *Surface Sci.* **17**, 132–60 (1969).

69.171 M. P. SEAH. Slow Electron Scattering from Metals. II. The Inelastically Scattered Primary Electrons. *Surface Sci.* **17**, 161–80 (1969).

69.172 M. P. SEAH. Slow Electron Scattering from Metals. III. The Coherently Elastically Scattered Primary Electrons. *Surface Sci.* **17**, 181–213 (1969).

69.173 M. P. SEAH and D. P. WOODRUFF. The Energy and Temperature Dependence of LEED Intensity Peak Widths. *Phys. Letters* **30A**, 250–1 (1969).

69.174 R. L. SCHWOEBEL. Step Motion on Crystal Surfaces. II. *J. Appl. Phys.* **40**, 614–18 (1969).

69.175 R. F. STEIGER, J. M. MORABITO, G. A. SOMORJAI and R. H. MULLER. A Study of the Optical Properties and of the Physical Adsorption of Gases on Ag Single Crystal Surfaces by LEED and Ellipsometry. *Surface Sci.* **14**, 279–304 (1969).

69.176 R. M. STERN and A. GERVAIS. Inner Potential Measurements in Electron Diffraction. *Surface Sci.* **17**, 273–97 (1969).

69.177 R. M. STERN, A. GERVAIS and M. MENES. Multiple Diffraction Origin of LEED Intensities. II. *Acta Cryst.* **A25**, 393–4 (1969).

69.178 R. M. STERN, J. J. PERRY and D. S. BOUDREAUX. LEED Dispersion Surfaces and Band Structure in Three-dimensional Mixed Laue and Bragg Reflections. *Rev. Mod. Phys.* **41**, 275–95 (1969).

69.179 R, M. STERN, H. TAUB and A. GERVAIS. Dynamical Origin of Three-dimensional LEED Intensities. *J. Vac. Sci. Technol.* **6**, 222–3 (1969).

69.180 R. M. STERN, H. TAUB and A. GERVAIS. Dynamical Interpretation of Three-dimensional LEED Intensities.

*The Structure and Chemistry of Solid Surfaces*, ed. G. A. Somorjai (John Wiley, N.Y., 1969), pp. 8–1 to 8–28.

69.181 D. J. STIRLAND. Epitaxy Modifications to Evaporated FCC Metals Induced by Electron Bombardment of Alkali Halide Substrates. *Appl. Phys. Letters* **15**, 86–8 (1969).

69.182 N. TAKAHASKI, H. TOMITA and S. MOTOO. Study of the Initial Oxidation of Cu by Diffraction of Slow Electrons. *C.R. Acad. Sci.* **269B**, 618–20 (1969).

69.183 Y. TAKEISHI and K. HIRABAYASHI. LEED Study of Surface Structures of Diamond Type Crystals. *Kotai Butsuri* **4**, 295–302 (1969).

69.184 P. W. TAMM and L. D. SCHMIDT. Interaction of $H_2$ with (100) W. I. Binding States. *J. Chem. Phys.* **51**, 5352–63 (1969).

69.185 H. TAUB and R. M. STERN. Origin of the Angular Dependence of Secondary Emission of Electrons. *J. Vac. Sci. Technol.* **6**, 237 (1969).

69.186 H. TAUB, R. M. STERN and V. F. DVORYANKIN. Temperature Dependence of Mean Free Path in Secondary Electron Emission. *Phys. Stat. Solidi* **33**, 573–7 (1969).

69.187 N. J. TAYLOR. Thin Reaction Layers and the Surface Structure of Si(111). *Surface Sci.* **15**, 169–74 (1969).

69.188 N. J. TAYLOR. Reply to Comments of H. E. Bishop and J. C. Rivière on "Auger Spectroscopy of Si". *Surface Sci.* **17**, 466–8 (1969).

69.189 N. J. TAYLOR. Resolution and Sensitivity Considerations of an Auger Electron Spectrometer Based on Display LEED Optics. *Rev. Sci. Instrum.* **40**, 792–804 (1969).

69.190 N. J. TAYLOR. Auger Electron Spectrometer as a Tool for Surface Analysis (Contamination Monitor). *J. Vac. Sci. Technol.* **6**, 241–5 (1969).

69.191 J. R. THOMPSON, J. C. DANKO, T. L. GREGORY and H. F. WEBSTER. Surface Characterization Studies on Chemically Vapor Deposited Tungsten. *I.E.E.E. Trans. on Electron Devices* **16**, 707–12 (1969).

69.192 J. C. TRACY and J. M. BLAKELY. A Study of Facetting of W Single Crystal Surfaces. *Surface Sci.* **13**, 313–36 (1969).

69.193 J. C. TRACY and J. M. BLAKELY. The Kinetics of Oxygen Adsorption on the (112) and (110) Planes of W. *Surface Sci.* **15**, 257–76 (1969).

69.194 J. C. TRACY and J. M. BLAKELY. Work Function and Surface Structure Correlations in the Adsorption of Oxygen on W Single Crystals. *The Structure and Chemistry of Solid Surfaces*, ed. G. A. Somorjai (John Wiley, N.Y., 1969), pp. 65–1 to 65–18.

69.195 J. C. TRACY and P. W. PALMBERG. Simple Technique for Binding Energy Determinations: CO on Pd(100). *Surface Sci.* **14**, 274–7 (1969).

69.196 J. C. TRACY and P. W. PALMBERG. Structural Influences on Adsorbate Binding Energy. I. CO on (100) Pd. *J. Chem. Phys.* **51**, 4852–62 (1969).

69.197 C. W. TUCKER. Iodine Faceting of the Ni(210) Surface. *The Structure and Chemistry of Solid Surfaces*, ed. G. A. Somorjai (John Wiley, N.Y., 1969), pp. 58–1 to 58–11.

69.198 H. VAN CAN. Reactivity of Metal Surfaces. *Bull. Soc. Chim. Fr.* **6**, 1901–7 (1969).

69.199 K. K. VIJAI and P. F. PACKMAN. Gas Adsorption Studies on the (100) Plane of V by LEED. *J. Chem. Phys.* **50**, 1343–9 (1969).

69.200 R. F. WALLIS, B. C. CLARK, R. HERMAN and D. C. GAZIS. Theoretical Temperature Dependence of the Mean-square Displacements and Velocities of Surface Atoms of FCC Crystals, *Phys. Rev.* **180**, 716–21 (1969).

69.201 C. M. K. WATTS. LEED From Crystal Surfaces II. A Clean Surface. *J. Phys.* C. **2**, 966–71 (1969).

69.202 R. E. WEBER and A. L. JOHNSON. Determination of Surface Structures Using LEED and Energy Analysis of Scattered Electrons. *J. Appl. Phys.* **40**, 314–18 (1969).

69.203 R. E. WEBER and W. T. PERIA. Work Function and Structural Studies of Alkali-covered Semiconductors. *Surface Sci.* **14**, 13–38 (1969).

69.204 W. H. WEBER and M. B. WEBB. Inelastic Scattering in LEED from Ag. *Phys. Rev.* **177**, 1103–10 (1969).

69.205 P. S. P. WEI, A. Y. CHO and C. W. CALDWELL.

Instrumental Effects of the Retarding Grids in a LEED Apparatus. *Rev. Sci. Instrum.* **40**, 1075–9 (1969).

69.206 J. R. WOEFE and H. W. WEART. Surface Self-diffusion on Ni(111) by Radioactive Tracers. *The Structure and Chemistry of Solid Surfaces*, ed. G. A. Somorjai (John Wiley, N.Y., 1969), pp. 32–1 to 32–38.

69.207 D. P. WOODRUFF and M. P. SEAH. The Temperature Dependence of the Energy of LEED Intensity Peaks and Its Effect on the Surface Debye Temperature. *Phys. Letters* **30A**, 263–4 (1969).

69.208 R. WÜBBENHORST, K. HARTIG AND R. NIEDERMAYER. Anisotropic Film Growth and Pseudomorphism of Ag on a Ge(110) Surface. *J. Vac. Sci. Technol.* **6**, 865–70 (1969).

69.209 G. K. ZURYANOV. Diffraction of Slow Electrons on CdS. *Vestn. Leningrad Univ. Fiz. Khim.* **2**, 69–72 (1969).

## 1970

70.001 S. AKSELA, M. KARRAS, M. PESSA, E. SUONINEN. Study of the Electron Optical Properties of an Electron Spectrograph with Coaxial Cylindrical Electrodes. *Rev. Sci. Instrum.* **41**, 351–5 (1970).

70.002 S. ANDERSSON. Resonances in LEED From the Cu(001) Surfaces. *Surface Sci.* **19**, 21–8 (1970).

70.003 E. BAUER. Interaction of Slow Electrons with Surfaces. *J. Vac. Sci. Technol.* **7**, 3–12 (1970).

70.004 H. E. BISHOP and J. C. RIVIÈRE. Characteristic Ionization Losses Observed in Auger Emission Spectroscopy. *Appl. Phys. Letters* **16**, 21–3 (1970).

70.005 L. A. BRUCE. Comment on "A Study of the Optical Properties and of the Physical Adsorption of Gases on Ag Single Crystal Surfaces by LEED and Ellipsometry" by R. F. Skiger, J. M. Morabito, G. A. Somorjai and R. H. Muller. *Surface Sci.* **20**, 187–9 (1970).

70.006 B. D. CAMPBELL and W. P. ELLIS. Auger Electron Studies of $UO_2$ Surfaces. *J. Chem. Phys.* **52**, 3303–4 (1970).

70.007 J. M. CHARIG and D. K. SKINNER. Auger Electron Spectroscopy of Ni Deposits on the Si(111) Surface. *Surface Sci.* **19**, 283–90 (1970).

70.008 J. CHEN and C. A. PAPAGEORGOPOULOS. Ditungsten Carbide Overlayer on W(112). *Surface Sci.* **20**, 195–200 (1970).

70.009 G. DALMAI-IMELIK and J. BERTOLINI. Étude de l'adsorption de l'éthylène sur la face (100) du Ni par la diffraction des électrons de faible énergie. *C.R. Acad. Sci.* **270C**, 1079–81 (1970).

70.010 T. H. DISTEFANO and D. T. PIERCE. Energy Resolution of the Photoemission Analyzer. *Rev. Sci. Instrum.* **41**, 180–8 (1970).

70.011 G. J. DOOLEY and T. W. HAAS. Some Further Studies of Gas Adsorption on the Mo(100) Surface. *J. Chem. Phys.* **52**, 461–2 (1970).

70.012 G. J. DOOLEY and T. W. HAAS. Scattering of Low-Energy Electrons From Hydrogen-covered Mo(100) Surfaces. *J. Chem. Phys.* **52**, 993–6 (1970).

70.013 G. J. DOOLEY and T. W. HAAS. Chemisorption on Single Crystal Mo(112) Surfaces, *J. Vac. Sci. Technol.* **7**, 49–52 (1970).

70.014 G. J. DOOLEY and T. W. HAAS. Some Properties of the Re(0001) Surface. *Surface Sci.* **19**, 1–8 (1970).

70.015 M. DRECHSLER. Comments on "Simple Technique for Binding Energy Determinations: CO on Pd(100)" by J. C. Tracy and P. W. Palmberg. *Surface Sci.* **20**, 179–80 (1970).

70.016 C. B. DUKE, J. R. ANDERSON and C. W. TUCKER. The Inelastic Collision Model II. Second-order Perturbation Theory. *Surface Sci.* **19**, 117–58 (1970).

70.017 C. G. DUNN and L. A. HARRIS. Auger Electron Analysis of Electropolished High-purity Al. *J. Electrochem. Soc.* **117**, 81–2 (1970).

70.018 W. P. ELLIS and B. D. CAMPBELL. Secondary-electron

Energy Distribution Studies of $UO_2$ Surfaces. *J. Appl. Phys.* **41**, 1858–60 (1970).

70.019 L. G. FEINSTEIN. LEED Secondary Bragg Peaks for Ni(111). *Surface Sci.* **19**, 366–70 (1970).

70.020 L. G. FEINSTEIN, E. BLANC and D. DUFAYARD. LEED Study of the Epitaxy of Ag on Ni(111). *Surface Sci.* **19**, 269–82 (1970).

70.021 L. G. FEINSTEIN and O. MASSENET. The Multiple Diffraction Origin of Extra Reflections in Electron Diffraction from an Epitaxial System. *Surface Sci.* **20**, 437–40 (1970).

70.022 G. GAFNER. Determination of Various Mo(110) and W(110) Surface Structures from LEED Results using a Multiple-scattering Method. *Surface Sci.* **19**, 9–20 (1970).

70.023 R. L. GERLACH, J. E. HOUSTON and R. L. PARK. Ionization Spectroscopy of Surfaces. *Appl. Phys. Letters* **16**, 179–81 (1970).

70.024 J. T. GRANT and T. W. HAAS. Auger Studies of Cleaved (111) Si Surfaces. *J. Vac. Sci. Technol.* **7**, 77–9 (1970).

70.025 T. W. HAAS and J. T. GRANT. Chemical Effects on the KLL Auger Electron Spectrum from Surface Carbon. *Appl. Phys. Letters* **16**, 172–3 (1970).

70.026 T. W. HAAS, J. T. GRANT and G. J. DOOLEY. Some Problems in the Analysis of Auger Electron Spectra. *J. Vac. Sci. Technol.* **7**, 43–5 (1970).

70.027 T. W. HAAS, J. T. GRANT and G. J. DOOLEY. Auger Electron Spectroscopy of Transition Metals. *Phys. Rev.* B **1**, 1449–59 (1970).

70.028 W. HAIDINGER and S. C. BARNES. LEED Investigations of Clean and Au-stabilised Si Surfaces. *Surface Sci.* **20**, 313–25 (1970).

70.029 R. C. HENDERSON, W. J. POLITO and J. SIMPSON. Observation of SiC with Si(111)–7 Surface Structure Using HEED. *Appl. Phys. Letters* **16**, 15–18 (1970).

70.030 M. HENZLER. LEED-Investigation of Step Arrays on Cleaved Ge(111) Surfaces. *Surface Sci.* **19**, 159–71 (1970).

70.031 D. A. HUCHITAL and J. D. RIGDEN. High Sensitivity

Electron Spectrometer. *Appl. Phys. Letters* **16**, 348–51 (1970).

70.032 P. J. Jennings. Spin-polarisation and Relativistic Corrections in LEED. *Surface Sci.* **20**, 18–26 (1970).

70.033 K. Kambe. A Multiple-scattering Theory of LEED Intensities. *Surface Sci.* **20**, 213–19 (1970).

70.034 H. Kanter. Slow-Electron Mean Free Paths in Al, Ag and Au. *Phys. Rev.* B **1**, 522–36 (1970).

70.035 E. Kerre and P. Phariseau. A Green Function Approach of LEED. I. Application to Simple Semi-infinite Crystals with Muffin-tin Potential. *Physica* **46**, 411–37 (1970).

70.036 A. Klopfer. Desorption Durch Elektronenbeschuss Von Molybdän Mit Adsorbierten Gasen: $H_2$, $O_2$, $H_2O$. *Surface Sci.* **20**, 129–42 (1970).

70.037 N. C. MacDonald. Auger Electron Spectroscopy in Scanning Electron Microscopy: Potential Measurements. *Appl. Phys. Letters* **16**, 76–80 (1970).

70.038 E. Margot, J. Oudar and J. Bénard. Étude par la diffraction des électrons de faible énergie de l'adsorption du soufre sur la surface (100) du fer. *C.R. Acad. Sci.* **270C**, 1261–4 (1970).

70.039 J. A. D. Matthew. A Temperature Dependent Contribution to Auger Electron Energy Distributions. *Surface Sci.* **20**, 183–6 (1970).

70.040 C. S. McKee and N. W. Roberts. LEED and Auger Electron Spectroscopy. *Chem. Brit.* **6**, 106–10 (1970).

70.041 A. J. Melmed and J. J. Carroll. Ellipsometry, LEED and FEM Study of Evaporated Epitaxial Films of Fe on (011) W. *Surface Sci.* **19**, 243–8 (1970).

70.042 W. M. Mularie and T. W. Rusch. Inelastic Effects in Auger Electron Spectroscopy. *Surface Sci.* **19**, 469–74 (1970).

70.043 R. G. Musket and J. Ferrante. Auger Electron Spectroscopy Study of Oxygen Adsorption on W(110). *J. Vac. Sci. Technol.* **7**, 14–17 (1970).

70.044 K. L. Ngai, E. N. Economou and **M. H.** Cohen. Theory of Surface Plasmon Excitation in LEED and in Photoemission. *Phys. Rev. Letters* **24**, 61–3 (1970).

70.045 M. PERDEREAU and J. OUDAR. Structure, Méchanisme de Formation et Stabilité de la Couche d'Adsorption du Soufre sur le Nickel. *Surface Sci.* **20**, 80–98 (1970).

70.046 M. PESSA, S. AKSELA and M. KARRAS. New Fine Structure in Electron-excited Auger Spectra from Solid Surfaces. *Phys. Letters* **31A**, 382–3 (1970).

70.047 J. H. POLLARD. A Correlation of Auger Emission Spectroscopy, LEED and Work Function Measurements for the Epitaxial Growth of Th on a W(100) Substrate. *Surface Sci.* **20**, 269–84 (1970).

70.048 J. W. T. RIDGWAY and D. HANEMAN. The Diffusion of Fe and Ni to Si Surfaces. *Phys. Stat. Solidi* **38**, K31–3 (1970).

70.049 G. J. RUSSELL. Observations of Cleaved and Oxygen Exposed Surfaces of Si and Ge by RHEED Diffraction. *Surface Sci.* **19**, 217–29 (1970).

70.050 H. Z. SAR-EL. Criterion for Comparing Analyzers. *Rev. Sci. Instrum.* **41**, 561–4 (1970).

70.051 E. N. SICKAFUS. Sulfur and Carbon on the (110) Surface of Ni. *Surface Sci.* **19**, 181–97 (1970).

70.052 R. F. STEIGER, J. M. MORABITO, G. A. SOMORJAI and R. H. MULLER. Reply to the Comments of L. A. Bruce on "A Study of the Optical Properties and of the Physical Adsorption of Gases on Ag Single Crystal Surfaces by LEED and Ellipsometry". *Surface Sci.* **20**, 190–1 (1970).

70.053 D. TABOR and J. WILSON. The Amplitude of Surface Atomic Vibrations in the (100) Plane of Nb. *Surface Sci.* **20**, 203–8 (1970).

70.054 P. W. TAMM and L. D. SCHMIDT. Interaction of $H_2$ with (100) W. II. Condensation. *J. Chem. Phys.* **52**, 1150–60 (1970).

70.055 H. TOKUTAKA and J. A. D. MATTHEW. Inner Potential Calculations for Alkali Halides and Their Relation to (00) Intensity-Voltage Measurements. *Surface Sci.* **19**, 427–34 (1970).

70.056 J. C. TRACY and G. K, BOHN. Auger Electron Spectrometer Preamplifier. *Rev. Sci. Instrum.* **41**, 591–2 (1970).

70.057 J. C. TRACY and P. W. PALMBERG. Reply to the Comments of M. Drechsler on "Simple Technique for Binding Energy Determinations: CO on Pd (100)". *Surface Sci.* **20**, 181–2 (1970).

70.058 J. J. UEBBING. Auger Electron Spectroscopy of Contaminated GaAs Surfaces. *J. Vac. Sci. Technol.* **7**, 81–3 (1970).

70.059 J. J. UEBBING. Use of Auger Electron Spectroscopy in Determining the Effect of Carbon and Other Surface Contaminants on GaAs–Cs–O Photocathodes. *J. Appl. Phys.* **41**, 802–4 (1970).

70.060 J. J. UEBBING and N. J. TAYLOR. Auger Electron Spectroscopy of Clean GaAs. *J. Appl. Phys.* **41**, 804–9 (1970).

70.061 P. S. P. WEI. Studies of Si and its Oxygen Adsorption by Low Energy Electron Scattering. *Surface Sci.* **20**, 157–62 (1970).

70.062 D. P. WOODRUFF and B. W. HOLLAND. Time Reversal Symmetry in LEED. *Phys. Letters* **31A**, 207–8 (1970).

# AUTHOR INDEX

# MATERIAL INDEX

# SUBJECT INDEX

67.009; 67.021; 67.023; 67.036; 67.037; 67.039; 67.045; 67.046; 67.054; 67.055; 67.065; 67.066; 67.068; 67.103; 68.017; 68.034; 68.035; 68.047; 68.059; 68.060; 68.067; 68.075; 68.078; 68.079; 68.080; 68.088; 68.089; 68.099; 68.107; 68.115; 68.118; 68.124; 68.125; 69.008; 69.016; 69.024; 69.029; 69.032; 69.037; 69.045; 69.046; 69.047; 69.048; 60.064; 69.067; 69.072; 69.089; 69.091; 69.095; 69.098; 69.099; 69.102; 69.107; 69.119; 69.120; 69.128; 69.129; 69.130; 69.131; 69.143; 69.152; 69.153; 69.154; 69.155; 69.176; 69.177; 69.178; 69.179; 69.180; 69.201; 70.003; 70.016; 70.022; 70.032; 70.033; 70.035; 70.044; 70.055; 70.062

# SURFACE IONIZATION AND ITS APPLICATIONS

N. I. Ionov
*A. F. Ioffe Physico-Technical Institute,*
*Academy of Sciences of the U.S.S.R., Leningrad, K-21, U.S.S.R.*

## CONTENTS

## 1. Introduction

In the same issue of *The Physical Review* in 1923, Kingdon and Langmuir[1] and Ives[2] reported the discovery of a new phenomenon, which acquired the name of "surface ionization". This phenomenon consisted in a pronounced desorption from the surface of a heated tungsten filament of cesium atoms in the form of positive ions. An electric current passed through the vacuum diode filled with cesium vapors, if the heated tungsten filament was rendered positive with respect to the surrounding cylindrical electrode. The current increased with increasing cesium vapor pressure in the diode and dropped to zero when the cesium vapor condensed out.

Further studies have shown the surface ionization (SI) phenomenon to be a common feature, so that atoms of virtually all elements, as well as many thermally stable molecules, become ionized in thermal desorption from the surface of high melting-point solids. However, the degree of SI $\alpha$, which is the ratio of the flux of ionized desorbed particles $\nu_+$ to that of neutral desorbed particles $\nu_0$ of the same chemical composition, varies for different cases within a wide range from $\alpha \ll 1$ to $\alpha \gg 1$.

Morgulis[3] suggested in 1934 that atoms possessing electron affinity should form negative ions in SI. Indeed, in 1935 Sutton and Mayer[4] observed SI of the iodine atoms on tungsten involving the formation of negative $I^-$ ions.

Up to now, several hundred publications on SI have appeared in the literature. Investigations were conducted with the view of both studying the characteristics of the phenomenon proper and finding possible applications in electronics, as well as using it as a method for the study of the physicochemical processes on solid surfaces. A detailed description of these investigations, as of the beginning of 1968, was given in the recently published monograph by Zandberg and Ionov[5]. The purpose of the present review is to acquaint, in a "broad-brush" fashion, the physicists, chemists and engineers with the principal features of SI and the role of this phenomenon in the operation of many types of electronic and ionic devices, as well as to consider cases of the most advantageous application of SI in physicochemical studies. Specific problems of a more narrow scope, and works which are now mostly of historical interest, will not be treated in the review. Therefore, the list of references at the end does not cover all available publications on SI and its applications.

## 2. Quantitative Characteristics of Surface Ionization

Figure 1 shows schematically the surface of an adsorbent with a layer of adsorbed particles of concentration $N$. At the temperature $T$, the adsorbent surface emits the fluxes: $\nu_0$ of neutral particles, $\nu_+$ of positive and $\nu_-$ of negative ions. To ensure a complete collection of ions from the surface, an accelerating electric field of the appropriate sign is applied, which blocks out the ion fluxes of the opposite sign emitted

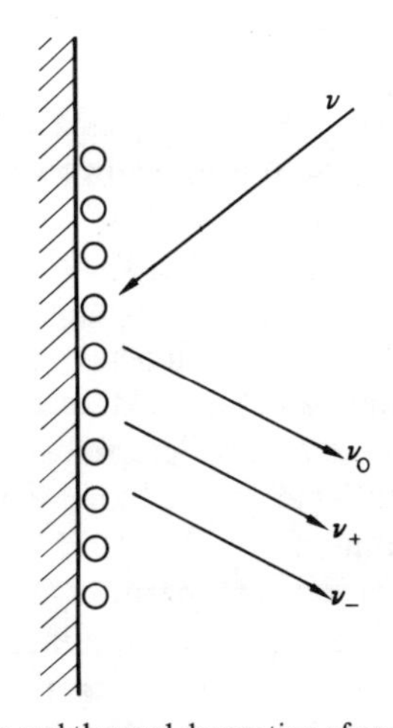


FIG. 1. Adsorption and thermal desorption of particles.

from the surface. In this case, the degree of ionization is defined as the ratio $\alpha = \nu_+/\nu_0$ for positive and $\alpha = \nu_-/\nu_0$ for negative ions. The major part of SI experiments are carried out under steady-state conditions, where the flux of particles desorbing from the surface $(\nu_0 + \nu_+)$ or $(\nu_0 + \nu_-)$ is equal to that of particles $\nu$ adsorbing on this surface from the surrounding space, i.e.,

$$\nu = \nu_0 + \nu_\pm. \tag{2.1}$$

The quantity measured experimentally is the ionic current from the surface, which is determined by the ionization coefficient $\beta$ and, in the steady state, by the particle flux $\nu$, so that

$$j = e\nu\beta \tag{2.2}$$

where $j$ is the ionic current density, $e$ the ionic charge and

$$\beta = \frac{\nu_\pm}{\nu} = \frac{\nu_\pm}{\nu_0 + \nu_\pm} = \frac{1}{1 + \alpha^{-1}}. \tag{2.3}$$

Expressions (2.1) to (2.3) are valid for SI involving the formation of both positive and negative ions.

## 3. $\alpha(T, \varepsilon)$-Function

In this and the next sections we will consider SI involving the formation of positive ions.

It is known, experimentally, that the magnitude of the ionic current $j$ from the surface and, hence, the magnitudes of $\alpha$ and $\beta$ for each adsorbent-ionizing atom pair, depend on the temperature $T$ of the system and the strength $\varepsilon$ of the field accelerating ions away from the surface. The function $\alpha(T, \varepsilon)$ is determined theoretically for a homogeneous surface over which the adsorption properties and the work function $\varphi$ are the same.

From a consideration of thermodynamic equilibrium between the gas and the adsorbed phases, Kingdon and Langmuir[6] derived an expression for $\alpha(T)$. Pyatigorskii[7] generalized the thermodynamic derivation to obtain the function $\alpha(T, \varepsilon)$. The advantage of the thermodynamic derivation lies in its being of a general nature, since it can readily be applied to the description of ionization on both metals and semiconductors, provided the conditions for thermodynamic equilibrium in the adsorbent-adsorbate system are satisfied.

A statistical method was employed by Dobretsov[8, 9] and Zandberg[10] to derive $\alpha(T, \varepsilon)$, for the case of ionization on metals, using the Sommerfeld model for the metal-adsorbent system. Following the ideas expressed by Ansel'm[11], Gurney[12] and Grover[13], the adsorbed atom and the adsorbent are considered as making up a single quantum-mechanical system. The ground level of the valence electrons of the atoms, located at an equilibrium distance $x_{eq}$ (Fig. 2) from the metal

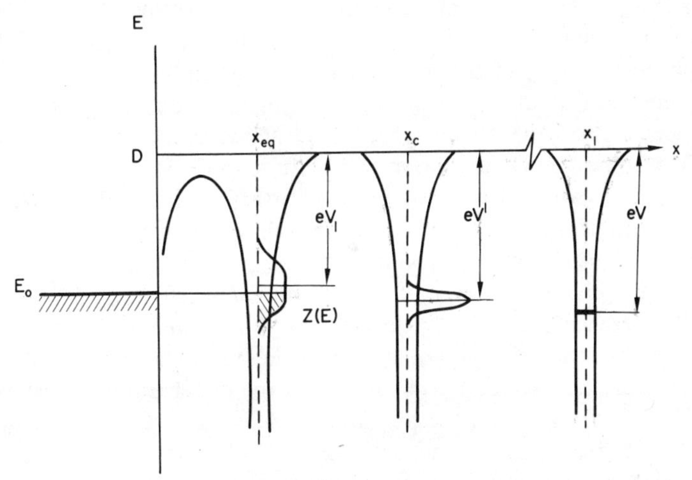


FIG. 2. Transformation of the energy level of the valence electron of an adsorbing atom.

surface, broadens to form a band $Z(E)$, whose central level $eV_1$ is, generally speaking, shifted relative to the free atom level $eV$. The valence electrons of the adsorbed atoms form a common system with the conduction band electrons of the metal, their distribution over the energy states of the system being described by the Fermi function. The level $eV_1$ of the adsorbed atoms does not coincide with the Fermi level

$E_0$ of the metal, the adsorbed atomic charge $e'$ averaged over time and volume of the atom not being equal to the elementary charge $e$, i.e. $0 < e' < e$. In the case of electropositive adsorbates, this effective positive charge $e'$ produces, when combined with its electric image, a double-surface layer lowering the work function $\varphi$ of the surface. In the case of electronegative adsorbates, whose atoms possess electron affinity, Fig. 2 retains its meaning if one replaces the levels $eV$ by the affinity levels $eS$. Here the effective charge of the adsorbed atoms will be negative, its magnitude again lying within $0 < e' < e$.

Now let us remove an atom from the surface (desorb it) along the $x$-axis normal to it (Fig. 2). The potential barrier between the levels of the atoms and the metal, which is transparent for electrons at $x = x_{eq}$, broadens in the course of desorption, its transparency for electrons decreasing sharply and becoming practically zero at $x = x_c$. The magnitude of $x_c$ is of the order of the atomic radius. The single adsorbent-atom quantum-mechanical system now splits into two independent and comparatively weakly interacting subsystems. The band $Z(E)$ becomes much narrower and shifts to $eV'$. The critical charge exchange distance $x_c$ is naturally a statistical quantity. However, its spread should be small, owing to a strong dependence of the transparency of the barrier on its width.

Figure 2 illustrates a case of a zero external accelerating field $\varepsilon = 0$. In the presence of a field, the levels $V_1$, $V'$ and $V$ should be calculated from the external field potential $e\varepsilon(x)$, which results in their additional shift with respect to the Fermi level $E_0$ of the metal. Since the external field $\varepsilon$ should also affect the potential in the surface layer of the metal and the atomic potential, and hence the potential barrier between them, some dependence of $x_c$ on $\varepsilon$ should exist.

In desorption, at a distance $x_c$ from the surface no more electrons are transferred from the atomic levels $eV'$ to the levels of the metal, so that the desorbing particle moves further away with the same charge (0 or $+e$) it possessed at the distance $x_c$. The probability of ionization in the flux of desorbing particles at $x = \infty$, i.e. the quantity $\alpha(T, \varepsilon)$, is determined by the ionization probability in the flux at $x = x_c$, described by the Fermi distribution, and by the ratio of the probabilities that the desorbing particles have a kinetic energy sufficient to perform work against the forces binding the ion and atom with the adsorbent on the way from $x_c$ to $x = \infty$. The adsorbate–adsorbent system is assumed to be in thermal equilibrium.

Consider now a kinetic derivation of $\alpha(T, \varepsilon)$. It should be stressed that all three methods, viz. the thermodynamical, statistical and kinetic methods, yield identical expressions for $\alpha(T, \varepsilon)$. We will start from the above concept of the adsorbate–adsorbent system, assuming in addition that the system is in thermal equilibrium, and that the adsorbent surface remains homogeneous, irrespective of the adsorbate concentration on it. We will regard the thermal desorption of particles in the neutral and ionic states as representing bimolecular reactions $AM \rightarrow [A + M]$ and $AM \rightarrow [A^- + M^+]$, their rates being determined by the following kinetic formulae:

$$\nu_+ = CN \exp\left(-\frac{E_+}{kT}\right) \tag{3.1}$$

$$\nu_0 = DN \exp\left(-\frac{E_0}{kT}\right). \tag{3.2}$$

Here $\nu_0$ and $\nu_+$ are the desorption rates (flux densities of the desorbing particles) of the atoms and ions; $N$ is the concentration of the adsorbed atoms residing on the surface in the same states of partial ionization: $T$ is the temperature of the adsorbent–adsorbate system; $E_0$ and $E_+$ are the desorption activation energies of the atoms and ions; $C$ and $D$ are factors which depend weakly upon $T$ as compared with the exponentials.

Figure 3 presents schematically the dependence of the potential energy of the adsorbent–adsorbed particle system upon the separation $x$ between them. In this graph, $E$ is the desorption activation energy, $E'$ the adsorption activation energy and $l$ the isothermal heat of desorption. In accordance with the theory of absolute reactions rates, the point $x'$ of the peak of the curve corresponds to the system being in the state of an activated transition complex $[A\text{-}M]^{\ddagger}$. The particle that transferred from the adsorbed into the activated complex state will desorb, since it will not return back to the surface from the point $x'$, even if it has a very small momentum in the $x$-direction. We assume here that $x_c < x'$, i.e. in the activated complex state, the charge state of the particles is already quite definite.

From the condition of equilibrium between the adsorbed particles and the particles residing in the activated complex state, it follows that, in (3.1) and (3.2), $C = Q^{\ddagger}_{+}/Q$ and $D = Q^{\ddagger}_{0}/Q$, where $Q$ is the sum of the states of particles adsorbed on the surface, and $Q^{\ddagger}_{+}$ and $Q^{\ddagger}_{0}$ are those of the

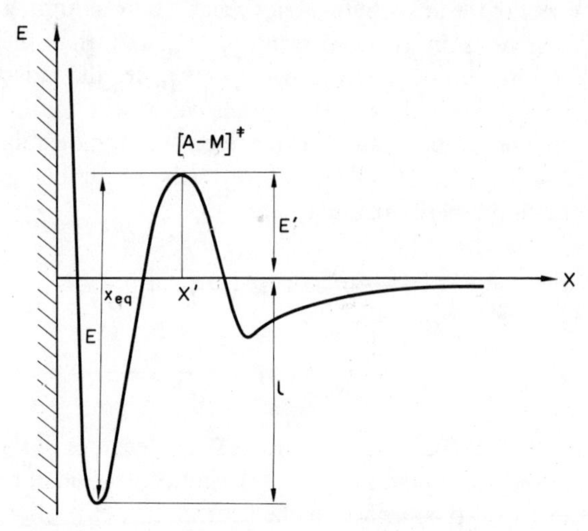


FIG. 3. Potential-energy variation of an $[A + M]$ system in the desorption of an $M$ particle in the direction of the $x$-axis.

ion and atom in the activated complex state. Substituting these quantities in (3.1) and (3.2) yields (see also Fig. 3):

$$\alpha = \frac{Q^{\ddagger}_{+}}{Q^{\ddagger}_{0}} \exp\left[\frac{1}{kT}(E'_{0} - E'_{+} + l_{0} - l_{+})\right]. \tag{3.3}$$

When in the activated complex state, the particles are only weakly bound with the adsorbent surface, so that the quantities $Q^{\ddagger}_{+}$ and $Q^{\ddagger}_{0}$ may be justifiably replaced by the sum of states of the free ion $Q_{+}$ and atom $Q_{0}$.

The existence of an activation barrier for adsorption $E'_{0}$ and $E'_{+}$ results in a partial reflection of the particle flux adsorbed on or desorbed from the surface. With a Maxwellian distribution in energy, exp

$(-E'_0/kT) \sim (I - R_0)$ and $\exp(-E'_+/kT) \sim (I - R_+)$, where $R_0$ and $R_+$ are the reflection coefficients of the atoms and ions, respectively, (3.3) reduces to:

$$\alpha = A\left(\frac{1 - R_+}{1 - R_0}\right)\exp\left[\frac{1}{kT}(l_0 - l_+)\right] \tag{3.4}$$

where we have introduced a quantity $A = Q^{\ddagger}_{+}/Q^{\ddagger}_{0} = Q_+/Q_0$ representing the statistical sum ratio of the ionic and neutral states of desorbing particles. For atoms and atomic ions with the same masses, the quantity $A$ is equal to the ratio of the statistical sums over the electronic states of the ion and atom. For molecular particles, $A$ is determined by the ratio of the total sums of states over all degrees of freedom (vibrational, rotational and electronic states of the shells) except for the translational motion, since the ion and the neutral particle have identical masses.

The quantities $l_+$ and $l_0$ depend on surface concentration of the adsorbed particles (surface coverage $\theta$), as well as on the strength $\varepsilon$ of the accelerating field. Turning to the consideration of the dependences $l_+(\varepsilon)$ and $l_0(\varepsilon)$, at $\theta = \text{const}$, again assuming the surface to be homogeneous, we will use the potential curve method for the systems $[A + M]$ and $[A^- + M^+]$[5, 14]. Two particular cases will be studied (Fig. 2), viz. $V' < \varphi$ and $V' > \varphi$.

*(i) $V' < \varphi$ (Fig. 4) case*

We calculate the energy from that of the $[A^- + M^+]$ system with the $M^+$ ion at $x = \infty$ from the surface, and the field $\varepsilon = 0$. As the $M^+$ ion slowly approaches the surface, it experiences the following attractive forces: (*a*) the polarization image force $F_p = e^2/4x^2$: (*b*) the van der Waals force $F_v$ attracting the ion to the surface and the force $F_\theta$ of interaction of the ion with the surface electric dipoles of the adsorbed particles $M$. (Both these forces decrease with increasing $x$ much faster than $F_p$ does): (*c*) the force with which a nonuniform electric field pulls the ion to the surface. The field acting on the ion near the surface is actually a sum of the external field $\varepsilon$ and the field responsible for all electric interactions of the ion with the surface. The work done by all these forces, shown in Fig. 4 by the broken line, is equal to $\lambda_{+0}$, along the distance $x = \infty$ to $x_c$. Along the distance from $x_c$ to $x_{eq}$, $M^+$ and $A^-$

are acted on by the exchange forces (the chemical binding forces), whose work is denoted here by $q_0$. Thus, the heat of ion desorption from the surface, which equals the depth of the potential curve for the system $[A^- + M^+]$ at the point $x_{eq}$, is

$$l_{+0} = q_0 + \lambda_{+0}. \tag{3.5}$$

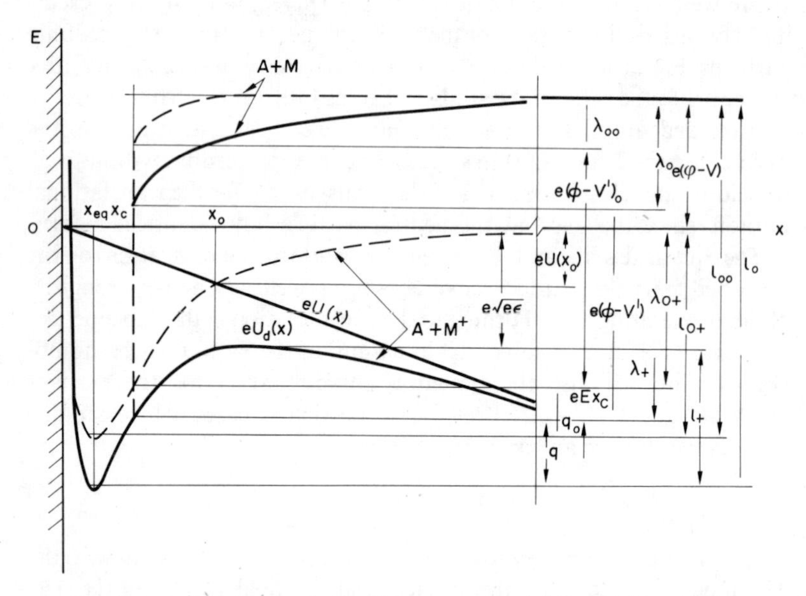


FIG. 4. Potential-energy variation of the systems $[A + M]$ and $[A^- + M^+]$ in the presence of an external accelerating electric field $\varepsilon$ (solid curves) and for $\varepsilon = 0$ (broken curves), $V' < \varphi$[5, 14].

Here, and in what follows, the subscript "0" refers to zero field, $\varepsilon = 0$.

Consider now the potential curve for the system $[A + M]$. In order to go over from the $[A^- + M^+]$ potential curve to the curve for $[A + M]$ at $x = \infty$, an electron must be transferred from the Fermi level of the metal to the level $eV$ of the $M^+$ valence electron. The energy required for this is $e(\varphi - V)$. As a particle $M$ moves slowly towards the surface $A$, the electric forces acting between $M$ and $A$ (the van der Waals force and the force exerted on a neutral dipole molecule by an inhomogeneous field) do the work $\lambda_{00}$ along the distance from $x = \infty$ to $x_c$. Since $V' < \varphi$, at the point $x_c$ an electron of the $M$ atom transfers to a lower level $e\varphi$ of the metal, the potential energy of the $[A + M]$ system decreases stepwise by $e(\varphi - V')$, which means that the state of the system at the point $x_c$ moves from the potential curve $[A + M]$ to the curve $[A^- + M^+]$. Along the distance from $x_c$ to $x_{eq}$, the system $[A^- + M^+]$ performs the work $q_0$, exactly as in the first case. This agrees with the earlier statement on the indistinguishability of the states of adsorbed particles on the surface, i.e. in the region $x \leqslant x_c$. The depth of the potential curve for the system $[A + M]$, as follows from Fig. 4, is

$$l_{00} = q_0 + \lambda_{00} + e(\varphi - V'). \tag{3.6}$$

*(iii) $V' > \varphi$ (Fig. 5) case*

The graphs of the potential energy at $\varepsilon = 0$ (broken curves) for the systems $[A + M]$ and $[A^- + M^+]$ are drawn similar to the first case. From their consideration, the following expressions for the heats of desorption can be derived:

$$l_{+0} = q_0 + \lambda_{+0} + e(V' - \varphi) \tag{3.7}$$

and

$$l_{00} = q_0 + \lambda_{00}. \tag{3.8}$$

From (3.5), (3.6), (3.7) and (3.8), as well as from the graphs of Figs. 4 and 5, we can obtain the well-known *Schottky relation*:

$$l_{00} - l_{+0} = \lambda_{00} - \lambda_{+0} + e(\varphi - V') = e(\varphi - V). \tag{3.9}$$

Since in all cases the work $\lambda_{+0}$, required to carry ions from $x_c$ to $x = \infty$ at $\varepsilon = 0$, is apparently larger than the corresponding work for the atoms $\lambda_{00}$, it follows from (3.9) that $V' < V$, which means that the mean level of the atomic valence electrons rises, whereas the atomic ionization potential decreases near the adsorbent surface (at $x = x_c$).

Consider now the effect of the external field $\varepsilon$ on the quantities $l_0$ and $l_+$. In most experiments, the accelerating electric field is inhomogeneous, so that $\varepsilon$ increases with decreasing $x$. Therefore, if the constant or induced dipole moments of the atoms $M$ are not zero, the work

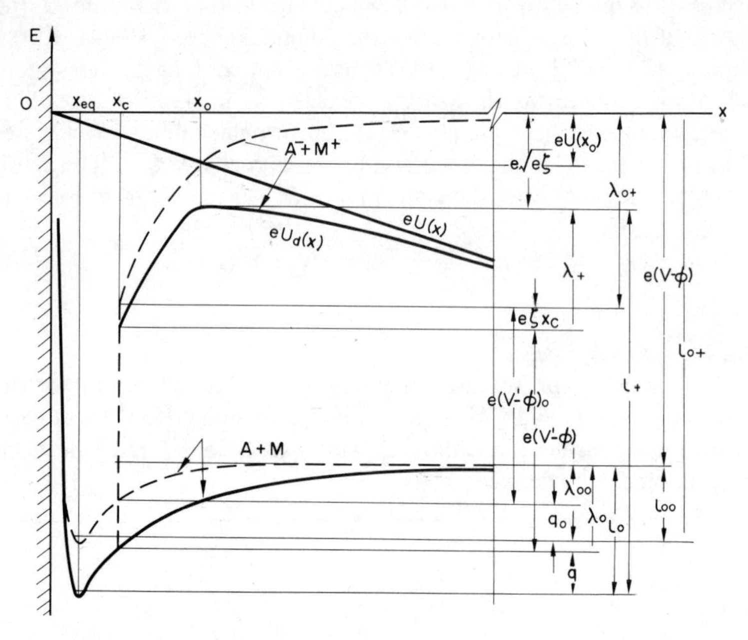


FIG. 5. Potential-energy variation of the systems $[A + M]$ and $[A^- + M^+]$ in the presence of an external accelerating electric field $\varepsilon$ (solid curves) and for $\varepsilon = 0$ (broken curves), $V' > \varphi$[5, 14].

performed against the forces of interaction between the atom and adsorbent, along the distance from $x_c$ to $x = \infty$ in the external field, increases to $\lambda_0$. The potential curve for the system $[A + M]$ lies below the curve for $\varepsilon = 0$ (the solid curves in Figs. 4 and 5). The chemical binding energy of the adsorbed particles also probably changes in the field $\varepsilon$ from $q_0$ to $q$.

The field $\varepsilon$ affects more strongly the work of desorption of ions $\lambda_+$. The potential curve for the system $[A^- + M^+]$ in the field $\varepsilon$ should be constructed by taking into account the changes in both the forces of interaction of the ion with the surface and the potential energy of the ion produced by the external field $\varepsilon$. Since the force exerted on the ion by the external field is directed away from the surface, the potential energy curves with external field (solid lines in Figs. 4 and 5) have a peak at $x = x_0$. At $x < x_0$, the force acting on the ion is directed towards, and at $x > x_0$ away from, the surface. It should also be borne in mind that, in the graphs, the potential of the external field on the surface $A$ is assumed to be zero, whereas at $x = x_c$ it is $e\varepsilon x_c$.

Confining ourselves to the simplest case of the ion being acted upon only by the image force along the distance from $x_c$ to $x = \infty$, we obtain that, at $x_c$, this force is equal in magnitude and opposite in sign to the force $e\varepsilon$ exerted on the ion by the external field, i.e. $e\varepsilon = e^2/4x_0^2$, whence $x_0 = \frac{1}{2}\sqrt{(e/\varepsilon)}$. The lowering of the potential curve for $[A^- + M^+]$ at $x_0$ in this case is

$$e\varepsilon x_0 + e^2/4x_0 = e\sqrt{(e\varepsilon)}, \tag{3.10}$$

i.e. it is equal to the Schottky lowering of the image-force barrier by the field $\varepsilon$. Figure 6 shows graphically the dependence of $x_0$ and $\Delta l_+ = e\sqrt{(e\varepsilon)}$ on lg $\varepsilon$ (V/cm) for the case of only the Coulomb image force acting between the ion and the surface.

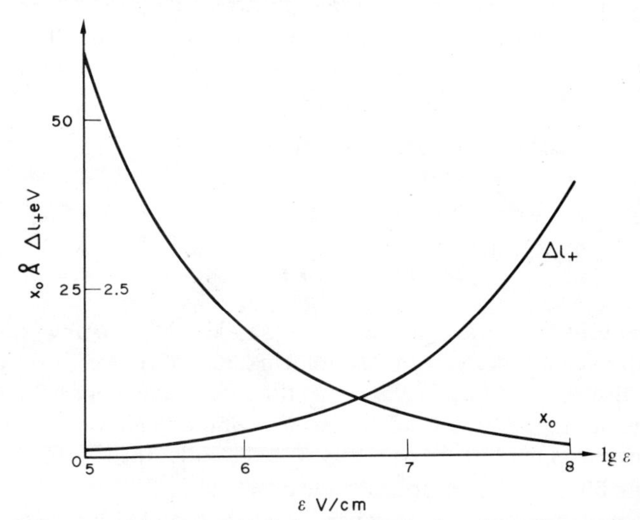


FIG. 6. The curves of $x_0(\varepsilon)$ and $\Delta l_+(\varepsilon)$ for the case of only an image force acting between ion and surface ($\varepsilon$ in V/cm).

Bearing this in mind, we readily obtain from the potential diagrams of Fig. 4, for the case $\varepsilon > 0$ and $V' < \varphi$, the following relations:

$$l_0 = l_{00} + (q - q_0) + e\varepsilon x_c \qquad (3.11)$$
$$l_+ = l_{+0} + (q - q_0) + e\varepsilon x_c - e\varepsilon x_0 - e^2/4x_0$$

and from Fig. 5 for $V' > \varphi$:

$$l_0 = l_{00} + (q - q_0) + (\lambda_0 - \lambda_{00}) \qquad (3.12)$$
$$l_+ = l_{+0} + (q - q_0) + (\lambda_0 - \lambda_{00}) - e\varepsilon x_0 - e^2/4x_0.$$

From (3.11), (3.12) and (3.9), and the fact that $x_0 = \frac{1}{2}\sqrt{(e/\varepsilon)}$, comes

$$l_0 - l_+ = l_{00} - l_{+0} + e\sqrt{(e\varepsilon)} = e(\varphi - V + \sqrt{(e\varepsilon)}). \qquad (3.13)$$

Substituting (3.13) into (3.4) yields an expression for the degree of ionization called the *Saha–Langmuir formula*, which is valid for any $V$ and $\varphi$:

$$\alpha(T, \varepsilon) = A\left(\frac{1 - R_+}{1 - R_0}\right)\exp\left[\frac{e}{kT}(\varphi - V + \sqrt{(e\varepsilon)})\right]. \qquad (3.14)$$

Equations (3.11) to (3.14) are applicable only for the case $x_0 > x_c$, i.e. for the fields $\varepsilon < \varepsilon_c = e^2/4x_c^2$. The field $\varepsilon_c$ cancels out the image force along the distance from $x = \infty$ to $x_c$. For fields $\varepsilon > \varepsilon_c$, we have to replace $x_0$ by $x_c$ in (3.11) and (3.12), obtaining

$$l_0 - l_+ = l_{00} - l_{+0} + e^2/4x_c + e\varepsilon x_c = e(\varphi - V + e/4x_c + \varepsilon x_c) \qquad (3.15)$$

and

$$\alpha(T, \varepsilon) = A\left(\frac{1 - R_+}{1 - R_0}\right)\exp\left[\frac{e}{kT}\left(\varphi - V + \frac{e}{4x_c} + \varepsilon x_c\right)\right] \qquad (3.16)$$

In deriving expressions for $\alpha(T, \varepsilon)$, we did not place any constraints on the magnitude of $V$ for the ionizing atoms. Hence, (3.14) or (3.16) can also be used to describe ionization, for example, of helium atoms in the ionic projectors. In this case, to obtain high values of $\alpha$ at the projector point, we must have a sufficiently high field strength $\varepsilon$ (probably $\varepsilon > \varepsilon_c$), so as to make the exponent in (3.14) or (3.16) positive. It should, however be remembered that (3.14) and (3.16) were obtained under the assumption that the only force acting between the adsorbent surface and the ion at $x > x_c$ is the polarization image force. In a more rigorous derivation of the expression for $\alpha(T, \varepsilon)$[5, 14] one should also take into account the above-mentioned forces $F_v$ and $F_\theta$, as well as a possible deviation of the polarization interaction of the ion with the metal surface from the Coulomb law for the image force at distances $x$ from the surface comparable with the metal lattice constant. The work done by the force attracting an ion possessing a dipole moment should be considered only along the distance from $x_0$ to $x_c$, since at $x = \infty$ and $x = x_0$ (the maxima of the potential curve) the field acting on the ion is zero[7].

A quantity which is more important than $\alpha$, from the practical point of view, is the ionization coefficient $\beta$ which, according to (2.2), determines in the steady-state the magnitude of the ionic current collected from the surface, as well as its dependence on $T$ and $\varepsilon$. In accordance with (2.3) and (3.4), for $\varepsilon < \varepsilon_c$, we have

$$\beta = \left\{1 + \frac{(1 - R_0)}{A(1 + R_+)} \exp\left[\frac{e}{kT}(V - \phi - \sqrt{(e\varepsilon)})\right]\right\}^{-1}. \tag{3.17}$$

Figure 7 presents graphs of $\ln \beta = f(T)$ for various values of the exponent in (3.17), $\delta = V - \varphi - \sqrt{(e\varepsilon)}$, expressed in volts. In the calculations, we assumed

$$\frac{1 - R_0}{A(1 - R_+)} = 1.$$

From (3.17) and these graphs, it follows that, while the magnitude of $\beta$, and accordingly of the ionic current $j$, do not depend on the adsorbent temperature $T$ at $\delta = 0$, they increase with increasing $T$ for $\delta > 0$ and decrease for $\delta < 0$. According to (3.17), at the given values of $\varphi$ and $V$, the dependence of the quantities $\beta(T)$ and $j(T)$ on T, with increasing field strength $\varepsilon$, may go over from an increasing to a decreasing function,

In experiments on SI, one ordinarily studies the dependence of the observed ionic currents on surface temperature $T$ and the accelerating electric field $\varepsilon$, at a constant atomic flux $\nu$ to the surface. From these experimental dependences, one can calculate the dependence on $T$ and $\varepsilon$ of the degree $\alpha$, and coefficient $\beta$, of surface ionization. The equation for the ionic current density $j(T, \varepsilon)$ can be obtained by substituting the expression for $\beta$ from (3.17) into (2.2):

$$d = ev\left\{1 + \frac{1 - R_0}{A(1 - R_+)}\exp\left[\frac{e}{kT}(V - \phi - \sqrt{(e\varepsilon)})\right]\right\}^{-1}. \tag{3.18}$$

The magnitudes of the work function $\varphi$ of the high-melting point metals are confincd to a comparatively narrow range of 4–6 volts. The atomic ionization potentials $V$ (the first number in each square), see also

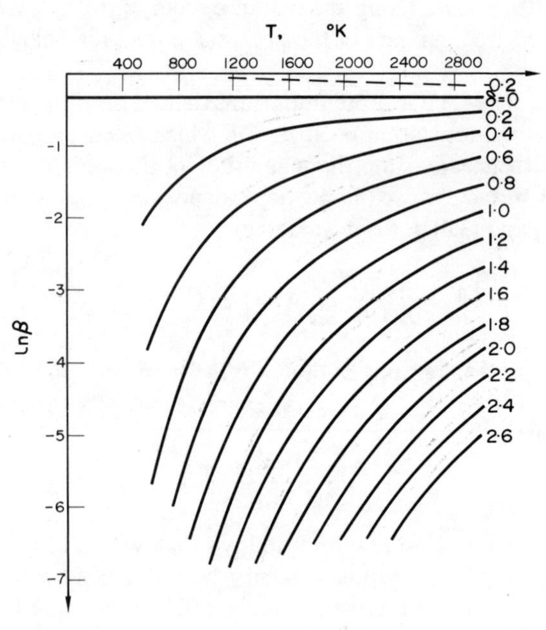


FIG. 7. The graphs of ln $\beta$ as a function of $T$ for different values of $\delta = V - \varphi - \sqrt{(e\varepsilon)}$ expressed in volts.

Table 4), are listed in Table 1.[15] The second number is the most reliable value of the electron affinity $S$ (see also Table 5). The same table contains the ground state scheme for the outer electrons of the atoms needed to determine the statistical weight $g_0$ of these states. As mentioned earlier, in the case of atoms and atomic ions, the magnitude of $A$ in (3.18) is determined by the statistical sum ratio $Q_e$ over the electron states of the ion and atom which, as is well known, is

$$Q_e = g_0 + \sum_i g_i \exp(-\Delta E_i/kT), \tag{3.19}$$

where $g_i$ is the statistical weight of the $i$th excited state of the ion or atom and $\Delta E_i$ is the excitation energy of the ion or atom from the ground state into the $i$th excited state. Table 2 contains the values of $g_i$ and $\Delta E_i$ for the levels of the atoms and atomic ions of the alkali and alkali-earth metals lying most closely to the ground state. It follows from this table that, for the alkali metals, the first excited states of the ions and atoms lie sufficiently far from the ground states, so that they may be neglected for all temperatures attainable in experiment ($T < 3000$°K). For the alkali metals, $A = g_{+0}/g_{00} = \frac{1}{2}$. For the alkali-earths, and the majority of other elements with excited states of the valence electrons close to the ground state, one should, in (3.17) and (3.18), strictly speaking, take into consideration the temperature dependence of the factor $A$.

When $e(V - \varphi - \sqrt{(e\varepsilon)}) \gg kT$, and $A$ is of the order of unity, the second term in the denominator of (3.18) is much greater than unity, so that for this particular case (3.18) becomes

$$j = evA\left(\frac{1 - R_+}{1 - R_0}\right)\exp\left[\frac{e}{kT}(\varphi - V + \sqrt{(e\varepsilon)})\right]. \tag{3.20}$$

The dependence of $\ln j$ on $1/T$, at fixed $v$, $\varepsilon$ and $A$, is linear, which has been confirmed experimentally in many cases.

If $V < \varphi + \sqrt{(e\varepsilon)}$ and the exponents in (3.18) are negative, then the magnitude of current $j$ decreases with increasing $T$, as is shown schematically in Fig. 8. The magnitude of $\beta$, in this case, is close to unity (the second term in the denominator of (3.18) is less than unity) and the magnitude of $j$ depends only weakly on $T$. The degree of ionization $\alpha$ may also vary over a wide range. However, the quantities $\beta$ and $j$, for $\alpha \gg 1$, are not very sensitive to a change in $\alpha$ and, therefore, the determination of $\alpha(T)$ from the experimental relationships $j(T)$ is difficult. According to (3.18), at $V < \varphi + \sqrt{(e\varepsilon)}$, the maximum current density $j$ can be obtained at $T = 0$. On the other hand, if one reduces $T$ starting from a high value (Fig. 8), then at some value $T_0$ the magnitude of $\varphi$ will decrease, and the current $j$ will drop sharply to almost zero,

TABLE

| | I | II | III | IV | V |
|---|---|---|---|---|---|
| 1 | $1s^1$<br>**H** 13,595<br>0.75416 | | | | |
| 2 | $2s^1$<br>**Li** 5.390 | $2s^2$<br>**Be** 9.320 | $2s^22p^1$<br>8.296 **B** | $2s^22p^2$<br>11.256 **C**<br>1.25 ± 0·03 | $2s^22p^3$<br>14.53 **N** |
| 3 | $3s^1$<br>**Na** 5.138 | $3s^2$<br>**Mg** 7.644 | $3s^23p^1$<br>5.984 **Al** | $3s^23p^2$<br>8.149 **Si** | $3s^23p^3$<br>10.484 **P** |
| 4 | $4s^1$<br>**K** 4.339 | $4s^2$<br>**Ca** 6.111 | $3d^14s^2$<br>**Sc** 6.54 | $3d^24s^2$<br>**Ti** 6.82 | $3d^34s^2$<br>**V** 6.74 |
| | $3d^{10}4s^1$<br>7.724 **Cu**<br>1.5 ± 0.5 | $3d^{10}4s^2$<br>9.391 **Zn** | $4s^24p^1$<br>6.00 **Ga** | $4s^24p^2$<br>7.88 **Ge** | $4s^24p^3$<br>9.81 **As**<br>$\lesssim 2$ |
| 5 | $5s^1$<br>**Rb** 4.176 | $5s^2$<br>**Sr** 5.692 | $4d^15s^2$<br>**Y** 6.38 | $4d^25s^2$<br>**Zr** 6.84 | $4d^45s^1$<br>**Nb** 6.88 |
| | $4d^{10}5s^1$<br>7.574 **Ag**<br>2.0 ± 0.2 | $4d^{10}5s^2$<br>8.991 **Cd** | $5s^25p^1$<br>5.785 **In** | $5s^25p^2$<br>7.342 **Sn** | $5s^25p^3$<br>8.639 **Sb**<br>$\lesssim 2$ |
| 6 | $6s^1$<br>**Cs** 3.893 | $6s^2$<br>**Ba** 5.210 | Rare earths | $5d^26s^2$<br>**Hf** 7 | $5d^36s^2$<br>**Ta** 7.88 |
| | $5d^{10}6s^1$<br>9.22 **Au**<br>2.8 ± 0.1 | $5d^{10}6s^2$<br>10.43 **Hg** | $6s^26p^1$<br>6.105 **Tl** | $6s^26p^2$<br>7.415 **Pb** | $6s^26p^3$<br>7.287 **Bi** |
| 7 | $7s^1$<br>**Fr**<br>3.98 ± 0.1 | $7s^2$<br>**Ra** 5.277 | $6d^17s^2$<br>**Ac**<br>6.86 ± 0.6 | $6d^27s^2$<br>**Th**<br>6.95 ± 0.06 | $5f^26d^17s^2$<br>**Pa** |
| Rare earths | $5d^16s^2$<br>**La** 5.61 | $4f^26s^2$<br>**Ce**<br>5.60 ± 0.05 | $4f^36s^2$<br>**Pr**<br>5.49 ± 0.04 | $4f^46s^2$<br>**Nd**<br>5.51 ± 0.02 | $4f^56s^2$<br>**Pm** |
| | $4f^96s^2$<br>**Tb**<br>5.98 ± 0.02 | $4f^{10}6s^2$<br>**Dy**<br>5.80 ± 0.02 | $4f^{11}6s^2$<br>**Ho**<br>6.19 ± 0.02 | $4f^{12}6s^2$<br>**Er**<br>6.08 ± 0.03 | $4f^{13}6s^6$<br>**Tm**<br>6.14 ± 0.06 |

| VI | VII | VIII | | | |
|---|---|---|---|---|---|
| | | $1s^2$<br>**He** 24.581 | | | |
| $^2 2p^4$<br>.614 **O**<br>48 ± 0.10 | $2s^22p^5$<br>17.418 **F**<br>3.448 ± 0.005 | $2s^22p^6$<br>**Ne** 21.559 | | | |
| $^2 3p^4$<br>.357 **S**<br>07 ± 0.07 | $3s^23p^5$<br>13.01 **Cl**<br>3.613 ± 0.003 | $3s^23p^6$<br>**Ar** 15.755 | | | |
| $3d^54s^1$<br>r 6.764 | $3d^54s^2$<br>**Mn** 7.432 | | $3d^64s^2$<br>**Fe** 7.87 | $3d^74s^2$<br>**Co** 7.86 | $3d^84s^2$<br>**Ni** 7.633 |
| $^2 4p^4$<br>75 **Se**<br>2.0 | $4s^24p^5$<br>11.84 **Br**<br>3.363 ± 0.003 | $4s^24p^6$<br>**Kr** 13.996 | | | |
| $4d^55s^1$<br>o 7.10 | $4d^55s^2$<br>**Tc** 7.28 | | $4d^75s^1$<br>**Ru** 7.364 | $4d^85s^1$<br>**Rh** 7.46 | $4d^{10}$<br>**Pd** 8.33 |
| $^2 5p^4$<br>)1 **Te**<br>2.0 | $5s^25p^5$<br>10.454 **I**<br>3.063 ± 0.003 | $5s^25p^6$<br>**Xe** 12.127 | | | |
| $5d^46s^2$<br>7.98<br>1.0 | $5d^56s^2$<br>**Re** 7.87<br>0.65 | | $5d^66s^2$<br>**Os** 8.7 | $5d^76s^2$<br>**Ir** 9 | $5d^96s^1$<br>**Pt** 9.0 |
| $^2 6p^4$<br>43 **Po** | $6s^26p^5$<br>9.2 ± 0.4 **At** | $6s^26p^6$<br>**Rn** 10.746 | | | |
| $5f^36d^17s^2$<br>6.08 ± 0.08 | | | | | |
| $4f^66s^2$<br>n<br>5.70 ± 0.02 | $4f^76s^2$<br>**Eu**<br>5.68 ± 0.03 | $4f^75d^16s^2$<br>**Gd** | | | |
| $4f^{14}6s^2$ | $5d^16s^2$<br>**Lu**<br>5.41 ± 0.02 | | | | |

TABLE 2

| Element | Atom | | | Ion | | |
|---|---|---|---|---|---|---|
| | Levels | $g$ | $\Delta E$, eV | Levels | $g$ | $\Delta E$, eV |
| **Li** | $2s$ $^2S_{\frac{1}{2}}$ | 2 | 0.00 | $1s^2$ $^1S_0$ | 1 | 0.00 |
| | $2p$ $^2P^0_{\frac{1}{2}}$ | 2 | 1.85 | $2s$ $^3S_1$ | 3 | 59.01 |
| **Na** | $3s$ $^2S_{\frac{1}{2}}$ | 2 | 0·00 | $2p^6$ $^1S_0$ | 1 | 0·00 |
| | $3p$ $^2P^0_{\frac{1}{2}}$ | 2 | 2·10 | $2p^5(^2P^0_{\frac{3}{2}})3s$ | 5 | 32.8 |
| **K** | $4s$ $^2S_{\frac{1}{2}}$ | 2 | 0.00 | $3p^6$ $^1S_0$ | 1 | 0.00 |
| | $4p$ $^2P^0_{\frac{1}{2}}$ | 2 | 1.61 | $4p^5(2P^0_{\frac{3}{2}})4s$ | 5 | 20.14 |
| **Rb** | $5s$ $^2S_{\frac{1}{2}}$ | 2 | 0.00 | $4p^6$ $^1S_0$ | 1 | 0.00 |
| | $5p$ $^2P^0_{\frac{1}{2}}$ | 2 | 1.56 | $4p^5(^2P^0_{\frac{3}{2}})4d$ | 3 | 15.67 |
| **Cs** | $6s$ $^2S_{\frac{1}{2}}$ | 2 | 0.00 | $5p^6$ $^1S_0$ | 1 | 0.00 |
| | $6p$ $^2P^0_{\frac{1}{2}}$ | 2 | 1.38 | $5p^5(^2P^0_{\frac{3}{2}})6s$ | 5 | 13.31 |
| **Mg** | $3s^2$ $^1S_0$ | 1 | 0.00 | $3s$ $^2S_{\frac{1}{2}}$ | 2 | 0.00 |
| | $3p$ $^3P^0_0$ | 1 | 2.71 | $3p$ $^2P^0_{\frac{1}{2}}$ | 2 | 4.42 |
| **Ca** | $4s^2$ $^1S_0$ | 1 | 0.00 | $4s$ $^2S_{\frac{1}{2}}$ | 2 | 0.00 |
| | $4p$ $^3P^0_0$ | 1 | 1.88 | $3d$ $^2D_{\frac{3}{2}}$ | 4 | 1.69 |
| **Sr** | $5s^2$ $^1S_0$ | 1 | 0.00 | $5s$ $^2S_{\frac{1}{2}}$ | 2 | 0.00 |
| | $5p$ $^3P^0_0$ | 1 | 1.77 | $4d$ $^2D_{\frac{3}{2}}$ | 4 | 1.80 |
| **Ba** | $6s^2$ $^1S_0$ | 1 | 0.00 | $6s$ $^2S_{\frac{1}{2}}$ | 2 | 0.00 |
| | $5d$ $^3D_1$ | 3 | 1.12 | $5d$ $^2D_{\frac{3}{2}}$ | 4 | 0.60 |
| | $^3D_2$ | 5 | 1.14 | $^2D_{\frac{5}{2}}$ | 6 | 0.70 |
| | $^3D_3$ | 7 | 1.19 | $6p$ $^2P^0_{\frac{1}{2}}$ | 2 | 2.51 |
| | $5d$ $^1D_2$ | 5 | 1.41 | | | |
| | $6p$ $^3P^0_0$ | 1 | 1.52 | | | |
| | $^3P^0_1$ | 3 | 1.57 | | | |
| | $^3P_2$ | 5 | 1.67 | | | |
| | $6p$ $^1P^0$ | 3 | 2.24 | | | |

because of increasing coverage $\theta$ of the surface by absorbed electropositive atoms (with low values of $V$). This has been experimentally established in the first experiments on the surface ionization of cesium ($V = 3.88$ volts) on tungsten ($\varphi = 4.5$ volts). If one now gradually increases $T$ from low values ($j = 0$), then the current $j$ will start to increase sharply, reaching a maximum value at a temperature $T_0' > T_0$ (the direction of variation of $T$ is shown in Fig. 8 by arrows). One also observes a hysteresis of the temperature dependence of the current $j(T)$, seen for the first time in the surface ionization of cesium on tungsten.[16]

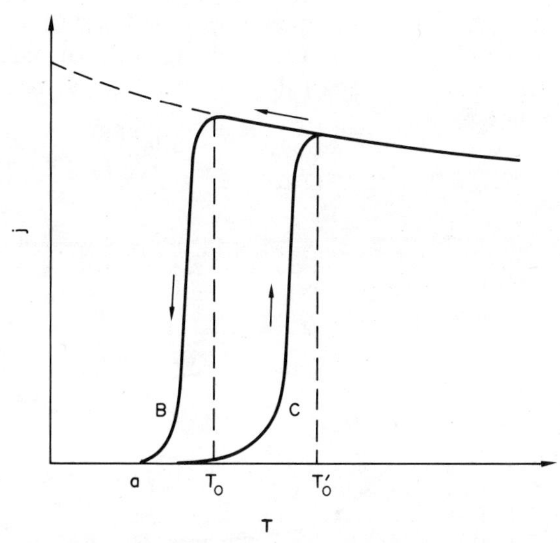


FIG. 8. The graph of the ionic current $j(T)$ in the ionization temperature threshold region. Arrows indicate the sequence of measurements.

In the following we will call the low temperature decrease of the ionic current the temperature threshold, and the temperatures $T_0$ and $T'_0$ the first and the second threshold temperatures.

## 4. Temperature and Field Thresholds of Surface Ionization

As mentioned earlier, the temperature dependence of the ionic current, in the threshold region, is due to the dependence of the metal work function $\varphi$, and the heats of desorption $l_+$ and $l_0$ of the ions and atoms, on the concentration $N$ of adsorbed atoms, or the coverage $\theta = N/N_0$ ($N_0$ is the concentration of atoms in the monolayer) of the surface by the adsorbate. Comparatively reliable data on the functions $l_+(\theta)$, $l_0(\theta)$ and $\varphi(\theta)$ have been obtained for cesium adsorbed on the surface of polycrystalline tungsten[17, 18]. These functions are shown graphically in Fig. 9. In the region of small $\theta < 0.2$, the functions $l_+(\theta)$, $l_0(\theta)$ and $\varphi(\theta)$ are approximately linear, so that the quantity $l_+$

increases with increasing $\theta$, the quantities $\varphi$ and $l_0$ decreasing. Confining ourselves to low values of the accelerating electric field $\varepsilon < 10^7$ V/cm, the functions $l_+(\theta, \varepsilon)$ and $l_0(\theta, \varepsilon)$ for $\theta < 0.2$ may be written as

$$\begin{aligned} l_+ &= l^0_{+0} - e\sqrt{(e\varepsilon)} + a_1\theta \\ l_0 &= l^0_{00} - a_2\theta. \end{aligned} \tag{4.1}$$

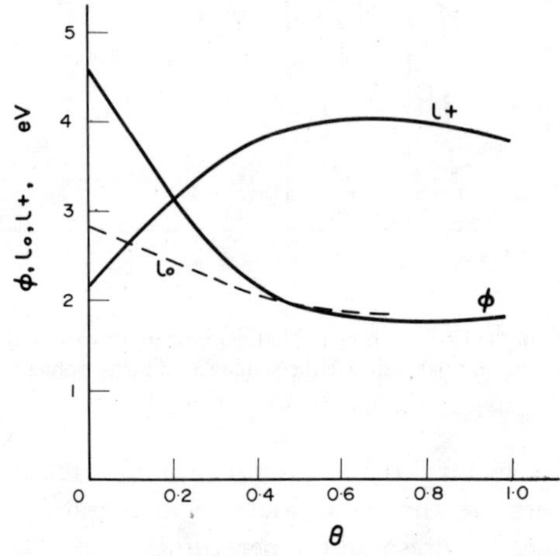


FIG. 9. The curves of $\varphi(\theta)$, $l_0(\theta)$ and $l_+(\theta)$ for cesium adsorbed on tungsten.

Here $l^0_{+0}$ and $l^0_{00}$ are heats of desorption at $\varepsilon = 0$ (the subscript) and $\theta = 0$ (the superscript) and $a_1$ and $a_2$ are constant positive coefficients.

We again assume that the adsorbent surface remains homogeneous at all values of $\theta$ and $\varepsilon$ with respect to $l_+$, $l_0$ and $\varphi$. The analysis is performed for steady-state SI, $\nu = \nu_0 + \nu_+ =$ const. Using (3.1) and (3.2), for the rates of desorption of ions and atoms, and taking (4.1) for $l_+$ and $l_0$, yields the condition for the steady state, namely

$$\nu = \nu_0 + \nu_+ = N_0\theta\left[C\exp\left(-\frac{l^0_{+0} - e\sqrt{(e\varepsilon)} + a_1\theta}{kT}\right) + D\exp\left(-\frac{l^0_{00} - a_2\theta}{kT}\right)\right] = f(\theta) \tag{4.2}$$

which enables all the experimentally observed features of surface ionization in the temperature threshold region to be explained.

A qualitative explanation for the formation of thresholds in the graphs $j(T)$ and the hysteresis phenomenon has been given by Becker[19] on the basis of an analysis of the experimental relationships $f(\theta)$. A consistent analysis of (4.2) for the case $a_2 = 0$ has also been carried out.[20–22] We will perform this analysis, taking the linear dependence of $l_0$ on $\theta$[5, 23] also into account.

Figure 10 presents a graph of $f(\theta)$ (solid line) in units of $CN_0$ and the graphs of

$$\exp\left(-\frac{l_+^0 + a_1\theta}{kT}\right) \text{and} \frac{D}{C}\exp\left(-\frac{l_0^0 - a_2\theta}{kT}\right)$$

(broken lines). These graphs represent, in relative units, the dependence

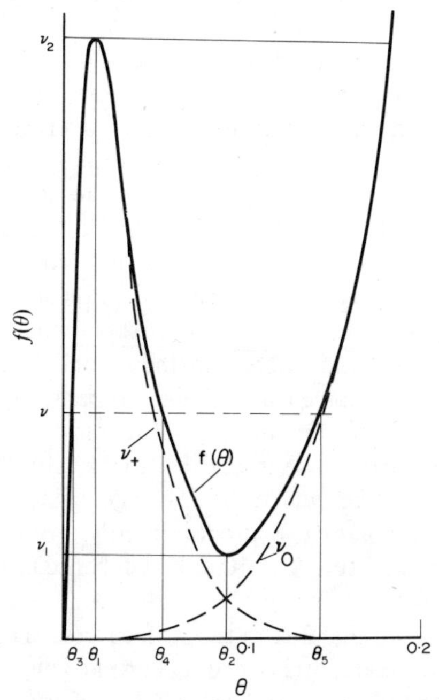


FIG. 10. The relationship $f(\theta) = \nu_+(\theta) + \nu_0(\theta)$ of the rate of desorption of cesium from tungsten surface at $T = 1000°$K.

on $\theta$ of the rates of desorption $f(\theta)$ of adsorbed particles, as well as of ions $\nu_+$ and atoms $\nu_0$, separately. The values assumed in calculations are $\varepsilon = 0$, $T = 1000°K$, $D/C = 2$, (the statistical weight ratio of the ground states of the cesium atom and ion). In addition, for cesium on tungsten, the graphs of Fig. 9 yield $l^0_{+0} = 2.1$ eV, $l^0_{00} = 2.8$ eV, $a_1 = 5$ eV and $a_2 = 2$ eV. Curves similar to those of Fig. 10 should be obtained in all cases, provided $l^0_{00} > l^0_{+0}$, and the heat of ionic desorption $l_+$ increases with increasing $\theta$. As seen from Fig. 10, the rate of desorption $\nu_+(\theta)$ of cesium ions increases sharply with increasing $\theta$, passes through a maximum at a coverage $\theta \lesssim 0.02$, and drops practically to zero at $\theta < 0.2$. The rate of desorption $\nu_0(\theta)$ of cesium atoms is small at $\theta < 0.05$, increasing sharply with increasing $\theta$. The desorption isotherms $f(\theta)$ have two extremums at the coverages $\theta_1$ and $\theta_2$.

If the atomic flux impinging on the surface is constant in time, $\nu = \text{const.}$ (Fig. 10), then the steady-state condition $\nu = f(\theta)$ will be fulfilled at three values of surface coverage $\theta_3$, $\theta_4$ and $\theta_5$: $\theta_3$ and $\theta_5$ are stable. The coverage $\theta_4$ is unstable, since any fluctuation in $\theta$ (or $T$) at this point will produce either a sharp increase in the rate of ionic desorption $\nu_+$, with a transition to a stable coverage $\theta_3$ or, vice versa, a sharp decrease of $\nu_+$ with a transition to a stable coverage $\theta_5$. At intermediate surface coverages $\theta_3 < \theta < \theta_5$, the steady-state condition $\nu = f(\theta)$ is not fulfilled, and $\nu = f(\theta) + dN/dt = \text{const.}$ One observes here a time dependence of the ionic current from the surface[24, 25].

For $\nu < \nu_1$ and $\nu > \nu_2$ (Fig. 10), the desorption isotherm $f(\theta)$ has only one stable state of coverage for each, the first of them corresponding predominantly to ionic desorption, and the second to the desorption of atoms. Hence, if, at a given adsorbent temperature, we increase the flux $\nu$ from $\nu = 0$, the ionic current will increase almost as $f(\theta)$. Now the coverage will also increase, from $\theta = 0$ to the value $\theta_1$ at the maximum of the isotherm $f(\theta)$, after which the coverage $\theta$ will start to increase up to the value corresponding to predominantly atomic desorption. For the flux $\nu_2$, the temperature $T = 1000°K$ will hence be the first threshold temperature $T_0$.

If we reduce the atomic flux to the surface from the value $\nu > \nu_2$, the coverage $\theta$ will decrease, and the ionic current will again be low, until at $\theta \leqslant \theta_0$, when the coverage of the surface changes stepwise to the value $\theta_3$, where the desorption will become predominantly ionic. The ionic current will increase for $\nu \leqslant \nu_1$ from a low value to the maximum $\nu_+ \cong f(\theta)$. The temperature $T = 1000°K$ of the adsorbent surface will be equal to that of the second threshold $T_0'$ for the atomic flux $\nu = \nu_1$.

From the above discussion, it follows that the threshold temperatures $T_0$ and $T_0'$ depend on the magnitude of the flux $\nu$. We turn now to the elucidation of the quantitative relationships between $T_0$, $T_0'$, $\nu$ and $\varepsilon$.

When studying the functions $j(T)_\varepsilon$ and $j(\varepsilon)_T$ experimentally, from which we can derive the functions $\alpha(T)$ and $\alpha(\varepsilon)$, the atomic flux to the surface remains constant, so that $T$ at $\varepsilon$ = const, or $\varepsilon$ at $T$ = const, is varied. Consider the function $j(T)_\varepsilon$. Fig. 11 presents schematically a

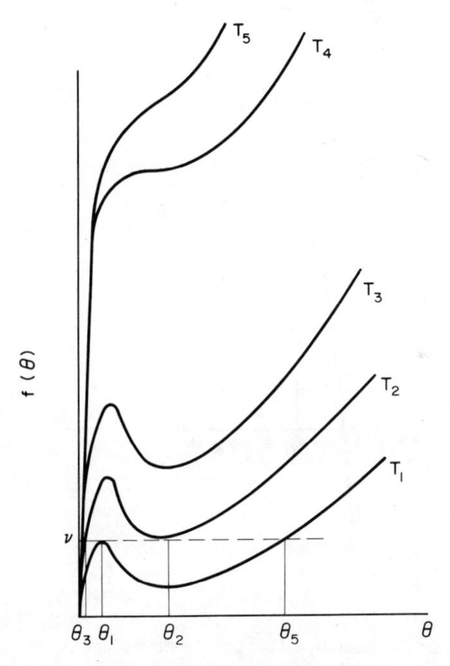


FIG. 11. Schematic representation of the family of desorption isotherms $f(\theta)$.

family of the desorption isotherms $f(\theta)$ for a number of values of $T$ in the ascending order from $T_1$ to $T_5$. The broken horizontal straight line indicates the constant flux level $\nu$. If, at fixed values of $\nu$ and $\varepsilon$, the temperature is gradually reduced down to $T_1$, then, as specified above, the first threshold in the function $j(T)$ will be at $T_1 = T_0$, when $\theta = \theta_1$ and the maximum of the isotherm $f(\theta)$ occurs. If, however, we start to increase the temperature from values $T < T_1$, then the second temperature threshold will be observed at $T'_0 = T_2 > T_0$, where $\theta = \theta_2$. and the minimum of the isotherm $f(\theta)$ is located. To obtain the values $\theta_1$ and $\theta_2$, we have to find the extrema of $f(\theta)$ in (4.2) from the condition $df(\theta)/d\theta = 0$, which yields

$$(a_1\theta - kT)\exp\left(-\frac{a_1\theta}{kT}\right) = (a_2\theta + kT)\exp\left(\frac{a_2\theta}{kT}\right)\frac{1}{\alpha(T, \varepsilon)}$$

or

$$y(x) = \frac{x-1}{ax+1}\exp\left[-(a+1)x\right] = \frac{1}{\alpha(T, \varepsilon)} = \frac{D}{C}\exp\left(\frac{l_+^0 - l_0^0}{kT}\right) \tag{4.3}$$

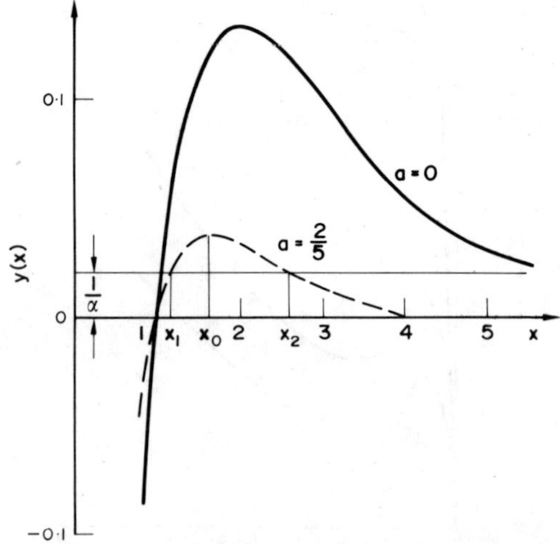


FIG. 12. The graphs of equation (4.3) for the determination of $\theta_1$ and $\theta_2$ at threshold temperatures $T_0$ and $T'_0$.

where the following notations are used: $x = a_1\theta/kT$, $a = a_2/a_1$, $\alpha(T, \varepsilon)$ – the degree of ionization of atoms at $\varepsilon$ = const and $\theta = 0$. From the transcendental equation (4.3) we readily obtain the extrema values $\theta_1$ and $\theta_2$ for the given values of $T$.

Presented in Fig. 12 are graphs of the function $y(x)$ for the two values $a = a_2 = 0$ and $a = 2/5$. The points $x_1$ and $x_2$ of the intercept of the graphs $y(x)$ with the straight line $\alpha^{-1} = 1/\alpha$, at the given values of $a$ and $T$. correspond to the extreme values of coverage $\theta_1$ and $\theta_2$. The point $x_0$ at the maximum of the curve $y(x)$ corresponds to the isotherm where the extrema coverages $\theta_1$ and $\theta_2$ coincide (the inflection point on the isotherm $T_4 = T_c$ in Fig. 11). The curves $j(T)$ can obviously exhibit temperature hysteresis only at $T < T_c$ ($T_c$ is the temperature at which the hysteresis disappears). The point $x_0$ can be determined as the maximum of the function $y(x)$ from the equation

$$ax_0^2 - (a - 1)x_0 = 2. \tag{4.4}$$

At $a = a_2 = 0$ (the heat of atomic desorption $l_0$ does not depend on $\theta$) $x_0 = 2$, at $a = 2/5$, $x_0 = 1.6$ and at $a = 1$, $x_0 = \sqrt{2}$. For values $\alpha^{-1} > y(x_0)$ (see Fig. 12) the extrema values $\theta_1$ and $\theta_2$ do not exist and, hence, the ionization curves $j(T)$ will not exhibit hysteresis. According to (4.3), the condition for the existence of hysteresis may be written as

$$\frac{ax_0 + 1}{x_0 - 1} \exp [(a + 1)x_0] < \alpha(T, \varepsilon). \tag{4.5}$$

Since, by (3.14), the magnitude of $\alpha(T, \varepsilon)$ increases with increasing $\varepsilon$, it follows from (4.5) that, if at $\varepsilon = 0$ the condition for the existence of hysteresis is not fulfilled, then, at sufficiently high values of $\varepsilon$, the hysteresis may appear.

With sufficient accuracy the extrema values $\theta_1$ and $\theta_2$ can be found with some simplifying assumptions.

(a) At the point $\theta_1$ of the isotherm maximum, desorption is predominantly ionic. At $\alpha = \infty$ (only ionic desorption, see Fig. 12) $x_1 = 1$ and $\theta_1 = kT_0/a_1$. For $\alpha \gg 1$,

$$\theta_1 = \frac{kT_0}{a_1}(1 + \delta), \tag{4.6}$$

where $\delta = x_1 - 1$ and $T_0$ is the first threshold temperature.

(b) At the minimum of $f(\theta)$, the rates of ionic and atomic desorption are approximately equal, i.e. from (4.2)

$$N_0\theta_2 C \exp\left[\frac{1}{kT'_0}(-l^0_+ - a_1\theta_2)\right]$$

$$\cong N_0\theta_2 D \exp\left[\frac{1}{kT'_0}(-l^0_0 + a_2\theta_2)\right],$$

whence

$$\theta_2 \cong \frac{1}{a_1 + a_2}[l^0_0 - l^0_+ + kT'_0 \ln(C/D)]. \quad (4.7)$$

Using expressions (4.6) and (4.7) for $\theta_1$ and $\theta_2$, and knowing the ionic desorption rate entering (4.2), we can establish the relationship between the threshold temperatures $T_0$ and $T'_0$ and the quantities $\nu$ and $\varepsilon$.

(i) *First thresholds*

Let us write an expression for the ratio of the ionic current densities at the surface $j_1 = e\nu_{+1}$ at fixed atomic flux $\nu_1$ and field strength $\varepsilon_1$, and $j = e\nu_+$ at any other values of the flux $\nu$ and field $\varepsilon$ at surface coverages $\theta'_1$ and $\theta_1$, viz.

$$\frac{j}{j_1} = \frac{\theta_1 \exp[(1/kT_{01})(l^0_{+0} + a_1\theta'_2 - e\sqrt{(e\varepsilon_1)})]}{\theta'_2 \exp[(1/kT_0)(l^0_{+0} + a_1\theta_1 - e\sqrt{(e\varepsilon)})]}$$

$$\cong \frac{T_0}{T_{01}} \frac{\exp[(1/kT_{01})(l^0_{+0} - e\sqrt{(e\varepsilon_1)})]}{\exp[(1/kT_0)(l^0_{+0} - e\sqrt{(e\varepsilon)})]}, \quad (4.8)$$

or

$$\frac{1}{T_0} = \frac{1}{T_{01}} \cdot \frac{l^0_{+0} - e\sqrt{(e\varepsilon_1)}}{l^0_{+0} - e\sqrt{(e\varepsilon)}} + \frac{k}{l^0_{+0} - e\sqrt{(e\varepsilon)}} \ln\left(\frac{j_1 T_0}{j\, T_{01}}\right). \quad (4.9)$$

Expressions (4.8) and (4.9) establish the relationship between fixed values of $j_1$, $T_{01}$ and $\varepsilon_1$ and any other values of $j$, $\varepsilon$ and $T_0$.

Consider two practically interesting cases:

(a) $\varepsilon = \varepsilon_1 =$ const., so that (4.9) takes the form

$$\frac{1}{T_0} = \frac{1}{T_{01}} + \frac{k}{l^0_{+0} - e\sqrt{(e\varepsilon_1)}} \ln\left(\frac{j_1 T_0}{j\, T_{01}}\right) \quad (4.10)$$

which expresses the relationship between the threshold temperature $T_0$ and the ionic current density $j$ at this temperature.

(b) $j = j_1 = \text{const}; e\sqrt{(e\varepsilon_1)} = 0$, for which (4.9) becomes

$$\frac{1}{T_0} = \frac{1}{l^0_{+0} - e\sqrt{(e\varepsilon)}}\left[\frac{l^0_{+0}}{T_{01}} + k\ln\left(\frac{T_0}{T_{01}}\right)\right]. \tag{4.11}$$

This expression yields the dependence of $T_0$ on the accelerating field $\varepsilon$, at a fixed value of ionic current density $j$. It follows from (4.10) and (4.11) that the threshold temperature $T_0$ increases with respect to the fixed value $T_{01}$ with increasing current $j$ (or the atomic flux $v$ impinging on the surface), and decreases with increasing $\varepsilon$. Hence, a shift of the temperature threshold $T_0$ with increasing $j$ can be compensated by a corresponding increase of the field strength $\varepsilon$, and vice versa. The relationship between the mutually compensating quantities $j$ and $\varepsilon$ is obtained from (4.10) and (4.11) at $T_0 = T_{01}$ and $e\sqrt{(e\varepsilon_1)} = 0$:

$$j = j_1 \exp\left(\frac{e\sqrt{(e\varepsilon)}}{kT_0}\right). \tag{4.12}$$

*(ii) Second thresholds*

As mentioned earlier, in the region of the second temperature thresholds, the surface coverage is $\theta_2$ (4.7), and $v_+ \cong v_0$ or $j \cong \frac{1}{2}ev$. Having written an expression for the ionic current ratio at $\theta = \theta_2$ in deriving (4.8), we obtain

$$\ln\left(\frac{j_1}{j}\right) = \ln\left[\frac{l^0_{00} - l^0_{+0} + e\sqrt{(e\varepsilon_1)}}{l^0_{00} - l^0_{+0} + e\sqrt{(e\varepsilon)}}\right] + \frac{1}{a_1 + a_2}\left[\frac{a_1 l^0_{00} + a_2(l^0_{+0} - e\sqrt{(e\varepsilon)})}{kT'_0} - \frac{a_1 l^0_{00} + a_2(l^0_{+0} - e\sqrt{(e\varepsilon_1)})}{k\,T'_{01}}\right] \tag{4.13}$$

which provides a relationship between fixed values of $j_1$, $\varepsilon_1$ and $T'_{01}$, and the value of $T'_0$ at any other values of $j$ and $\varepsilon$ in the second threshold region.

From (4.13) we may again obtain equations for two particular cases: (a) $\varepsilon = \varepsilon_1 = \text{const}$. In this case, (4.14) yields

$$\frac{1}{T'_0} = \frac{1}{T'_{01}} + \frac{k(a_1 + a_2)}{a_1 l^0_{00} + a_2(l^0_{+0} - e\sqrt{(e\varepsilon)})} \ln\left(\frac{j_1}{j}\right). \tag{4.14}$$

(b) $j = j_1$ = const and $e\sqrt{(e\varepsilon_1)} = 0$. Here (4.13) gives

$$\frac{1}{T'_0} = \frac{a_1 l^0_{00} + a_2 l^0_{+0}}{a_1 l^0_{00} + a_2(l^0_{+0} - e\sqrt{(e\varepsilon)})} \cdot \frac{1}{T'_{01}} + \frac{k(a_1 + a_2)}{a_1 l^0_{00} + a_2(l^0_{+0} - e\sqrt{(e\varepsilon)})} \ln\left(\frac{l^0_{00} - l^0_{+0} + e\sqrt{(e\varepsilon)}}{l^0_{00} - l^0_{+0}}\right) \tag{4.15}$$

Since the thresholds exist only under the condition $l^0_{00} > l^0_{+0}$ or $\alpha > 1$, both terms on the right-hand side of (4.15) increase with increasing $\varepsilon$ and, hence, the second threshold temperatures shift to smaller $T'_0$.

It should be stressed once more that the above equations describing the relationship between the quantities $T_0$ and $T'_0$ and $j$ (or $v$) and $\varepsilon$ are valid for comparatively low fields $\varepsilon < 10^7$ V/cm, for which the magnitude of $l_+$ decreases with increasing $\varepsilon$ following the Schottky law (4.1), and the dependence of $l_0$ on $\varepsilon$ may be neglected. At higher $\varepsilon$, one should take into consideration additional dependences of $l_+$ and $l_0$ on $\varepsilon$.

*(iii) Temperature hysteresis*

Expressions (4.10) and (4.14), as well as (4.11) and (4.15) enable us to determine the difference between $T_0$ and $T'_0$, i.e. to find the magnitude of temperature hysteresis for the above two cases.

(a) $\varepsilon = \varepsilon_1$ = const. By neglecting in (4.10), the small term $[l^0_{+0} - eV(e\varepsilon)]^{-1} \times k \ln (T_0/T_{01})$ we obtain

$$\frac{1}{T_0} - \frac{1}{T'_0} = \frac{1}{T_{01}} - \frac{1}{T'_{01}} - \frac{k}{l^0_{+0} - e\sqrt{(e\varepsilon_1)}} \times \left[\frac{l^0_{00} - l^0_{+0} + e\sqrt{(e\varepsilon_1)}}{l^0_{00} - (a_2/a_1)(l^0_{+0} - e\sqrt{(e\varepsilon_1)})}\right] \ln\left(\frac{j}{j_1}\right). \tag{4.16}$$

It follows from (4.16) that the difference $(1/T_0 - 1/T'_0)$ decreases with the increase of ionic current $j$, which means that the temperature hysteresis decreases. It disappears completely when $j = j_0$, at which $T_0 = T'_0$. From (4.16) we find the value $j_0$ meeting this condition is given by

$$\frac{1}{T_{01}} - \frac{1}{T'_{01}} = \frac{k}{l^0_{+0} - e\sqrt{(e\varepsilon_1)}} \left[ \frac{l^0_{00} - l^0_{+0} + e\sqrt{(e\varepsilon_1)}}{l^0_{00} - (a_2/a_1)(l^0_{+0} - e\sqrt{(e\varepsilon_1)})} \right] \ln\left(\frac{j_0}{j_1}\right) \tag{4.17}$$

(b) $j = j_1 = \text{const}$, $e\sqrt{(e\varepsilon_1)} = 0$. The quantity $\varepsilon$ is variable. Equations (4.11) and (4.15) yield

$$T'_0 - T_0 = \frac{a_1 l^0_{00} + a_2(l^0_{+0} - e\sqrt{(e\varepsilon)})}{a_1 l^0_{00} + a_2 l^0_{+0}} \cdot T'_{01} - \frac{l^0_{+0} - e\sqrt{(e\varepsilon)}}{l^0_{+0}} \cdot T_{01}. \tag{4.18}$$

Both terms on the right-hand side of (4.18) decrease with increasing field; however, the second term decreases faster than the first. Hence, the temperature hysteresis increases with increasing $\varepsilon$, at $\nu = \text{const}$.

*(iv) Field thresholds and field hysteresis*

According to (4.2), the rate of particle desorption from the surface depends also on the field strength $\varepsilon$. In comparatively weak fields, $\varepsilon < 10^7$ V/cm, this field essentially affects the rate of ionic desorption $\nu_+(\theta)$, at each value of $\theta$, and practically does not affect the atomic desorption rate $\nu_0(\theta)$. At a constant adsorbent surface temperature $T$, the desorption rate $f(\theta)$ will, by (4.2), depend on the field $\varepsilon$. In other words, if we plot the functions $f(\theta)$, for a given value of $T$ and different constant values of $\varepsilon$, we will obtain a family of curves $f(\theta)_\varepsilon$, similar to that of the isotherms in Fig. 11.

Repeating the analysis performed earlier for the temperature threshold of ionization and the temperature hysteresis, we conclude that the graphs of the ionic current $j(\varepsilon)_T$, for fixed values of $T$, should also exhibit field thresholds $\varepsilon_0$ and $\varepsilon'_0$, as well as field hysteresis. However, since we confine ourselves, in this section, to a consideration of comparatively weak fields, the surface temperature should lie only slightly below the first threshold temperature ($T \lesssim T_0$) at $\varepsilon \simeq 0$. Otherwise, the threshold values $\varepsilon_0$ and $\varepsilon'_0$ could exceed the specified field limit $\varepsilon < 10^7$ V/cm.

Let the first threshold temperature, at a given atomic flux $\nu$ incident on the adsorbent surface and $\varepsilon \simeq 0$, be $T_0$. Suppose that a sufficiently high electric field $\varepsilon > \varepsilon_0 < 10^7$ V/cm has been created near the surface and the adsorbent temperature set at $T \lesssim T_0$. If we now gradually decrease the magnitude of $\varepsilon$, then, at $\varepsilon = \varepsilon_0$, we will obtain the first field threshold on the ionization curves $j(\varepsilon)$. At $v$ = const, the ionic currents $j(T_0)$ at $\varepsilon \simeq 0$ and $j(\varepsilon_0)$ at $T \leqslant T_0$ are approximately equal, and the adsorbent surface coverages corresponding to them are, by (4.6), $\theta_1 = kT_0/a_1\,(1+\delta)$ and $\theta_1' = kT/a_1\,(1+\delta')$. The quantities $\delta$ depend on $T$ and $\varepsilon$ and, generally speaking, differ somewhat from one another (see Fig. 12). From the equality of ionic currents $j(T_0) = j(\varepsilon_0)$, we obtain

$$N_0 \frac{kT_0}{a_1}(1+\delta)\exp\left[-\frac{l_{+0}^0 - kT_0(1+\delta)}{kT_0}\right] = N_0 \frac{kT}{a_1}(1+\delta')\exp\left[-\frac{l_{+0}^0 - e\sqrt{(e\varepsilon_0)} - kT(1+\delta')}{kT}\right],$$

hence, to small correction terms

$$e\sqrt{(e\varepsilon_0)} = l_{+0}^0\left(1 - \frac{T}{T_0}\right). \tag{4.19}$$

The second field threshold we obtain at $\varepsilon = \varepsilon_0'$, by increasing the field strength from $\varepsilon \simeq 0$ at $T \lesssim T_0$. The relation between the second threshold temperature $T_0'$ and $\varepsilon_0'$ can be derived from the approximate equality of currents, $j(T_0')$ at $\varepsilon \simeq 0$ and $j(\varepsilon_0')$ at $T \lesssim T_0$, at surface coverages $\theta_2$ and $\theta_2'$ determined by (4.7), i.e.

$$e\sqrt{(e\varepsilon_0')} = \left(l_{+0}^0 + \frac{a_1}{a_2} l_{00}^0\right)\left(1 - \frac{T}{T_0'}\right). \tag{4.20}$$

The magnitude of the field hysteresis at $T \lesssim T_0$, via (4.19) and (4.20), follows from the expression

$$e\sqrt{(e\varepsilon_0')} - e\sqrt{(e\varepsilon_0)} = l_{+0}^0\left(\frac{1}{T_0} - \frac{1}{T_0'}\right)T + \frac{a_1}{a_2} l_{00}^0\left(1 - \frac{T}{T_0'}\right). \tag{4.21}$$

Figure 13 presents the graphs of (4.19) and (4.20), for ionization on a homogeneous surface of atoms with the desorption characteristics $l_{+0}^0 = 2.1$ eV, $l_{00}^0 = 2.8$ eV, $a_1 = 5$ eV and $a_2 = 2$ eV, which corresponds to the desorption of cesium atoms from polycrystalline tungsten. The intercept of the straight line $\sqrt{\varepsilon_0'} = f(T/T_0')$ on the ordinate axis is determined by (4.20) for $T = T_0$, and is equal to the field hysteresis at $\varepsilon_0 = 0$. The magnitude of the field hysteresis, as seen from Fig. 13, increases with decreasing surface temperature $T$.

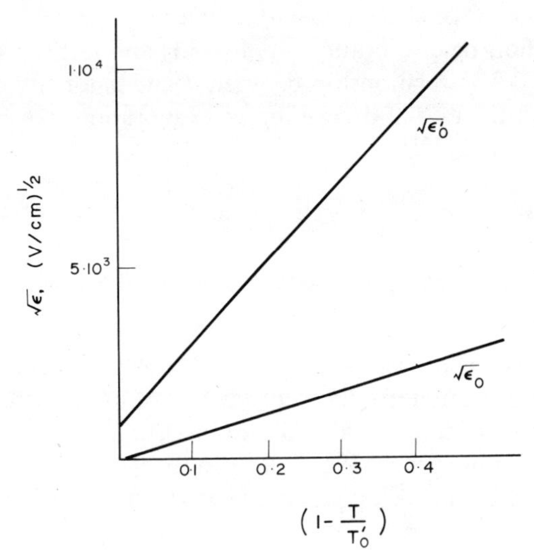


FIG. 13. The plots of (4.19) and (4.20) for the determination of the threshold values of $\varepsilon_0$ and $\varepsilon_0'$.

As mentioned earlier, at low adsorbent temperatures, a sharp decrease in the ionic current is observed when the isothermal heat of ionic desorption is less than that for atoms ($l^0_{+0} < l^0_{00}$), and decreases with increasing surface coverage $\theta$. The case mentioned, corresponds to the values of $\alpha > 1$ and, by (3.18), to a slow decrease of the ionic current $j$ with temperature at high $T(\theta \approx 0)$. For $\alpha < 1$, if the surface work function decreases with increasing surface coverage $\theta$, the ionic current will also decrease with decreasing $T$, but faster than it would follow from (3.20). This occurs, for example, in ionization on high-melting-point metals of atoms of alkali-earth and rare-earth elements, reducing the surface work function in adsorption.

In conclusion to this section, we will touch on the problem of ionic adsorption and evaporation by the electric field $\varepsilon$. As is well known, at sufficiently high values of $\varepsilon$ (comparable to the atomic fields), one observes ionic desorption of atoms adsorbed on the surface or of the adsorbent atoms proper. This phenomenon is revealed ordinarily at low surface temperatures, when thermal desorption or evaporation of atoms does not take place ($\nu_0 = 0$). However, at high $\varepsilon$, the rate of ionic desorption or evaporation $\nu_+$, by (3.1) and (3.12), can be quite high at any $T$. The relationship between $T$ and $\varepsilon$, for the given value $\nu_+ =$ const of the rate of desorption or evaporation, can be obtained from (3.1) and (3.12)[26].

## 5. Experimental Techniques Used in the Study of Surface Ionization

The principal relationships to be tested in experiments on steady-state surface ionization are the dependences of ionic current $j(T)$ and $j(\varepsilon)$ on the temperature of the emitting surface and the accelerating electric field on this surface. We shall consider briefly the main requirements imposed on the techniques used in these measurements.

### (i) *Ionizing surface*

To ensure a comprehensive test of the theory developed in §§3 and 4, the ionizing surface should be homogeneous. This means that, at all points of such a surface, all the emission and adsorption characteristics ($\varphi$, $l^0_{+0}$, $l^0_{00}$) should be exactly the same. As an approximation to such a perfect surface may serve single crystal faces, and, probably, surfaces of molten substances with a total absence of foreign particles adsorbed on them. Materials with such surfaces should be specially prepared and a means of checking their homogeneity in experiments should be provided.

In reality, the majority of materials are composed of fine crystals. On the surfaces of these materials emerge various faces of microcrystals with different face orientation relative to the geometrical surface of the macroscopic specimens. It is well known that the work function and the adsorption characteristics are essentially different on different faces of single crystals. Indeed, the work function on different faces of a tungsten single crystal varies within 1 eV, from 4.3 eV to 5.3 eV. By the Schottky formula (3.9), this implies a variation of $l^0_{+0}$ and $l^0_{00}$ within the same range. Hence, pure surfaces of polycrystalline materials are essentially inhomogeneous (inherent inhomogeneity). The inhomogeneity of the surfaces of pure materials is augmented by various imperfections in the lattice structure.

Due to a difference in the adsorption characteristics of the single crystal faces, the adsorption of foreign atoms will result in the adsorption inhomogeneity, which depends on many factors, being imposed on the inherent inhomogeneity. A nonuniform distribution of the adsorbate over a homogeneous surface will likewise render this surface inhomogeneous.

In practice, the most efficient method of cleaning a surface from adsorbed particles is high-temperature heating of specimens in high vacuum. Hence, high-melting non-volatile materials are preferable as emitters of ions. For such materials, the function $j(T)$ can be measured over a sufficiently broad surface temperature range.

The shape of the ionizing surface may be arbitrary. It may be directly heated filaments and ribbons, or surfaces heated by radiation or electron bombardment. The temperature of that area on the surface, from which the ionic current to be measured is collected, should be uniform and determined as precisely as possible. The methods of surface temperature measurement are well known. The optical pyrometer and the thermocouple techniques are the most widely used methods in SI experiments. In the latter case, the thermocouples should be made of a sufficiently high-melting-point material, which will not chemically react with the material of the emitter. The thermocouple should be in a good thermal contact with the surface, while, at the same time, affecting as little as possible the surface temperature near the contact through heat conduction.

The directly heated filaments and ribbons should be sufficiently long, so that at any temperature the central portion, from which the ionic current is collected, will have the same temperature. As is well known, because of heat removal to the current-carrying electrodes, the temperature of filaments and ribbons decreases from center to the ends. Special heaters at the ends are used sometimes,[27] to produce a more uniform temperature distribution over the surface of the filaments and ribbons.

To ensure uniform conditions of current collection from the filament surface, the ionic collector is made up of three cylinders coaxial with the filament: the central cylinder, which collects the current from the uniformly heated central part of the filament, is used in measurements, the other two being guard cylinders, which collect current from the nonuniformly heated end parts of the filament.

To measure the ionic current dependence on the field strength $j(\varepsilon)$, one has to use thin filaments. To maintain proper centering of filaments, in the course of experiments, the filaments should be stringed, preferably with properly chosen weights. To measure $j(\varepsilon)$ over broad ranges of variation of $\varepsilon$, needle points and sharpened ribbon edges (blades) are used by some workers. However, in these cases, the determination of the field strength $\varepsilon$ at the surface, from the measured emitter-to-collector potential difference, is much less reliable than with thin filaments.

Some means of checking the condition of the ionizing surface are desirable in any SI experiment.

### (*ii*) *Atomic flux to the surface*

All experiments on SI should be carried out in a sufficiently high vacuum, to prevent the residual gas from affecting the properties of the ionizing surface, should it become adsorbed on it. Only ultra-high vacuum meets these requirements. The requirements of the vacuum can be lowered, if ionization is done at high emitter temperatures, at which equilibrium coverages of the surface by adsorbed gases are very small, or if one uses high-speed methods of measurements.

The atoms and molecules to be ionized can arrive at the surface from the surrounding volume (the "vapor method"). In order to ensure a constant flux $\nu$ of particles to the surface, the vacuum chamber should be thermostatically reliable, since the vapor pressure of substance to be ionized depends very strongly on the thermostat temperature, particularly if this substance is in thermal equilibrium with its vapor in all the volume of the instrument. The vapor method provides a uniform arrival of particles over the surface area. However, it can be used only with volatile substances such as, for instance, the alkali metals. The alkali metals, adsorbing on a surface, essentially lower its work function, which can give rise to spurious currents (the photocurrent, secondary emission current) that interfere with measurements of the SI currents. Furthermore, the adsorption of alkali metals on the surface of insulators results in the appearance of leakage currents, whose effect is particularly important in high-field experiments.

A more universal method of producing fluxes of particles on a surface is the molecular-beam approach. It can be applied to both easy- and hard-volatile substances. The evaporator, containing the substance under study, can be heated to above 2000°K, it being easy to ensure the constancy of this temperature, i.e. the constant flux density of particles in the beam. A well-collimated beam produces lesser contamination of the instrument walls and can be shut off at will. The latter is particularly important, since it provides a possibility of isolating the ionic current, associated with the beam incident on a surface, from the background produced by the photocurrent from the collector and by the considerable thermionic emission of ions, mainly those of alkali metals, from the emitter. The drawbacks of the molecular-beam method include the difficulty, and sometimes impossibility, of ensuring uniform arrival of particles over the surface area. A nonuniform flux of particles to the surface can result in a nonuniform coverage of the surface by the adsorbate and in the appearance of temperature-dependent migration fluxes on the surface.

The problem of the purity of the ionizing substances, with respect to the impurities and contamination, is very important. Whereas low-level impurities of hard-ionizing substances ($\alpha \ll 1$) do not contribute essentially to the ionic current from the surface in the ionization of easy-ionizing substances ($\alpha > 1$), the opposite situation is observed in the study of the SI of hard-ionizing substances. A very small impurity of, say, alkali metals in hard-ionizing solids can affect essentially the magnitude of ionic current from the surface and the principal characteristics of this current. Figure 14 presents the graphs of $j(T)$, calculated from (3.18), for the ionic currents of sodium ($V = 5.14$V) and potassium ($V = 4.34$ V) ionizing on a homogeneous surface with a work function $\varphi = 4.5$ eV in the surface ionization of the sodium atom flux and of the flux of potassium atoms one-hundredth of the first one in density. The broken lines in Fig. 14 represent, on a semilog scale, the variation of these currents with $T^{-1}$. The solid curve depicts the temperature dependence of the sum of the sodium and potassium ionic currents, i.e. it corresponds to the ionization of the atomic flux of sodium containing 1% of potassium atoms, and shows the complete distortion of the monoatomic $j(T)$ curves.

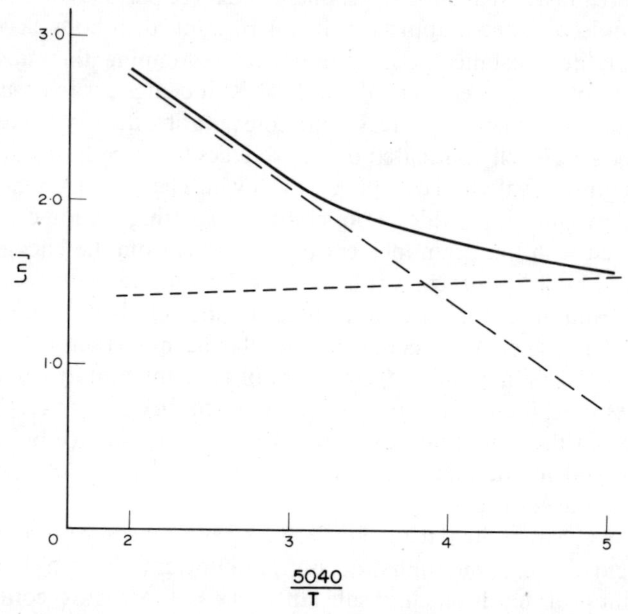


FIG. 14. The graphs of $\lg j = f(1/T)$ for the ionization of sodium with 1% of potassium on a homogeneous surface with $\varphi = 4.5$ V[35]

The required degree of purity of the ionizing substances can be evaluated by means of the graphs in Fig. 7. According to (2.2), in a combined ionization of two different substances on the same surface, the ionic currents desorbed from it will be equal, provided the corresponding fluxes and the ionization coefficients are related as $\nu_1\beta_1 = \nu_2\beta_2$. Therefore, assuming a 1% or lower contamination of the impurity ions in the total collected ionic current to be tolerable, it is required that the impurity in the incident flux $\nu_1$ be $\nu_2 < 0.01\,(\beta_1/\beta_2)\nu_1$. For example, in the surface ionization of magnesium atoms ($V = 7{\cdot}64$ V) on the surface with $\varphi = 5.4$ eV at $T = 2000°$K, the impurity of the alkali metal atoms in magnesium should be below $10^{-5}$%.

The problem of the chemical purity of the primary flux has another very essential aspect. All particles impinging with the flux on the surface undergo the adsorption process. The mean lifetime of particles in the adsorbed state $\tau$ is related to flux density $\nu$ and concentration $N$ of adsorbed particles on the surface by $N = \nu\tau$. According to (2.1), (3.1) and (3.2), the quantity

$$\tau = \left[C \exp\left(-\frac{E_+}{kT}\right) + D \exp\left(-\frac{E_0}{kT}\right)\right]^{-1} \tag{5.1}$$

at the given surface temperature $T$, is determined by the desorption characteristics. Therefore, at large $\tau$, even small impurity fluxes $\nu$ can produce large coverages, which will change the surface properties considerably. Moreover, chemical reactions can take place in the adsorbed layer, the products of these reactions desorbing thermally in the neutral and ionic form, with desorption characteristics different from those of the particles coming to the surface from the outside. This remark refers equally well to the impurities arriving at the surface both with the ionizing molecular beams and from the residual gas in the vacuum chamber.

The most advanced technique for the study of SI is the mass-spectrometer method. Most frequently used in these studies are static magnetic-sector mass spectrometers. The layout of a 90° mass spectrometer is shown in Fig. 15. The ionizing surface 1 (filament or ribbon) is placed at the focus $F_1$ of the mass spectrometer in front of accelerating electrode slit 2. The flux of ions, accelerated by a potential difference, passes through slit 2 and, shaped as a low-divergent ribbon, enters the homogeneous magnetic field 5, which separates it into beams with different $M/e$-ratio ($e$ being the ionic charge). At the focus $F_2$, is situated a second (exit) slit 3, which precedes the ion collector 4. As such, it can serve as the first dynode of a secondary electron multiplier, its final electrode being connected to the input of a d.c. amplifier. The sensitivity of the ionic flux detector is $10^{-18}$–$10^{-19}$ A per output meter division.

The molecular beam from evaporators 6 or 7 falls on the surface of emitter 1 through slits in the accelerating electrode and can be disrupted by shutters 8 and 9. Two evaporators are needed when studying combined ionization of two substances. Sometimes, when measuring the temperature and field dependences of the ionic current, the second evaporator is used to check the emitter position with respect to the mass-spectrometer entrance slit. An easy-ionizing element (for instance, cesium), with $\beta \approx 1$, whose ionic current (produced by the

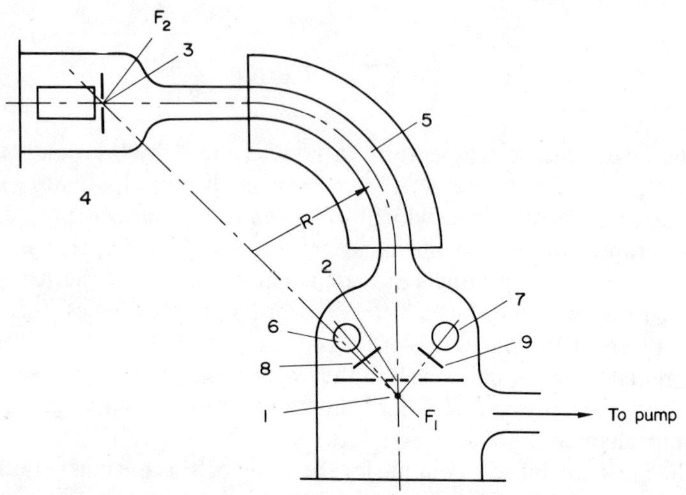


FIG. 15. The layout of the magnetic-sector mass spectrometer for the investigation of surface ionization.

ionization of atoms) is practically independent of $T$ and $\varepsilon$, is loaded for this purpose into one of the evaporators. The constancy of this current, at a constant flux of the cesium atoms, in the measurements of surface ionization of the molecular beam from the second evaporator, will indicate unchanging geometry of the experiment.

The accelerating electrode, whose shape can vary from case to case (a plane or a cylindrical surface), can be used as collector of the total current (both ionic and electronic) emitted by the surface, its measurement being necessary to check the condition of the surface in the course of experiment. The layout of one type of ion source is given in Fig. 16.

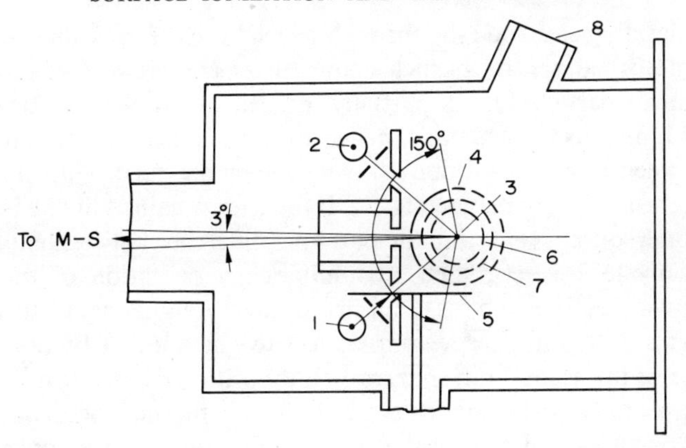


FIG. 16. The layout of a mass spectrometer ion source with a cylindrical electrode system[38].

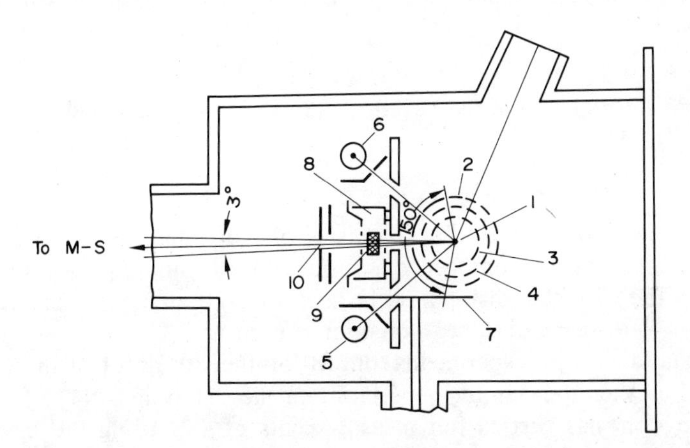


FIG. 17. The layout of a double ion source with surface ionization and with ionization of the neutral flux $\nu_0$ by electrons[28].

To detect and measure the flux of neutral particles desorbed from the surface, an additional ion source, with ionization of desorbed particles by electrons, was proposed[28, 29]. Figure 17 presents a layout of a combined ion source with surface ionization and ionization of the neutral flux by electrons. The ionizing surface 1 is situated in front of the accelerating electrode slit, through which both the ionic flux and the flux of desorbed neutral particles enter the space between electrodes. The neutral particle flux is partially ionized by an electron beam 9, and the ions thus formed are expelled by a weak electric field into the gap between electrodes 8 and 10, where they are further accelerated toward the analyzing magnetic field. If the energy gained by the ions in the volume source is $eU_2$ and the potential difference between emitter 1 and electrode 2 is $U_1$, then a simultaneous operation of the two sources will result in the mass spectra of the "volume" and "surface" ions being shifted in energy with respect to one another by $eU_1$. The presence of the second ion source, with ionization by electron impact, enables a study to be made of not only the thermal desorption of neutral particles, but also of the composition and relative concentrations of residual gases in the mass spectrometer. Moreover, in the course of the experiments, the variation of this composition can be followed.

## 6. Experimental Test of the Functions $j(T)$ at $\varepsilon =$ Constant and $j(\varepsilon)$ at $T =$ Constant

### (*i*) *Homogeneous surfaces*

As mentioned earlier, the properties of individual faces of single crystals are close to homogeneous surfaces. Unfortunately, only a few tests of the $j(T)$ relationship have been carried out on individual faces, several[30–32] being of rather a qualitative nature.

Reynolds[33], in experiments on SI of the atoms of strontium ($V = 5.69$ V) and calcium ($V = 6.11$ V) on the $\langle 110 \rangle$ and $\langle 111 \rangle$ faces of tungsten, has carried out measurements of $j(T)$, using the mass-spectrometer technique in a sufficiently high vacuum (the residual gas pressure in the chamber was $\sim 10^{-8}$ torr). The atomic flux density $\nu$ was not determined, but was maintained constant in the measurements. The homogeneity of the surface structure was checked by the X-ray technique both before and after the experiments. Figure 18 represents a graph of the equation $\lg i = f(1/T)$ for the ionic current† of $Sr^+$ from the $\langle 110 \rangle$ face.

† Current $i =$ Current density $j \times$ Area $s$.

For $T > 2100°K$, the experimental graph is closely approximated by a straight line. Since the condition $e(V - \varphi) \gg kT$ is met in these experiments, the function $j(T)$ should be given by (3.20). In this case, the slope of the linear parts of the graph in Fig. 18 for $T > 2100°K$ is equal to the difference $e(V - \varphi)$. The values of the work function,

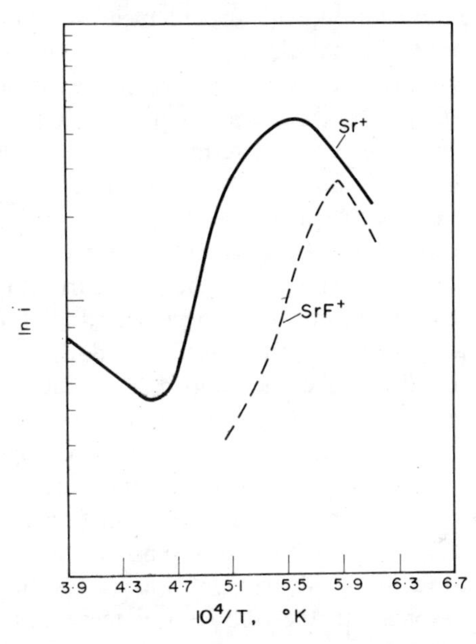


FIG. 18. The graphs of $\lg i = f(1/T)$ for $Sr^+$ and $SrF^+$ forming in surface ionization of the Sr atoms on the face (110) of tungsten[33].

calculated from the experimental curves $\ln i = f(1/T)$ obtained in the ionization of both strontium and calcium, turned out to be practically the same, viz. $\varphi_{(110)} = 5.38 \pm 0.03$ V and $\varphi_{(111)} = 4.49 \pm 0.04$ V. The same values for these faces were also obtained from the temperature dependences of the thermoelectronic current. For $T < 2100°K$, the curves $\ln i = f(1/T)$ (Fig. 18), deviate from the linear law, because, besides the current of $Sr^+$ ions, one observes at low temperatures an $SrF^+$ ionic current, which indicates a contamination of the tungsten surface by fluorine, with which Sr interacts chemically. Apparently, as the temperature decreases, the surface adsorbs impurities, which is accompanied by an increase in the work function $\varphi$ on the surface. It should be noted that these results clearly indicate the advantages of the mass-spectrometer method in the study of surface ionization.

For a complete experimental check of the principal formula (3.14) describing surface ionization, one should measure $\alpha(T)$. If (3.14) is valid, then $\ln \alpha = f(1/T)$ should be a straight line of the slope $(e/k)(\varphi - V + \sqrt{(e\varepsilon)})$. The intercept of this line on the axis $\ln \alpha$ should be equal to $\ln [(1 - R_+/1 - R_0)A]$. The function $j(T)$, (3.18) measured experimentally, can be used to determine $\beta(T) = (1/ev)j(T)$, if the absolute value of the atomic flux density $v$ to the surface is known. From $\beta(T)$, one can calculate $\alpha(T)$, using formula (2.3).

Such measurements have been carried out by Schroen[34]. He studied the temperature dependence of the ionic current in the surface ionization of potassium atoms on tungsten and platinum ribbons in a weak accelerating field ($e\sqrt{(e\varepsilon)} \simeq 0$). The ribbons were apparently strongly texturized, with one face emerging predominantly on the surface, so that their surface was nearly homogeneous. During the experiments, the work function of the surface was checked to be constant photo-electrically. High purity potassium was used and, since the potassium ionization potential is lower than the work function of the surfaces ($\alpha > 1$), the function $j(T)$ is only weakly sensitive to impurities. The residual gas pressure in the instrument was $\sim 3 \times 10^{-9}$ torr. The flux of potassium atoms $v$ to the ribbons was measured with a high-sensitivity quartz microbalance.

Figure 19 presents the graphs of $\ln \alpha = f(1/T)$, their slopes yielding the magnitude of $(\varphi - V)$. The intercept of the curves on the $\ln \alpha$-axis turns out to be $-(0.692 \pm 0.019)$ for tungsten and $-(0.736 \pm 0.048)$ for platinum. These values correspond, within the experimental errors, to $A = 0.5$ for the statistical weight ratio of the ground states of the potassium ion and the atom. Hence, it follows that, within the same accuracy limits, the reflection indices $R_+$ and $R_0$ in (3.14) and (3.18) are zero.

A highly sophisticated mass-spectrometer technique was used[27] to measure $\beta(T)$ in the surface ionization of gadolinium, europium and ytterbium atoms on the $\langle 112 \rangle$-face of tungsten. An essential drawback in these experiments arises from the fact that the flux $\nu$ to the surface was calculated from the dependence of the vapor pressure on the evaporator temperature, which was determined in other experiments, instead of being measured directly. Therefore, the quantitative disagreement between the experimental values of $\alpha(T)$ and (3.14), observed

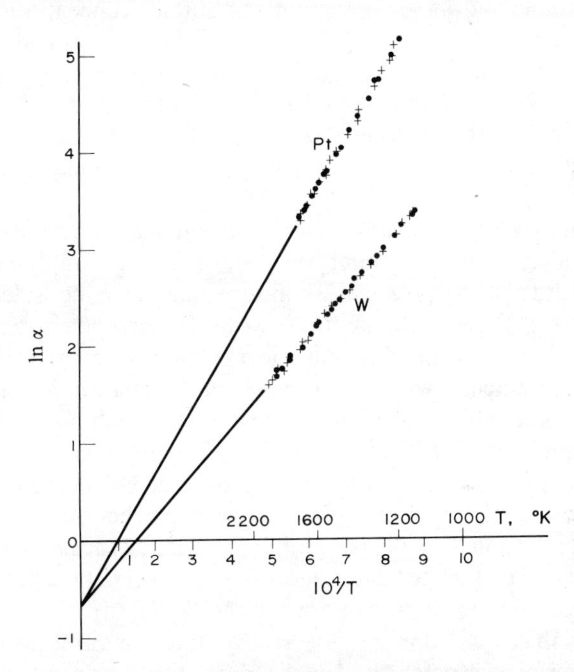


FIG. 19. The graphs of $\ln \alpha = f(1/T)$ for the surface ionization of potassium atoms on ribbons of Pt and W[34].

in these experiments, does not constitute, in our opinion, a convincing proof of the inapplicability of this formula to the surface ionization of rare-earth elements.

As far as we know, the function $j(\varepsilon)$ has not been tested experimentally for the case of ionization on homogeneous surfaces.

(*ii*) *Inhomogeneous surfaces*

In the majority of SI experiments, one uses polycrystalline surfaces of high-melting-point metals exhibiting a natural inhomogeneity with respect to the work function $\varphi$ and the desorption characteristics $l_0$ and $l_+$. Hence, in the steady state, with an atomic flux of constant density $\nu$ impinging on such a surface, the degree of ionization $\alpha$, the rate of atomic and ionic desorption and surface concentration of adsorbed particles $N$, in accordance with (3.1), (3.2), (3.4) and (5.1), will be different for different points on the surface. It may be added that on an inhomogeneous surface, for which the quantity $\varphi$ is a function of coordinates on the surface, there exists a contact field of patches affecting the work function of ion desorption $\lambda_+$ (and hence $l_+$) from the surface.

Thus, in order to calculate the ionic current density from the surface, one should know the coordinate dependence of $\varphi$, $l_+$ and $l_0$ on the surface, which is hardly possible. Within the qualitative consideration of the problem,[35] we will use the so-called sharp-patch diagram of Becker. We will assume the inhomogeneous surface to consist of a set of many patches, within each of which the quantities $\varphi$, $l_+$ and $l_0$ have constant values, changing stepwise at the patch boundaries. The surface temperature $T$ is assumed to be sufficiently high, so that, at the given particle flux $\nu$ to the surface, the coverage by adsorbed particles $\theta \simeq 0$ and the migration of particles between patches may be neglected [36, 37]. In addition, we will assume that the external accelerating electric field $\varepsilon$ is large enough to cancel out the contact field of the patches near the surface (the region of the normal Schottky effect).

Under these assumptions, the steady-state condition (2.1) should be satisfied for each patch, the degree of ionization $\alpha$ of desorbed particles being constant within each patch according to (3.14). If the surface consists of a set of $k$-kinds of patches, the total ionic current from such a patchy surface will be equal to the sum of the currents from all the patches, so that, from (3.18)

$$j = \frac{e\nu}{s} \sum_k s_k \left\{ 1 + \frac{1 - R_{0k}}{A(1 - R_{+k})} \exp \frac{e}{kT} \left[ V - \varphi_k - \sqrt{(e\varepsilon)} \right] \right\}^{-1} \tag{6.1}$$

where $s$ is the ion emitting area of the surface and $s_k$ the area occupied by patches of the $k$-kind with work function $\varphi_k$.

We will consider two limiting cases and, for simplicity, assume that $R_+ = R_0 = 0$.

(a) The second term in the denominator of each term in the sum (6.1) is smaller than unity. Since $A$ has the same value for all patches, and is of the order of unity, this condition is satisfied for each kind of patch, provided $V < \varphi_{min} + \sqrt{(e\varepsilon)}$ ($\varphi_{min}$ is the minimum value of the local work function on the surface). However, the exponential term is the greater, the smaller $\varphi_k$. In other words, the greatest contribution to the temperature dependence of the ionic current from the surface is due to patches with the lowest local work function. The same situation occurs in the case of the temperature dependence of the thermoelectronic current from inhomogeneous surfaces. The thermoelectronic work function $\varphi_R$ (the Richardson work function), determined from the slope of the Richardson plots

$$j_e = B^* T^2 \exp\left[-\frac{e(\varphi_R - \sqrt{(e\varepsilon)})}{kT}\right] = \frac{T^2}{s}\sum_k s_k B_k \exp\left[\frac{e}{kT}(-\varphi_k + \sqrt{(e\varepsilon)})\right] \tag{6.2}$$

turns out to be close to the lowest values $\varphi_{min}$ on the surface. Thus, if we use (3.18) for a description of the temperature dependence of ionic current from a patchy surface, as was done in the majority of earlier publications, the effective value $\varphi^*$, obtained from the experimental relationships $j(T)$ for the inhomogeneous surface, should lie close to the value $\varphi_R$. In this case, the real inhomogeneous surface is replaced by a fictitious homogeneous surface, for which an equation of the kind (3.18) agrees most closely with the experimental dependence $j(T)$, but with some effective values of the quantities $\varphi^*$ and $A^*$.

Such a case applies to surface ionization of the atoms of cesium, rubidium and potassium on high-melting metals.

(b) The second term in the denominator of each term in the sum (6.1) is much greater than unity, for all kinds of patches, provided

$$e(V - \varphi_{max} - \sqrt{(e\varepsilon)}) \gg kT. \tag{6.3}$$

For this case, (6.1) takes the form

$$j = (ev/s)\sum_k s_k A \exp\left[e(\varphi_k + \sqrt{(e\varepsilon)} - V)/kT\right]. \tag{6.4}$$

As follows from experiments on hard-ionizing elements ($\alpha \ll 1$), the experimental relationships $\ln j = f(1/T)$ may be closely approximated, within a broad temperature range, by straight lines with slopes corresponding to some effective values $\varphi^*$ and $A^*$, so that (6.4) may be replaced by an equation similar to (3.20), i.e.

$$j = e\nu A^* \exp\left[e(\varphi^* + \sqrt{(e\varepsilon)} - V)/kT\right] \tag{6.5}$$

where

$$A^* = (A/s \sum_k s_k \exp\left[e(\varphi_k - \varphi^*)/kT\right]. \tag{6.6}$$

In accordance with experiment, the dependence of $A^*$ on $T$ is comparatively weak, since (6.6) contains exponents $(\varphi_k - \varphi^*)$ with opposite signs. Since the exponents of all the terms in the sum (6.4) are negative, the largest contribution to the magnitude of $j$, and the temperature dependence $j(T)$, is due to patches with the largest $\varphi_k$. Therefore, the effective value $\varphi^*$, determined from the slope of the plots $\ln j = f(1/T)$ (6.5), should yield values close to $\varphi_{max}$, i.e. $\varphi_R < \varphi^* \lesssim \varphi_{max}$

If one studies surface ionization of atoms with various ionization potentials, on the same patchy surface, going over gradually from case (a) to case (b), then the values of $\varphi^*$, determined from the experimental plots $j(T)$, should gradually increase from the value $\varphi^* \gtrsim \varphi_{min}$ up to $\varphi^* \lesssim \varphi_{max}$, i.e. the dependence of $\varphi^*$ on $V$ should be observed. This dependence will not be observed, if the values of $V$ satisfy condition (6.3).

From the above qualitative consideration of surface ionization, one may deduce some important conclusions.

I. In all cases, according to (6.1) and (6.4), the magnitude of the ionic current collected from the surface is proportional to the incident particle flux $\nu$ to the surface.

II. The statistical sum ratio $A$ of the ionic and atomic states of particles cannot be determined from the function $j(T)$ for a patchy surface. Such attempts have been made in a number of previous investigations.

III. With increasing accelerating field $\varepsilon$, the ionic current $j$ increases according to the Schottky law, i.e.

$$j = j_0 \exp\left[e\sqrt{(e\varepsilon)}/kT\right] \tag{6.7}$$

where

$$j_0 = \frac{e\nu}{s} \sum_k s_k A \exp\left[\frac{e}{kT}(\varphi_k - V)\right]$$

only in the case of ionization of hard-ionizing elements.

An experimental check on the function $j(T)$ at $\varepsilon$ = const, for the surface ionization of hard-ionizing elements, was carried out mass spectrometrically and confirmed the validity of (6.5). For illustration, Fig. 20 presents plots of lg $i = f(1/T)$ for surface ionization of the atoms of indium ($V = 5.79$ V), calcium ($V = 6.11$ V), magnesium

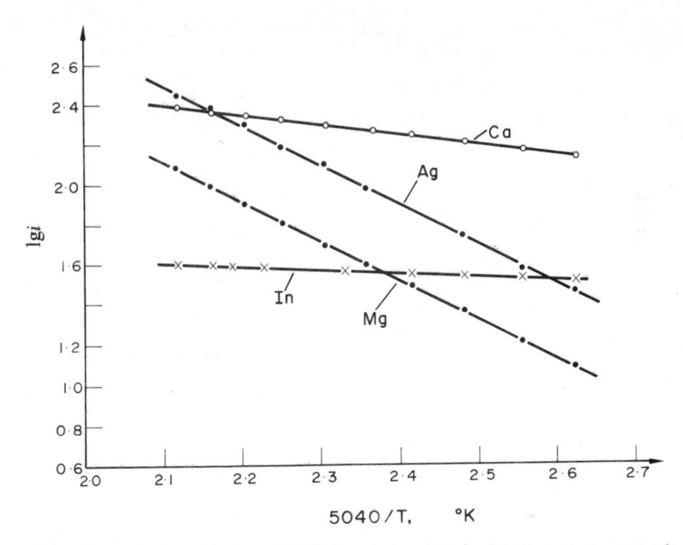


FIG. 20. The graphs of lg $i = f(1/T)$ for the surface ionization of Ca, Ag, Mg and In atoms on polycrystalline rhenium[38].

($V = 7.57$ V) and silver ($V = 7.64$ V) from the same polycrystalline rhenium filament[38]. From the slope of the graphs one can determine $\varphi^*$. In the case of the ionization of Ca, Mg and Ag, one obtains the same value $\varphi^* = 5.43 \pm 0.03$ V, whereas the graphs for In yield a lower value of $\varphi^*$. The latter follows from the fact that condition (6.3), for the validity of (6.4), is not fulfilled in the ionization of In atoms on polycrystalline Re. Measuring the temperature dependence of the thermoelectronic current to an auxiliary electrode of the ion source (see Fig. 16) from the same filament surface and plotting the Richardson graphs ln $i_e = f(1/T)$, one can determine the thermoelectronic work function $\varphi_R$. The difference between the work functions $\varphi^*$ and $\varphi_R$ represents a practical contrast of the surface with respect to the thermionic and thermoelectronic emissions. The same contrast can be measured by another method, namely plotting the temperature dependences of $i$ and $i_e$ obtained experimentally from the coordinates of $\ln (ii_e/T^2) = f(1/T)$[39]. In this case, according to (6.2) and (6.5), we have

$$\lg \frac{ii_e}{T^2} = \lg (B^*A^*ev) + \frac{5040}{T}(\varphi^* - \varphi_R - V + 2\sqrt{(e\varepsilon)}). \qquad (6.8)$$

At $e\sqrt{(e\varepsilon)} \simeq 0$, the slope of the graphs will be determined by the magnitude of $(\varphi^* - \varphi_R - V)$. This method may be used for an experimental verification of the practical contrast of an inhomogeneous surface.

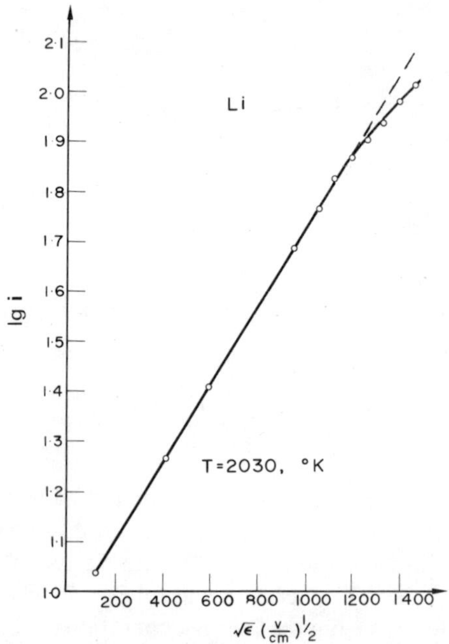


FIG. 21. The graph of $\lg i = f(\sqrt{\varepsilon}(\text{V/cm}))$ in the surface ionization of lithium atoms on polycrystalline tungsten at $T = 2030°\text{K}$[40].

The function $j(\varepsilon)$ at $T =$ const, for hard-ionizing elements, has also been measured by a mass spectrometer. Figure 21 presents a plot of $\lg i = f(\sqrt{\varepsilon})$ for the surface ionization of lithium atoms on a tungsten filament at $T = 2030°\text{K}$[40] in the range of $\varepsilon$ from $3.3 \times 10^4$ V/cm to

$2 = 10^6$ V/cm. Similar measurements have been made, under the same experimental conditions for the ionization of Na, Tl and Mg. In all cases, the plots of $\ln i = f(\sqrt{\varepsilon})$, in accordance with (6.7), have linear portions with slopes in the range 1.86–1.95, which coincide, within the experimental errors, with the theoretical value $e^{\frac{3}{2}}/k = 1.91$. However, in the case of the ionization of Li (Fig. 21), one observes a deviation from

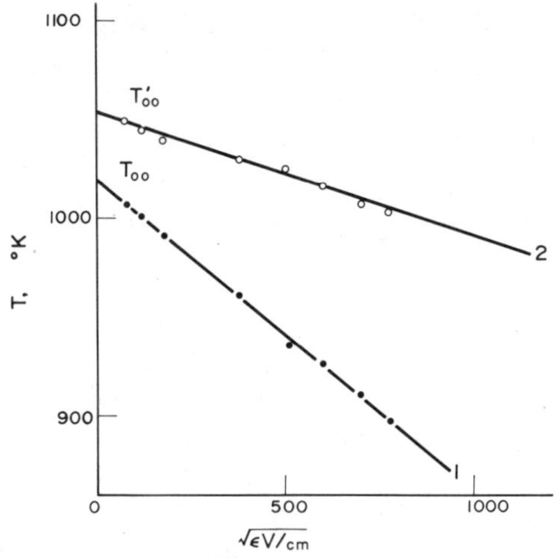


FIG. 22. The graphs of $T_0(\varepsilon)$ and $T'_0(\varepsilon)$ for the case of surface ionization of cesium on a tungsten filament[23].

linearity at $\varepsilon > 10^5$ V/cm. A similar deviation occurs for Na at still lower values of $\varepsilon$. Such deviations follow from the above analysis. With the increase of $\varepsilon$, the condition (6.3) ceases to be satisfied. This occurs, in ionization on a given surface, at values of $\varepsilon$ which are the lower, the smaller the difference $(V - \varphi_{max})$. If the condition (6.3) is satisfied, an experimental verification of (6.7), for ionization on inhomogeneous surfaces, would also indicate the validity of this equation for ionization on homogeneous surfaces; an assumption made in the derivation of (6.7).

In the surface ionization of easy-ionizing elements (for which the value of $\beta$ is close to unity) the ionic current, according to (6.1), depends only weakly on $\varepsilon$. The main effect of the accelerating field, according to (4.19) and (4.20), is revealed in a shift of the temperature thresholds of ionization. Presented in Fig. 22 are graphs of the threshold temperatures

$T_0$ and $T'_0$ vs. accelerating field $\varepsilon$ obtained in experiments on SI of cesium vapor on a tungsten filament[23]. In these experiments, a vacuum cylindrical diode with high-transmission suppressor grids was employed. With $\beta \simeq 1$, a low contamination of the cesium vapor with impurities does not affect the magnitude of the ionic current collected from the filament.

The plots in Fig. 22 confirm the linear dependence (4.19) and (4.20) between $\varepsilon_0$ and $\varepsilon'_0$, on the one hand, and the filament temperature $T$, on the other. It follows from the same plots that the magnitude of field hysteresis, i.e. the difference $\sqrt{\varepsilon'_0} - \sqrt{\varepsilon_0}$ in (4.21), increases with increasing filament temperature $T$. These graphs also agree with (4.18), namely, the temperature hysteresis of the ionization curves $j(T)$ increases with increasing $\varepsilon$. However, such an agreement may be only qualitative, since the experimental data obtained with the inhomogeneous surface were compared with the conclusions of § 4 valid for ionization on homogeneous surfaces.

Consider now the features of the ionization curves $j(T, \varepsilon)$, which may be expected in the ionization of easy-ionizing atoms on a patchy surface in the threshold region. We will again use the sharp-patch scheme. As the surface temperature decreases from high values of $T$, at which $\theta \simeq 0$, the surface concentration of adsorbed atoms at $\nu =$ const, according to (5.1), will increase on all patches. However, the concentration $N$, at the given temperature $T$, will be relatively higher on those patches for which $l_+$ is greater, since ions are predominantly desorbed. Thus a concentration gradient will exist between neighboring patches. If there is no migration of atoms between patches, then, with decreasing $T$, the coverage $\theta_1$ at the maximum of $f(\theta)$ (see Fig. 10), i.e. the first temperature threshold $T_0$ will occur, first on patches with $l_{+\max}$ and then subsequently on the other patches. It is assumed here that the areas of different kinds of patches on the surface are comparable. Hence, in the absence of migration, the decrease of ionic current in the first threshold region in ionization on patchy surfaces should have a step structure corresponding to different values of $l_+$ on the patches. In the absence of migration of atoms, the transition from a steady coverage $\theta_3$ to another $\theta_5$ (Fig. 10) will be hampered by a flux of migration to the adjacent patches, resulting in a smoothing-out of the step structure.

As the temperature $T$ increases from low values $T < T'_0$ corresponding to high coverages $\theta \gtrsim \theta_5$ (Fig. 10), the surface concentration of atoms will decrease, its value at a given atomic flux to the surface $\nu$ = const being determined, according to (5.1), by the rate of desorption (the lifetime $\tau$) first of atoms, and then of ions. It is clear that, as we approach the second threshold in the curve $i(T)$, the ionic desorption should start at the patches with the lowest values of $l_{0\min}$. In this case, one should also observe, in the absence of migration, a step structure in the threshold portions of the curves $i(T)$, which is smoothed out by the migration of adsorbed particles. The step structure in the region of the first thresholds on the curves $i(T)$ is ordinarily not revealed in experiments at a slow variation of $T$, which apparently indicates the essential role of migration. If, however, we increase the rate of variation of the filament temperature $dT/dt$, for instance, by successively switching on and off the heater current, then the curves $i(T)$ will exhibit the characteristic structure of Fig. 23 [41,42]. At a rapid variation of temperature, the relative role

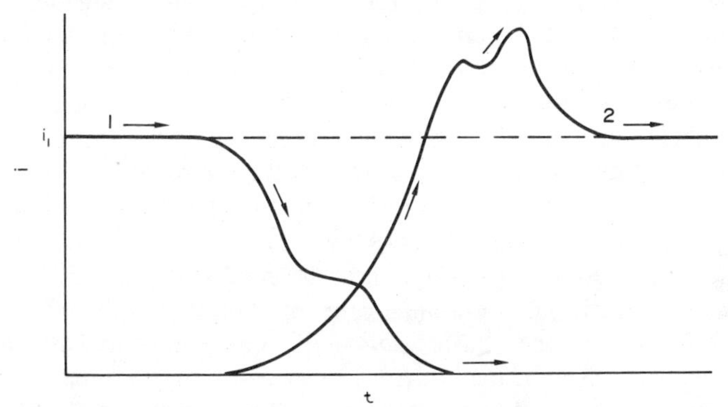


FIG. 23. The structure in the plots $i(T)$ revealed in a temperature flash[42]. Decreasing $T$ (curve 1): Increasing $T$ (curve 2).

of migration should decrease. The problems of kinetics of the surface ionization process at low temperatures have not been adequately studied up to now.

*(iii) Surface ionization on semiconductors*

As mentioned in (3.14) for the temperature dependence of ionization may also be derived from a consideration of the thermodynamic equilibrium between the gas and the phase condensed on the surface of a solid. Since, in such a consideration, one does not impose any constraints concerning the electrical properties of the solid adsorbent, it follows that (3.14) for $\alpha(T)$, and the corresponding expressions for the ionic current density $j(T)$ for both homogeneous (3.20) and inhomogeneous (6.1) surfaces, should be applicable to the description of SI semiconductors. The condition for the thermodynamic equilibrium in the adsorbed layer also includes the condition for the onset of charge equilibrium in the adsorbate–adsorbent system. However, if the levels $eV_1$ and $eV'$ of the atoms on the adsorbent surface (Fig. 2) fall into the band of forbidden energy states in a semiconductor, then the isoenergetic transition of a valence electron between the levels of the adsorbent and the adsorbed particle cannot take place. Such a transition may be realized only in an activated process via the levels in the conduction band of the semiconductor. An activated transition of this kind requires a definite time and, if it is much smaller than the lifetime of particles in the adsorbed state, then the condition for the onset of charge equilibrium on the surface will be satisfied, and (3.14) for $\alpha(T)$ will also be applicable to SI on semiconductors.

Zandberg and Paleev[43] carried out a thorough investigation of the function $j(T)$ for the surface ionization of the atoms of Cs, Na, K, Li and In on the $\langle 111 \rangle$ face of silicon and found it to correspond to (3.20). From the slope of the graphs $\ln j = f(1/T)$ for the ionic currents of $Na^+$, $Li^+$ and $In^+$ they obtained the same value $\varphi = 4.9$ V. In another investigation[44] the same value of $\varphi$ was found for the $\langle 111 \rangle$ face of silicon, from the measurement of the contact-potential difference by the retardation curve method (see § 7). However, the thermoelectronic work function obtained under the same conditions, from the slope of the Richardson plots[44] turned out to be equal to $\varphi_R = 4.04 \pm 0.05$, although for a homogeneous surface the values of $\varphi$ and $\varphi_R$ should coincide. The reason for such disagreement is still unclear.

The energy gap for silicon in the temperature range 1100–1600°K is about 1 eV, the time for the onset of charge equilibrium on the surface being apparently sufficiently short compared with the lifetime of atoms in the adsorbed state. It is still unclear whether (3.14), for $\alpha(T)$ in the surface ionization on semiconductors with a wider energy gap, will be valid.

As for the function $\alpha(\varepsilon)$, in the case of semiconductors, it may take on another form, because of specific differences between the semiconductors and metals, such as the presence of surface states, the penetration of external electric fields into the semiconductor, and the corresponding bending of the energy band and so on. No measurement of $j(\varepsilon)$ appears to have been made in the case of surface ionization on semiconductors.

## 7. Distribution of Ions in Initial Energies

All the theoretical methods (thermodynamic, statistical and kinetic) for obtaining the function $\alpha(T, \varepsilon)$ are based on the assumption that the particles incident from the gas phase on the surface of the ion emitter are adsorbed on this surface and, during adsorption, reach thermal and charge equilibrium with the adsorbent. This essential point can be tested experimentally by measuring the energy spectrum of particles desorbed in both the ionic and neutral states. The ionic energy spectrum may be measured by the well-known retardation field technique. However, at elevated temperatures, the surface simultaneously emits both ions and thermoelectrons, the current of the latter sometimes exceeding by far the ionic current. To separate the ionic and thermoelectronic currents, an intermediate grid, charged negatively with respect to both the filament and the collector, was introduced between the emitting filament and the surrounding collector of a cylindrical diode[45]. The electric field between the filament and the grid accelerated the ions and retarded the thermoelectrons. By varying the potential difference $U$ between the filament and the collector, the current–voltage characteristics of the ionic current (the so-called *retardation curves*) were obtained. The retardation curves for ions emitted by a heated tungsten filament (the thermionic emission), showed that the curves of $\ln i = f(U)$ have linear portions corresponding to the Maxwellian distribution of ions in initial energy.

A similar method was used[46] to obtain the retardation curves of potassium ions $K^+$ in SI on a polycrystalline tungsten filament of the atoms of potassium and molecules of potassium chloride and bromide impinging on the filament as molecular beams from the evaporator. By changing the grid potential in the same instrument, the retardation curves for thermoelectrons, for which the Maxwellian distribution in energy has also been proved, were obtained for the case of uniform electrodes[47]. The field at the filament surface, which accelerates the ions or thermoelectrons, was sufficient to cancel out the contact field of the patches.

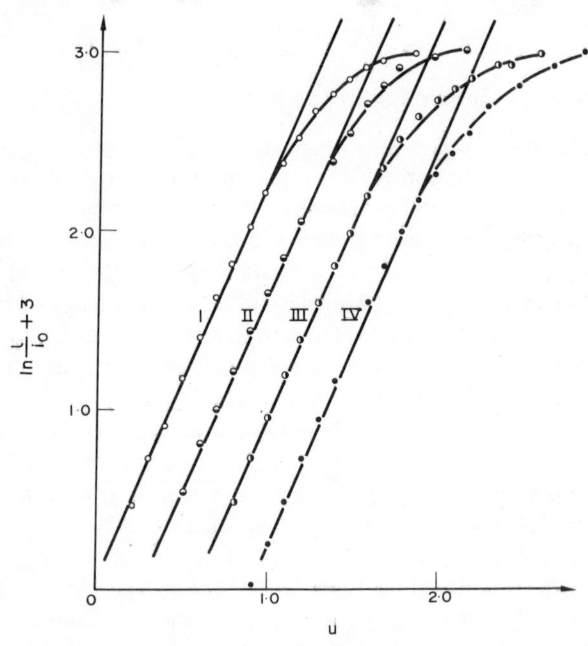


FIG. 24. The retardation curves for the positive ions $K^+$ formed by surface ionization on tungsten filament of the K atoms (curve I), KBr (curve II) and KCl (curve III) molecules. Curve IV is for thermoelectrons. Filament temperature $T = 2050°K$. The curves are shifted for the sake of comparison[46].

Figure 24 shows the $\lg(i/i_0) = f(U)$ ($i_0$ is the saturation current) retardation curves for ions in the ionization of K, KBr and KCl, as well as for thermoelectrons, obtained at the filament temperature $T = 2050°K$. All the curves have linear portions with the same slope corresponding, within experimental errors, to the filament temperature. Deviations from the linearity at low retardation potentials $U$ were ascribed to inhomogeneity of the emitter and collector surfaces. Moreover, the theoretical retardation curves for the Maxwellian distribution of particles in the field of a cylindrical condenser, in contrast to the plane condenser, deviate from the linear law at low retardation fields. The question of the shape of the retardation curves for the Maxwellian distribution of ions in energy, in the case of a non-uniform emitter and collector, has been considered in detail(48). The similarity of the retardation curves, for ions and thermoelectrons (Fig. 24), gives grounds to conclude(46) that the energy distribution of $K^+$ ions is Maxwellian and that the accommodation coefficient of the K atoms and KBr and KCl molecules on tungsten is unity, assuming the chemical bonding of the potassium atoms does not affect the ionic distribution in energy.

Since no experiments have been carried out with a uniform plane emitter and collector, it remains unclear whether or not the energy spectrum of the ions is Maxwellian at low energies. The distribution of thermoelectrons and $K^+$ and $Cl^-$ ions in the tangential components of the initial energy has been measured(49) by the technique proposed by Berry(50). Figure 25 shows the experimental layout. The beam of KCl molecules was fed from the evaporator 10 to the heated plane emitter 1. The ions emitted from the surface were accelerated and passed through a number of slits before entering the space between the plates of a plane deflecting condenser 5 as a ribbon-shaped beam. The beam broadened on the way from the source to the entrance slit into collector 7, due to the tangential components in the initial velocity. By varying the potential $U$ of the condenser 5, the current density distribution over the cross section of the ion beam was measured. The accelerating field near the emitter surface varied from 300 up to 1500 V/cm. This field cancelled out the contact field of the patches and, apparently, only weakly affected the tangential components of the ion velocity.

Displayed in Fig. 26 are the current distribution curves over the beam cross section as a function of $U$ for $K^+$ ions and thermoelectrons for two values of the potential difference accelerating the ions in the source, $U_1 = 200$V and 500V. According to theory(50), for a Maxwellian distribution of ions, we have $i/i_0 = \exp(-eU^2/TU_1)$, where $i_0$ is the current at $U = 0$. The plot of $\lg(i/i_0) = f(U^2/TU_1)$, constructed

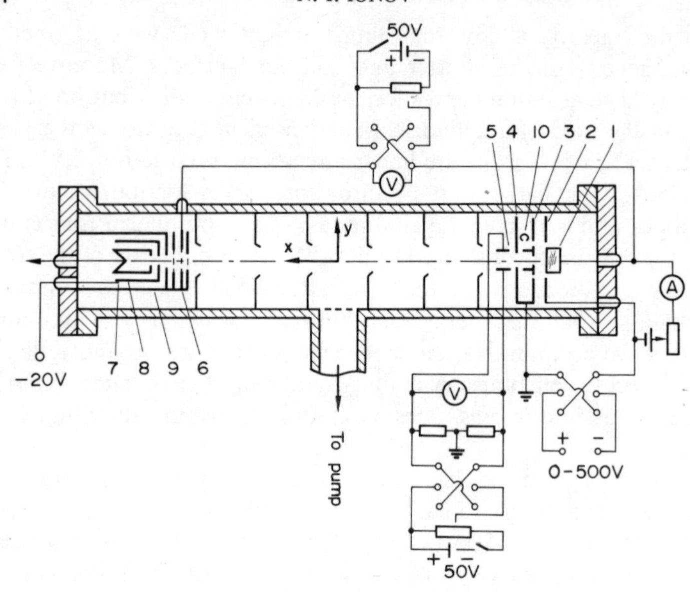


FIG. 25. The layout of the instrument for the investigation of the normal and tangential components of the initial ion velocities [49].

from the data of graphs in Fig. 26, is shown for the current of $K^+$ ions and thermoelectrons in Fig. 27. Irrespective of the magnitude of $U_1$, the experimental points for both the ions and the thermoelectrons lie along the straight line corresponding to the Maxwellian distribution law. The experimental results for the current of $Cl^-$ ions, formed under the same conditions through ionization of the KCl molecules, can also be fitted to the same straight line. In the latter case, the current from the emitter was separated from the thermoelectronic current by superimposing a weak magnetic field near the ion source. Using the same instrument, retardation curves, similar to those of Fig. 24, were also obtained for the normal components of the ions and the electrons, by employing a retarding field in the collector region (electrode 6). The tangential components of the initial velocities were investigated mass-spectrometrically.[51] The ions $Li^+$, $Rb^+$ and $Cs^+$, formed in SI on platinum, have been shown to have a Maxwellian distribution.

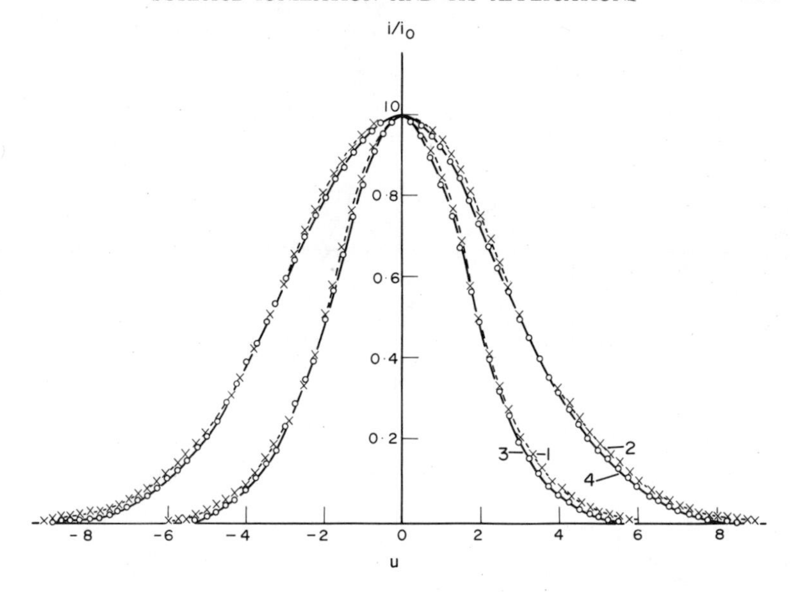


FIG. 26. The distribution curves of the tangential components of the initial velocities of thermoelectrons and $K^+$ ions from a tungsten emitter at $T = 2100°K$[49]. Curves 1 and 3 are for electrons and ions with 200 eV; curves 2 and 4 are for electrons and ions with 500 eV.

If thermal desorption occurs only in the form of neutral particles ($\beta = 0$, or a field retarding the ions is produced near the adsorbent surface), then, in the steady state, the desorbed flux $\nu_0$ should be equal to the flux $\nu$ incident on the surface. In the combined ion source of Fig. 17, the flux $\nu_0$ passing through the ionization chamber can be partially ionized by electron impact. The ionic current $i$ obtained will be proportional to the electronic current $i_e$ and the concentration $n$ of the desorbed particles in the intersection area of the beams $i_e$ and $\nu_0$. For Maxwellian distribution, and $\nu_0 = \nu = \text{const}$, $n \sim \nu_0/\sqrt{T}$ and, hence, $i \sim i_e \nu_0/\sqrt{T}$[52]. Recently[53], such a relationship for $i(T)$ was obtained experimentally, for the case of desorption of cesium and iridium chloride molecules, the temperature of particles in the flux being assumed to be equal to the adsorbent temperature. These experiments indicate that, in this case, the accommodation coefficient for the interaction of

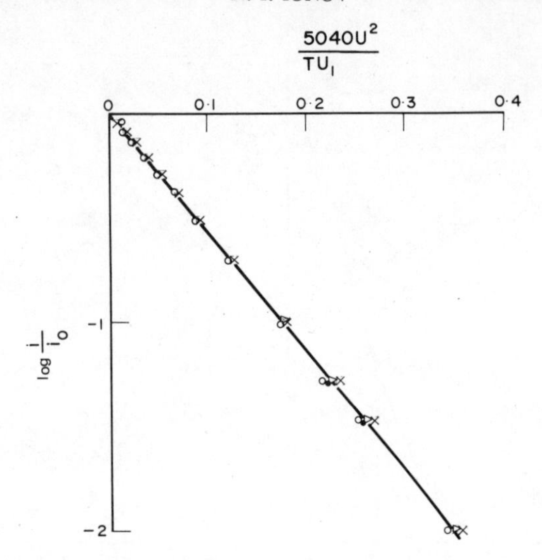


FIG. 27. The plot of lg $(i/i_0) = f(u^2/Tu_1)$ from data of Fig. 26[49]

CsCl molecules with Ir is unity. This method of experimentally checking the energy distribution of desorbed particles is simple and of a general nature.

## 8. Negative Surface Ionization

If particles adsorbed on a metal surface possess electron affinity, then they may desorb as negative ions. The rate of desorption will be determined by a formula similar to (3.1), in which the quantity $E_+$ will be replaced by the activation energy for the desorption of negative ion $E_-$. For the degree of ionization, we obtain an expression similar to (3.4), viz.

$$\alpha = A\left(\frac{1 - R_-}{1 - R_0}\right)\exp\left(\frac{l_0 - l_-}{kT}\right) \tag{8.1}$$

where $A$ is the statistical sum ratio for a particle in the negative ion and neutral states, $R_-$ is the reflection coefficient of negative ions from the surface and $l_-$ is the isothermal heat of negative ion desorption. For a homogeneous surface, the difference $l_0 - l_-$ obtained from the potential curves for the systems $[A + M]$ and $[A^+ + M^-]$, is similar to the curves for positive ions. Figure 28 presents such curves for the case $S < \varphi$ and $\varepsilon = 0$. These curves immediately yield the following expressions:

$$\left.\begin{aligned} l_{00} &= q_0 + \lambda_{00} \\ l_{-0} &= q_0 + \lambda_{-0} + e(\varphi - S') \\ l_{00} - l_{-0} &= \lambda_{00} - \lambda_{-0} + e(S' - \varphi) = e(S - \varphi) \end{aligned}\right\} \quad (8.2)$$

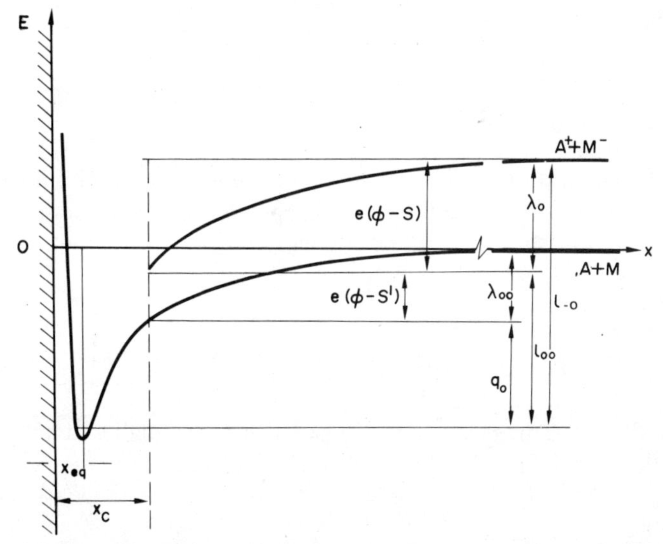


FIG. 28. The variation of the potential energy of the systems $[A + M]$ and $[A^+ + M^-]$ for $\varepsilon = 0$ and $S < \varphi$.

Here $\lambda_{00}$ and $\lambda_{-0}$ is the work of desorption of the neutral particles and the corresponding negative ions at $\varepsilon = 0$, respectively, and $S'$ is the position of the affinity level at $x = x_c$. As before, $\lambda_{-0} > \lambda_{00}$, so from (8.2) it follows that $S' > S$, i.e. as the particle approaches the metal surface its affinity level becomes lower.

In an accelerating electric field $\varepsilon$, the heat of ionic desorption $l_{-0}$ is reduced by $e\sqrt{(e\varepsilon)}$. It should be remembered that, in the case of negative ion emission, the field strength $\varepsilon$ cannot be high, since for $\varepsilon > 10^7$ V/cm the electron field-emission current will destroy the emitter. Therefore, the expression for the degree of ionization can be transformed from (8.1) and (8.2) to the form

$$\alpha(T, \varepsilon) = A\left(\frac{1 - R_-}{1 - R_0}\right)\exp\left[\frac{e}{kT}(S - \varphi + \sqrt{(e\varepsilon)})\right]. \tag{8.3}$$

Similar to (3.18) for the current of positive ions, one can, using (8.3), write an expression for the negative ion current from a homogeneous surface:

$$j = ev\left\{1 + \frac{1 - R_0}{A(1 - R_-)}\exp\left[\frac{e}{kT}(\varphi - S - \sqrt{(e\varepsilon)})\right]\right\}^{-1}. \tag{8.4}$$

The equation for the negative ion current from an inhomogeneous surface under the condition of compensation of the contact patch field by the external accelerating field, may be written out similar to (6.1), i.e.

$$j = \frac{ev}{s}\sum_k s_k\left\{1 + \frac{(1 - R_0)_k}{A(1 - R_-)_k}\exp\left[\frac{e}{kT}(\varphi_k - S - \sqrt{(e\varepsilon)})\right]\right\}^{-1}. \tag{8.5}$$

For all atoms, the magnitude of $S$ (see Table 1) is smaller than 4 eV and, for the case of SI on polycrystalline high-melting metals, the condition $e(\varphi_{\min} - S - \sqrt{(e\varepsilon)}) \gg kT$ is satisfied. Equation (8.5) now assumes the form

$$j = \frac{ev}{s}\sum_k s_k A\left(\frac{1 - R_-}{1 - R_0}\right)_k \exp\left[\frac{e}{kT}(S - \varphi_k + \sqrt{(e\varepsilon)})\right] =$$

$$= evA^* \exp\left[\frac{e}{kT}(S - \varphi_-^* + \sqrt{(e\varepsilon)})\right] \tag{8.6}$$

where

$$A^* = s^{-1} \sum_k s_k A \left( \frac{1 - R_-}{1 - R_0} \right)_k \exp \left[ \frac{e}{kT} (\varphi_-^* - \varphi_k) \right].$$

The quantities $A^*$ and $\varphi_-^*$ are effective characteristics of such a homogeneous surface, for which the magnitude of the negative ion current and its temperature dependence, in the given range of temperature variation, coincide with those of a real patchy surface. Just as in the case of thermoelectronic emission, the largest contribution to the total negative ion current, according to (8.6), is due to patches with the lowest work function $\varphi_{min}$. Therefore, the quantity $\varphi_-^*$ should be close to the Richardson work function $\varphi_R$, i.e. $\varphi_-^* \simeq \varphi_R$. As with the positive ions, we may draw some general conclusions from (8.6): (a) in the steady state, the current density $j$ of the negative ions from a patchy surface is proportional to the atomic flux $\nu$ incident on the surface; (b) in an accelerating electric field, the ionic current increases according to the Schottky law; (c) from the experimental relationship $\ln j = f(1/T)$ we can determine the effective values of the work function $\varphi_-^* \simeq \varphi_R$ and $A^*$.

Turning to a review of the experimental investigations of surface ionization involving the formation of negative ions, we will now specify some requirements of the techniques used. As mentioned earlier, in experiments on the ionization on pure high-melting metals, the quantity $\alpha \ll 1$ in almost all cases. To increase $\alpha$, and to insure an adequate surface purity, the experiments should be carried out at high temperatures $T$. However, this will cause an increase in the impurity ionic current and the thermoelectronic current, from which the current of the ions of interest should be separated. Therefore, in experiments on negative SI, satisfactory results can only be obtained with the mass-spectrometer technique.

The majority of investigations of the thermoemission of negative ions have been aimed at determining the quantity $S$ from the temperature dependence of the ionic current, rather than at testing (8.6) itself. The quantity $S$ for atoms and molecules, is as important a physicochemical quantity as the ionization potential $V$. However, in contrast to $V$, the quantity $S$ cannot be inferred from optical spectra. The method of surface ionization is one of the few direct methods of determining $S$, and we will consider this problem separately in § 11.

In the first experiments, where the formation of negative halogen ions on a tungsten filament surface was detected, the ions were separated from the electrons by a magnetic field in a magnetron with an additional grid[4, 54]. Later[55], the magnetron method was improved and used by some authors to determine the magnitude of $S$ for complex molecules. However, without a mass-spectrometer analysis of the ions desorbed from the surface, the interpretation of the experimental data on the ionization of complex molecules is ambiguous.

The initial attempt to test the function $i(T)$ was made by Ionov[56]. The ionization of iodine molecules $I_2$ on a polycrystalline tungsten filament was studied by the magnetron method. It was assumed that at the experimental temperature, the $I_2$ molecules dissociate completely on the surface of tungsten and desorb only in the form of atoms and negative atomic ions. Shown in Fig. 29 are the plots of $\lg i_e/T^2 i_- = f(1/T)$

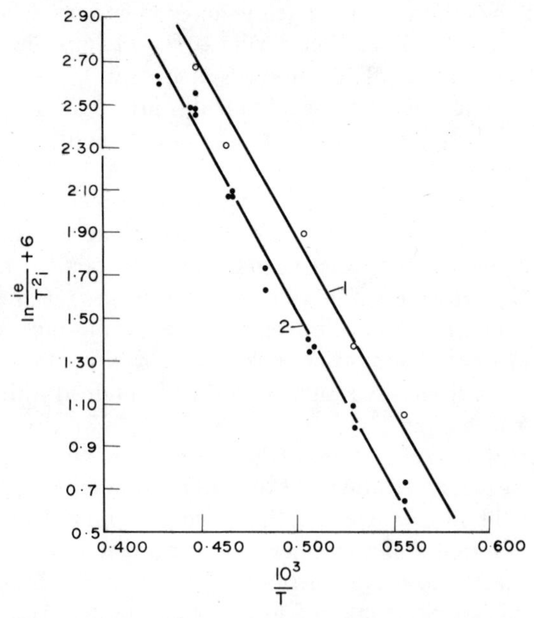


FIG. 29. The plots of $\lg (i_e/T^2 i_-) = f(1/T)$ for two values of the vapor pressure of $I_2$ in the magnetron: $0.3 \times 10^{-2}$ torr (curve 1) and $1 \times 10^{-2}$ torr (curve 2)[56].

for two values of the iodine vapor pressure in the magnetron. For an inhomogeneous surface, this relation is obtained from (6.2) and (8.6), assuming $\varphi_R \simeq \varphi^*$ thus,

$$\frac{i_e}{T^2 i_-} = \frac{B_R^*}{evA^*} \exp\left(\frac{eS}{kT}\right). \tag{8.7}$$

In a sufficiently broad temperature range, the plots of $\ln i_e/T^2 i_- = f(1/T)$ turn out to be linear, with slopes corresponding, according to (8.7), to the value of $S$ for the iodine atoms.

The relationship (8.7), and hence (8.6), was also checked for the emission of negative ions in SI on polycrystalline tungsten of the molecules of alkali halides[57] and the atoms of gold[58] and silver[59]. The same relationships have been confirmed for the ionization of iodine on textured W and Mo ribbon which closely resemble a homogeneous surface[60]. The distribution of negative ions in the initial kinetic energy, and the correspondence of this distribution to adsorbent temperature, has been measured[49]. We do not know of any experiments which test the dependence of negative ion currents on the accelerating electric field.

## 9. Surface Ionization of Molecules

Expression (3.14), for the degree of surface ionization, is also valid for the case of thermal desorption of not only atoms, but molecular particles as well. In the latter case, by the ionization potential $V$, one understands the quantity corresponding to a transfer of the most weakly bound electron from the molecule residing in the non-excited electronic and vibrational states to the ground state of the non-excited positive ion (the first adiabatic ionization potential). In addition, the quantity $A$, which enters (3.14) in the case of ionization of molecules, is equal to the ratio of the total sums of states for the molecule in the ionic and neutral states.

When in the adsorbed state, molecules can undergo various chemical transformations, in which the adsorbent can act as a catalyzer. On the surface of a pure metal in high vacuum, these are mainly dissociation reactions. On adsorbents of complex composition (for instance, high-melting metal oxides), as well as in a relatively poor vacuum, fluxes of different substances incident simultaneously on the same surface, there may occur any type of chemical reaction, both in the adsorbate layer, and between the adsorbate and the adsorbent. In this case, particles of different chemical composition can thermally desorb, their fluxes $\nu_k$ from the surface being dependent on the efficiency of the formation $\gamma_k$ of particles of the given $k$th kind on the surface. If a stationary flux of molecules $\nu$ impinges from the gas phase on the surface, then the flux $\nu_k = (\nu_{0k} + \nu_{+k})$, desorbed at the given temperature $T$, is related to the flux $\nu$ by the formula $\nu_k = \gamma_k \nu$. The quantities $\gamma_k$ and $\nu_k$ depend on $T$, $\varepsilon$ and the concentration of adsorbate particles on the surface; the latter depending, according to (5.1), on the magnitude of the incident flux $\nu$. The charge equilibrium in the desorbed fluxes $\nu_k$ is determined by the degree of ionization $\alpha_k(T, \varepsilon)$, in accordance with (3.14).

The ionic current $i_k$, for the given kind of desorbed particles, is

$$i_k(T, \varepsilon, \nu) = e\nu\gamma_k(T, \varepsilon, \nu)\, \beta_k(T, \varepsilon, \nu). \tag{9.1}$$

The quantities $\alpha_k$ and $\beta_k$, according to (3.4) and (3.11), may depend in their turn, via the quantities $l_0$, $l_+$ and $\varphi$, on the concentration of the adsorbate on the adsorbent surface, and eventually on the flux of particles to the surface. At low coverages (high temperatures $T$ and low fluxes $\nu$), at which the quantities $l_0$, $l_+$ and $\varphi$ characterize the pure adsorbate, the quantity $\beta_k$ in (9.1) will not depend on $\nu$. Thus, in the surface ionization of molecules, the ionic currents $i_k$ and their dependence on $T$ and $\varepsilon$, for each kind of desorbed ion, are determined, according to (9.1), not only by $\beta_k$ but also by the $\gamma_k$ characterizing the yield of chemical reactions on the surface. If, for instance, the quantity $\beta_k$ increases with $T$, then, depending on whether the quantity $\gamma_k$ increases or decreases in some temperature range $T$, the ionic currents $i_k$ will increase or decrease faster or slower than $\beta_k$. In the latter situation, the function $i_k(T)$ will pass through a maximum. The SI of molecules provides a means of investigating catalytic reactions on surfaces.

### (i) *Surface ionization of inorganic molecules*

The most well-studied reaction is the surface ionization of the molecules of alkali halides MX on high-melting metals. The effective ionization of MX molecules on tungsten has been observed[61] and the temperature dependence of the ionic current $i(T)$ of KI, KBr and KCl molecules measured[62, 63]. This dependence was shown to agree with (3.20) for $T > 1800°K$, thus implying total dissociation of the MX-molecules on the tungsten surface. In order to test the effect of the chemical bonding of alkali-metal atoms on $i(T)$, we studied[24], under the same conditions, the ionization of K atoms and KBr and KCl molecules on tungsten. Figure 30 shows the graphs of $i(T)$, which

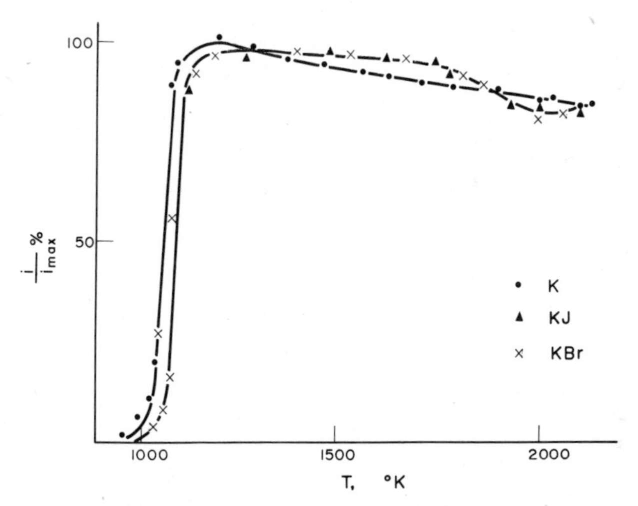


FIG. 30. The relationship $i/i_{max} = i(T)$ in the surface ionization of the K atoms and KI and KBr molecules on a tungsten filament [24].

practically coincide throughout the temperature range studied. The small differences between the K curve and those for KBr and KCl are accounted for by the change in the work function of tungsten, which results from the adsorption of the halogen atoms. Mass-spectrometer analysis of the desorbed ions has shown[57] that only $M^+$ ions are desorbed in the ionization of the MX molecules. A study has also been made[64, 65] of the ionization of CsCl and NaCl molecules on tungsten, in electric fields ε up to $7 \times 10^6$ V/cm, at which the temperature threshold of ionization dropped down to $T_0 \simeq 400°$K. A conclusion was made that, at a low coverage, total dissociation of the molecules occurs near room temperatures ($T \gtrsim T_0$).

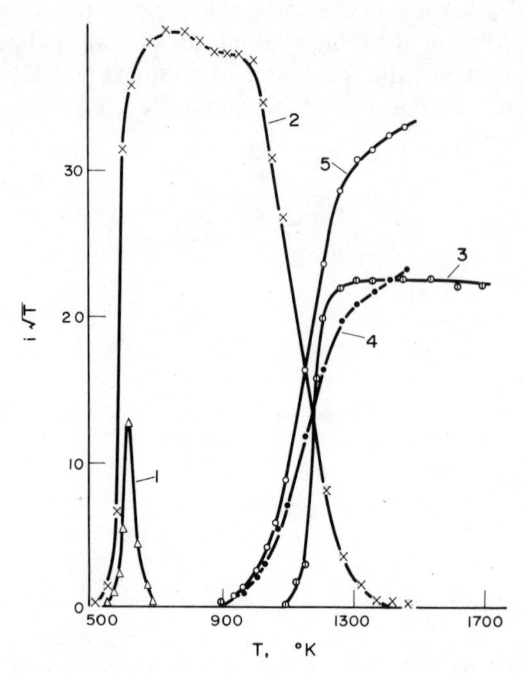


FIG. 31. The dependence of molecular, atomic and ionic fluxes desorbed from a tungsten ribbon on $T$ at a constant flux of the NaI molecules to the ribbon[29]. Curve 1—$Na_2I$; curve 2—NaI; curve 4—I; curve 5—Na. Curve 3—surface-ionization current.

The mass-spectrometer technique has been used[29] to study the products of thermal desorption on a tungsten ribbon surface at a constant incident flux $\nu$ of NaI and NaCl molecules at temperatures both above and below the temperature ionization threshold $T_0$. An ion source similar to that shown in Fig. 17, in which the desorbed flux of neutral particles is partially ionized by an electron beam, was used. Figure 31 presents the temperature dependence of the ionic currents obtained in the desorption of NaI molecules. For a flux of neutral particles from the ribbon, the efficiency of the ionization by electrons depends on the concentration $n$ of the desorbing particles in the ionization region, which is related to the flux by the formula $n \sim \nu/\sqrt{T}$. Therefore, along the ordinate axis of Fig. 31 are plotted the reduced currents $i\sqrt{T}$. It follows from the graphs that, at high ribbon temperatures, total dissociation of the molecules occurs, a fraction of the Na atoms desorbing from the surface in the form of ions $Na^+$ (curve 3). Moreover, desorption of Na (curve 5) and I (curve 4) atoms is observed. With a decrease of ribbon temperature and an increase of coverage, the atomic fluxes drop off and the desorption of the NaI molecules increases (curve 2). In the range $T = 500$–$700°K$, condensation of molecules on the ribbon begins, the concentration exceeds the monolayer value and desorption of dimer molecules is observed (ions $Na_2I^+$, curve 1). With the increase (decrease) of incident molecular flux $\nu$, all the curves in Fig. 31 shift towards higher (lower) $T$. Figure 31 illustrates, for the case of simple diatomic molecules, the complexity of the thermal desorption process. The temperature variation of the fluxes MX and $(MX)_2$ passes through a maximum, which indicates an increase of the rate of molecular dissociation with increasing $T$.

The kinetics of desorption and dissociation of MX molecules on metal surfaces have been studied[66, 67], showing the composition of particles desorbed from the surface to be determined by the rates of desorption of atoms and molecules, as well as by the rates of dissociation and association of particles on the surface. At a given flux of MX molecules to the surface, in some comparatively. narrow temperature range, transition occurs from almost total desorption in the form of $M$ and $X$ (or $X_2$) atoms (see Fig. 31), i.e. the temperature threshold of dissociation is observed. If the dissociation threshold of the given molecular flux $\nu$ to the surface lies near $T = T_0$, the ionization threshold for equal flux $\nu$ of M atoms to the same surface (see § 4), then the ionization curves $i(T)$ for the atoms M and molecules MX do not differ much from one another. This takes place in the ionization of alkali-halide molecules and alkali-metal atoms on tungsten.

In experiments on the ionization of alkali-metal chloride molecules on the surface of rhenium[68], it was observed that the dissociation threshold of molecules lies at higher temperatures than the threshold $T_0$ on the ionization curves for equal fluxes of the alkali-metal atoms, the difference increasing when going over from the ionization of CsCl to that of LiCl. Shown in Fig. 32 are the graphs of the ionic currents $i(T)$ for the separate ionization of equal fluxes of Li atoms (curve 1), LiCl molecules (curve 2) and combined Li + LiCl fluxes (curve 3). The threshold $T_D$ for the dissociation of LiCl molecules is seen to shift, with respect to the threshold $T_0$ for Li, by more than 400°K. It is essential, for

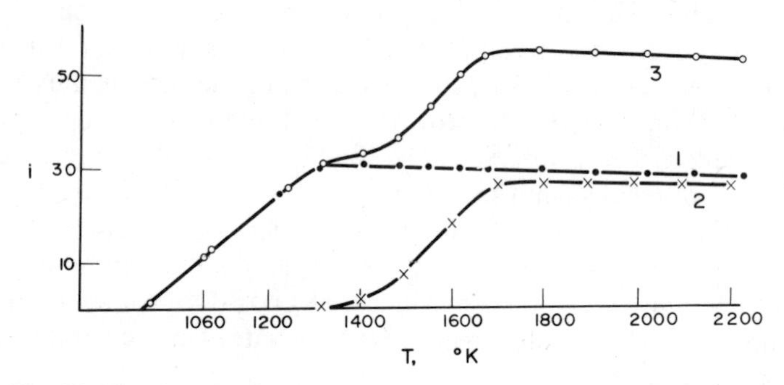


FIG. 32. The plot of $i(T)$ at simultaneous ionization of Li atoms and LiCl molecules on a rhenium filament [68]. Curves: 1—Li; 2—LiCl; 3—Li + LiCl.

the combined ionization of Li and LiCl (curve 3), that the ionization of Li atoms does not exhibit any changes in the properties of rhenium surface, which may be produced by the flux of LiCl molecules. Apparently, in the temperature range $T_D$–$T_0$, the coverage of the surface by MX, M and X particles remains low, the desorption of LiCl from rhenium occurring predominantly in the form of molecules. The difference in the ionization of MX molecules and M atoms on rhenium, according to (9.1), should thus be ascribed to the temperature dependence of the efficiency $\gamma(T)$ of molecular dissociation on the surface. The temperature dependence (Fig. 32) of the ionic current $i(T)$ of LiCl-ionization on rhenium was used[67] to calculate the heat of dissociation of LiCl molecules into atoms, which was found to coincide with the value obtained in other experiments.

From the study of the ionization on platinum of KI, KBr, KCl and KF molecules, it was found[69] that the ionization coefficient of these compounds on Pt is much smaller than its value for the ionization of K atoms. Similar results were obtained in the ionization of MX molecules on platinum and iridium[70,71]. However, if the surface is cleaned of carbon by means of prolonged heating of Pt and Ir in oxygen (at a pressure $\sim 10^{-6}$ torr), the ionization coefficient of MX becomes the same as on tungsten. It turns out that the presence of carbon on the surface of platinum metal shifts the dissociation threshold

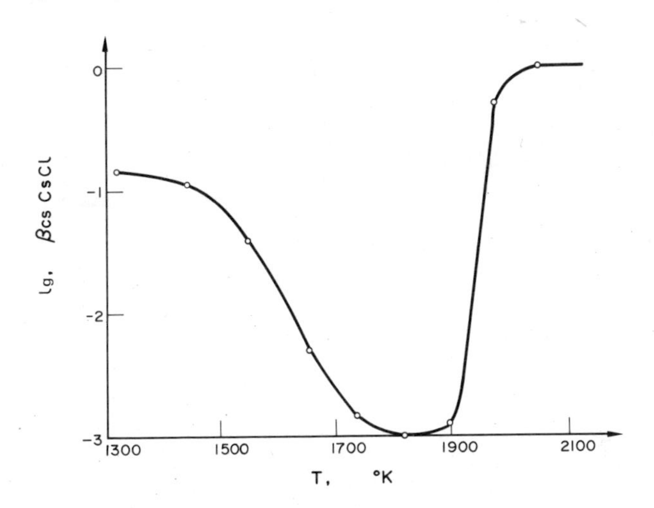


FIG. 33. Variation of the surface ionization coefficient $\beta$, for CsCl molecules, as a function of iridium surface coverage by carbon[72].

of the molecules towards higher temperatures (close to the melting point in the case of Pt). Zandberg and Tontegode[72] deposited carbon in a thick layer on the cold surface of Ir. Next, the temperature of iridium was increased gradually and the ionic current $i(T)$ was measured in ionization of a constant flux $\nu$ of CsCl molecules (Fig. 33). For $T < 1300$°K (the case of multilayer coverage of Ir by carbon), the value of $\beta$ corresponded to the ionization on a carbon filament. For $T > 1300$°K, carbon began to evaporate from Ir and the magnitude of $\beta$ dropped down by about $10^2$ times that at $T \simeq 1800°K$ (the case of monolayer carbon on the surface). For $T > 1800°K$, the magnitude of $\beta$ increased. reaching almost unity eventually. It has been shown[53] that, at temperatures below the ionization threshold, CsCl molecules desorbing from an Ir surface contaminated with carbon have a Maxwellian distribution in the initial energy, which corresponds to the surface temperature. As shown in the works cited, carbon adsorbed on the surface of the platinum group metals suppressed the reaction of thermal dissociation of the alkali-halide molecules.

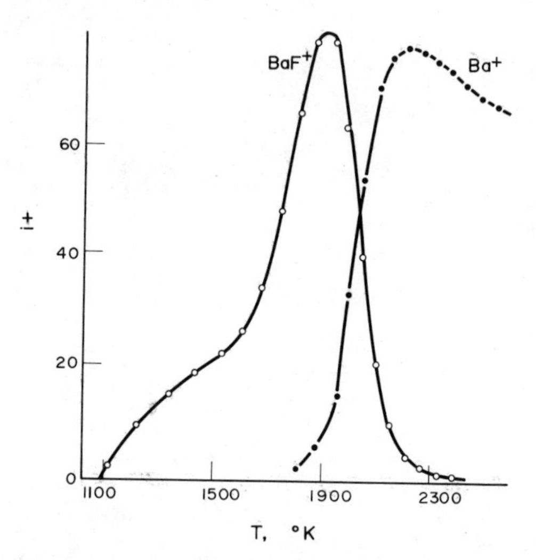


FIG. 34. The temperature variation of ionic currents in the ionization of $BaF_2$ molecules on tungsten[73].

Surface ionization of $BaF_2$ molecules has been studied[73] in a mass spectrometer (Fig. 15). Desorption of $Ba^+$ and $BaF^+$ ions was observed, and the temperature dependence of the ionic currents obtained is presented in Fig. 34. With the increase of the tungsten temperature, the predominant reaction at first is $BaF_2 \rightleftarrows BaF + F$, after which the reaction $BaF \rightleftarrows Ba + F$ occurs. In the same experiments, as well as those of Mittsev[74], were studied the combined ionizations Ca + LiF,

Ba + LiF, Ba + NaCl and Nd + LiF. Besides the ions $Li^+$, $Ba^+$, $Ca^+$, $Na^+$, $Nd^+$ were also observed the ions $BaF^+$, $CaF^+$, $BaCl^+$, $NdF^+$ and $NdF_2^+$, which were formed on the surface during the substitution reactions of the type Ba + LiF $\rightleftarrows$ BaF + Li.

Diverse chemical reactions, occurring in the adsorbed layers, are due to the presence of residual gases in the instrument, or to surface ionization on the metal oxide. The method of thermal desorption, used to study catalytic reactions in the adsorbed state, is most useful if one investigates, in addition to the fluxes of positive and negative ions from the surface, the fluxes of desorbed neutral particles.

### (ii) *Surface ionization of organic molecules*

It is known that, in mass spectrometers making use of thermionic sources, one observes a background spectrum in which mass lines of different intensity cover a broad mass range, with a separation of one mass unit[75]. This background interferes with mass spectrometer analysis of substances for low-level impurities, as well as with accurate determination of scarce isotopes. A suggestion was made that this background is produced by surface ionization of organic compounds contained in pump oil or the products of their cracking.

The first studies of the temperature dependence of ionic currents for some mass lines of the organic background in the spectrum of positive and negative ions have been carried out[76, 77]. Essentially, it was found that the positive ion current increases, if the tungsten surface is oxidized until it is covered with many layers of oxides. The temperature dependence of the current of both positive and negative ions for individual mass lines has a complicated character. Many dependences are bell-shaped which, according to (9.1), corresponds to complicated processes occurring on the surface. From these first experiments the following important conclusions have been drawn: (a) many organic molecules and radicals have a high thermal resistance on heated surfaces, and (b) they have rather low ionization potentials and possess a sufficiently high electron affinity.

In a number of subsequent experiments[78–84], Zandberg and co-workers have used a mass spectrometer to study surface ionization on tungsten and its oxides of a large number of pure organic substances, which belong to various classes of organic compounds. In these experiments, the residual gas pressure was $\sim 10^{-8}$ torr and the spectrum of the organic background was beyond the sensitivity of ionic current detection (currents below $10^{-18}$ A) and did not interfere with measurements. The vapour pressure of the substances in question in the mass spectrometer did not exceed $10^{-6}$ torr. Surface ionization of many oxygen- and nitrogen-containing organic compounds has been observed, the ion spectrum having a small number of mass lines. Molecular ions $M^+$ of the original molecules have also been observed impinging on the surface, as well as ions with masses M − 1, M + 1 and some others arising from reactions on the surface.

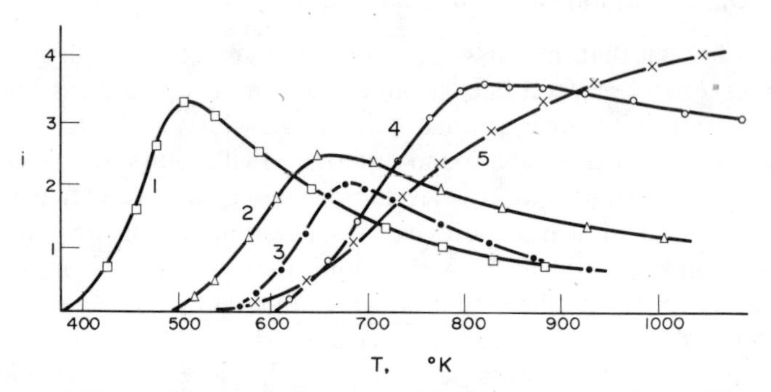


FIG. 35. The graphs of the ionic currents $i(T)$ with mass $M + 1$ (1—pyridine, 2—pyperidine, 3—amylamine) and $M - 1$ (4—pyperidine, 5—amylamine) formed in the ionization of pyridine, pyperidine and amylamine on oxidized tungsten[84].

It is interesting that some ions produce a noticeable current at surface temperatures as low as $T \simeq 400$°K. Figure 35 shows the curves of $i(T)$ for ions with mass M + 1 formed in ionization on oxydized tungsten of the molecules of pyridine, pyperedine, amylamine and ions with mass M − 1 in ionization of pyperedine and amylamine. From these curves can be obtained the ionization potentials of many compounds. Their values coincide with those found independently and published elsewhere. In some cases, metastable ions formed on the surface, which broke up on the way from the surface to the entrance into the analyzing magnetic field. The mean lifetimes of these metastable ions has been evaluated and the energy of the ruptured bonds determined. An analysis of the retardation curves for the ions of organic molecules and $Cs^+$, measured in the same conditions, gives grounds to conclude that the desorbing ions were in a thermal equilibrium with the surface. It is hoped that these interesting and promising investigations will yield a large amount of information on catalytic processes with organic substances and enable the structure and the energy characteristics of the molecules to be determined. These studies are also undoubtedly of considerable interest in analytic mass spectrometry

## 10. Surface Ionization in Evaporation of Metals and Impurities

Up to now we have considered the surface ionization of atoms and molecules impinging in a stationary flux on the surface from outside, namely, from a gas enclosing the adsorbent or a molecular beam from the evaporator. In this case, the mass flux of desorbed particles is always equal to that of particles adsorbed on the surface (the steady-state condition). We will now consider thermionic emission associated with the surface ionization of impurities present in the adsorbent and diffusing to the surface from the bulk, as well as that observed in the evaporation of the adsorbent proper. The charge equilibrium in the flux of particles evaporating from a homogeneous surface is determined by (3.14) for the degree of surface ionization. However, the net flux $\nu = \nu_0 + \nu_+$ of particles of each kind evaporating from the surface, in the ionic and neutral states, depends on the adsorbent temperature $T$. Therefore, the temperature dependence of the thermionic current will, in both cases, be described by (3.18), in which $\nu$ is temperature-dependent.

### (i) *Thermionic emission of impurities*

The emission of positive ions from the surface of heated metals is well known[85]. Mass-spectrometer analysis of the ions emitted showed that they were predominantly those of elements with low ionization potentials (the alkali and alkali-earth metals). The main contribution to the thermionic emission is due to the ions of sodium and potassium.

If the impurity concentration in the bulk of the metal is $n$, then, in desorbed from the surface, a concentration gradient $\partial n/\partial x$ will arise in the direction $x$ normal to the surface. The diffusion flux of the impurity atoms to the surface is

$$v = -D(\partial n/'\delta x) \tag{10.1}$$

where $D$ is the diffusion coefficient

$$D = D_0 \exp(-E_d/kT) \tag{10.2}$$

and $E_d$ is the activation energy of diffusion. Thus, in the case of the diffusion mechanism supplying the impurity atoms to the surface, the magnitude of the ionic current density from the surface is, by (3.18), (10.1) and (10.2),

$$j = -eD_0 \frac{\partial n}{\partial x} \exp\left(-\frac{E_d}{kT}\right)\left\{1 + A^{-1} \exp\left[\frac{e}{kT}(V - \varphi - \sqrt{(e\varepsilon)})\right]\right\}^{-1} \tag{10.3}$$

From this expression, in accordance with experiments, it follows that the ionic current density depends not only on $T$, but on the concentration and distribution of impurities over the bulk of the emitter as well. As the emitter is depleted of impurity atoms, the ionic current density should decrease with increasing collection time.

For easy ionizing impurities ($\beta \simeq 1$) eqn. (10.3) assumes the form

$$j = -eD_0 \frac{\partial n}{\partial x} \exp\left(-\frac{E_d}{kT}\right), \tag{10.4}$$

indicating that the temperature dependence of the ionic current density is determined only by the rate of atomic diffusion to the surface, being practically independent of the electric accelerating field $\varepsilon$. According to (10.4), from the temperature dependence $j(T)$ one can find the diffusion coefficient $D$, as well as the quantities $D_0$ and $E_d$[86].

For hard-ionizing impurities ($\beta \ll 1$), equation (9.3) yields

$$j = -eD_0 \frac{\partial n}{\partial x} A \exp\left\{\frac{1}{kT}[e(\varphi - V + \sqrt{(e\varepsilon)}) - E_d]\right\}$$
$$= j_0 \exp\left(\frac{e\sqrt{(e\varepsilon)}}{kT}\right), \tag{9.5}$$

$j_0$ being the ionic current density extrapolated to $\varepsilon = 0$. In this case, the dependence $j(T)$ is also determined by the activation energy of diffusion $E_d$. The dependence on $\varepsilon$ is determined by the Schottky law. The diffusion coefficient can be obtained from the difference between the slopes of the plots $\ln j = f(1/T)$ for the thermal emission of the impurity ions and the surface ionization, on the same surface, of the atoms impinging on it from the outside.

The thermionic emission of real polycrystalline solids is likewise determined by equations of the kind (10.4) or (10.5). However, the quantities $\varphi$ and $E_d$ are now averaged over the ionic current from the inhomogeneous surface. Moreover, real solids have various structural defects, such as contact planes of microcrystal faces, microcracks, dislocations, etc., which we will call micropores. The surfaces of these micropores can accumulate impurity atoms, which will emerge on to the surface, through pores opening on to it, at a greater rate than that of ordinary diffusion through the lattice. Closed micropores act as traps for the impurities, so that any process favouring their emergence to the surface (chemical and ionic etching, recrystallization, plastic and elastic deformations resulting in the emergence of dislocations to the surface[(89)]) should produce an increase of the mean ionic current from the surface. Thus, for instance, the emission of potassium ions from tungsten increases in an oxygen atmosphere. Oxidation of W and evaporation of oxides from the surface accelerates the opening of the micropores. Figure 36[(88)] presents the graphs of the temperature dependence of the $K^+$ ionic current and the rate of evaporation of tungsten oxides at various oxygen pressures. There is clearly a pronounced correlation between these graphs. At low average thermionic currents $i$, the emission exhibits spikes, due to the opening of pores. These current pulses of thermionic emission of the emitter material limit the sensitivity of surface ionization detectors to weak molecular beams, and probably represent one of the low-frequency sources of noise in electronic tubes (the flicker noise).

Various compounds have been proposed to produce stable and strong ionic emission, at comparatively low emitter temperatures, which contain ionizing elements in the form of their salts, oxides, or hydroxides at about 1% concentration. These compounds are used as coatings on the surface of ribbons made of W, Ta, Pt and other materials, which serve as heaters. One may also compact these compounds into pellets, which can be heated independently during operation of the source. Such mixtures are prepared on the basis of borax, iron oxides (Kunsmann emitter), tungsten powder (Koch emitter), aluminosilicate glasses, with various proportions of $Al_2O_3$ and $SiO_2$, etc. These mixtures emit many elements in the form of atomic ions and also, in

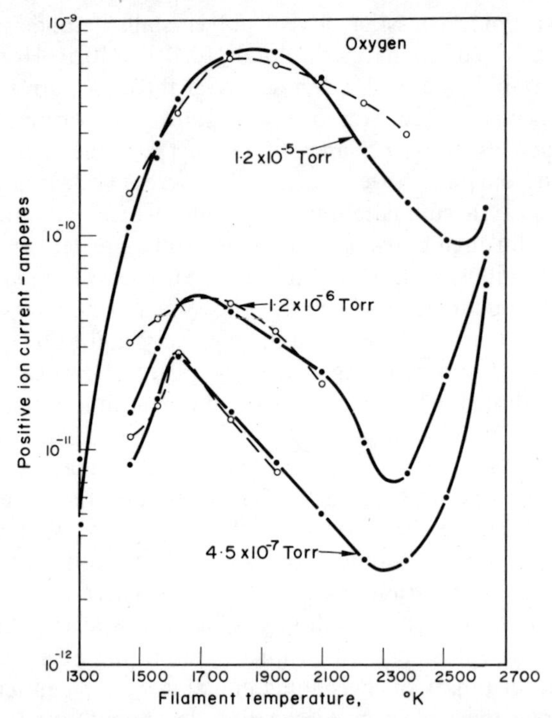


FIG. 36. The temperature dependence of the $K^+$ ionic current (solid curves) and the rates of evaporation of tungsten oxides (broken curves) at different oxygen pressures $p$[88].

many cases, ions of oxides of these elements. In emitters of this type, the charge equilibrium in the desorbed flux is undoubtedly determined by the mechanism of surface ionization, the ionizing particles being supplied from the mixture to the surface by diffusion.

*(ii) Thermionic emission in evaporation of metals*

If we assume that, in evaporation from a homogeneous metal surface, the charge equilibrium in the flux of evaporating particles is determined by (3.14), then the fluxes of the evaporating particles in the neutral and ionic states are described by (3.2) and (3.1). Since, in the course of evaporation of the metal, the surface concentration of atoms $N$ remains constant, we may derive the following expression for the ionic current density in evaporation from (3.1), namely,

$$j_{\pm} = eCN \exp(-l_{\pm}/kT) \tag{10.6}$$

which is accurate to the reflection index $R$.

In the evaporation of polycrystalline specimens, with inhomogeneous surfaces, the functions $j_{\pm}(T, \varepsilon)$ retain the form of (10.6), except that the quantities $l_{\pm}$ and $CN$ are replaced by their effective values $l^*_{\pm}$ and $(CN)^*$, which are averaged over the ion emission current, just as in the derivation of (6.5). The field $\varepsilon$ is assumed to be sufficient to cancel out the contact field of the patches (the region of the normal Schottky effect). With the ionic emission currents being small, the measurement of $j_{\pm}(T, \varepsilon)$ may only be carried out reliably by using the mass-spectrometer technique. This technique is necessary, not only to exclude emission of the impurity ions (since it can exceed that of the host ions), but also to take into account the composition of evaporating particles, which may be quite complicated.

The evaporation of polycrystalline W, Mo, Ta, and Re wires has been studied(28). The experimental layout is shown in Fig. 17. The apparatus enabled the composition of evaporating particles to be measured. The functions $i_{+}(T)$ and $\nu_0(T)$ (the neutral particle flux was ionized by an electron beam) were determined, and the work function $\varphi_R$ obtained from the Richardson plots of the thermoelectronic current. Similarly, the ionic work function $\varphi^*$ was found from the graphs $\ln i_{+} = f(1/T)$ of (6.5) for the surface ionization of hard-ionizing elements. All the measurements were carried out at a residual gas pressure of $p < 10^{-7}$ torr. For $p > 10^{-7}$ torr, the evaporation of metal oxides in neutral and ionic states was observed. Shown in Fig. 37 are the plots of $\lg (i_{+}\sqrt{T}) = f(1/T)$ for the evaporation of W, Re, Ta, and Mo ions. [Along the ordinate axis in Fig. 37 is plotted $\lg (i_{+}\sqrt{T})$, since in the original publication[28] the quantities $l_+^*$ were determined by the Clapeyron–Clausius formula for the equilibrium vapor pressure.] From the slope of the plots comes $l_+^*$. Similar curves have also been obtained for the quantities proportional to the fluxes of evaporating

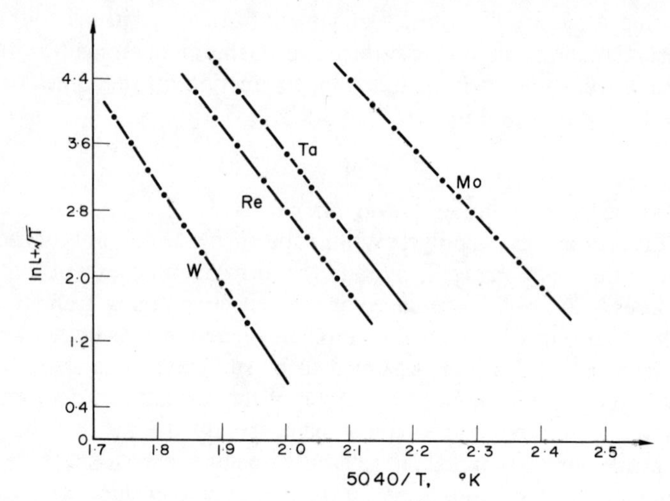


FIG. 37. The graphs of lg $(i_+\sqrt{T}) = f(1/T)$ in the evaporation of metals in the ionic state[28].

TABLE 3

| | Re | W | Mo | $T_a$ |
|---|---|---|---|---|
| $l_+^*$ (eV) | 10.2 ± 0.4 | 11.8 ± 0.4 | 8.3 ± 0.3 | 10.0 ± 0.4 |
| $l_0^*$ (eV) | 7.7 ± 0.4 | 8.6 ± 0.4 | 6.3 ± 0.3 | 7.1(5) ± 0.3 |
| $\varphi_R$ (V) | 4.93 ± 0.04 | 4.58 ± 0.03 | 4.33 ± 0.07 | 4.33 ± 0.03 |
| $\varphi^*$ (V) | 5.43 ± 0.03 | 5.14 ± 0.03 | 5.02 ± 0.05 | 4·78 ± 0.05 |
| $T$ | 2400–2750 | 2600–2900 | 2100–2400 | 2400–2650 |
| $j(a/\text{cm}^2)_T$ | $\sim 10^{-9}{}_{2430}$ | $\sim 5 \times 10^{-10}{}_{2700}$ | $\sim 10^{-8}{}_{2280}$ | $2 \times 10^{-11}{}_{2400}$ |

particles $\nu_0$: these give $l_0^*$. In Table 3 are given the values of $l_+^*$, $l_0^*$, $\varphi_R$ and $\varphi^*$, as well as the range of temperature variation in the measurements. The last column of the table presents the current densities of the metal ion emission. Measurements of $i_+(T)$ have been made for niobium[89] and W and Re[90]. Within experimental error, the $l_+^*$ results for W and Re agree with those obtained earlier[28]. Furthermore, the evaporation of W and Re in the form of negative ions $W^-$ and $Re^-$ made it possible to find $l_-^*$ for these metals[90].

A detailed study[28] has been performed of the applicability of the Schottky relation (3.9) to the case of evaporation of polycrystalline metals. This relation is shown to be only approximately valid for the quantities $l_+^*$, $l_0^*$ and $\varphi^*$. Although the function $j_\pm(\varepsilon)_T$, for homogeneous and patchy surfaces, should follow the Schottky law, it has not been tested experimentally, though in the determination of $l_+^*$, the correction $\varepsilon\sqrt{(e\varepsilon)}$ was taken into account[28].

## 11. Application of Surface Ionization in Physicochemical Studies

*(i) Determination of the ionization potentials of atoms and molecules*

The main method used in the determination of the atomic ionization potentials is that of optical spectroscopy. However, the spectra of many elements with unfilled inner electronic shells (rare-earth elements, uranium, thorium and transuranium elements) are complicated, and do not lend themselves easily to unambiguous interpretation. Therefore, until recently, the magnitude of $V$ has not been determined spectroscopically for a number of elements. The possibility of employing surface ionization, for the determination of $V$, follows from formula (3.20) for the ionization on homogeneous surfaces, and from (6.5) for that on inhomogeneous surfaces. If $\alpha \ll 1$, then, from the slope of the plots $\ln i = f(1/T)$ and $\nu = \text{const}$ and $A^*$ independent of $T$, we can derive the quantity $\varphi^* + \sqrt{(e\varepsilon)} - V$. Knowing the quantity $\varphi^* + \sqrt{(e\varepsilon)}$, we can find the value of $V$ (for homogeneous surfaces $\varphi^* = \varphi$).

The most accurate values of $V$ are obtained by measuring the temperature dependence of the ratio of ionic currents $i_1$ and $i_2$ in the simultaneous surface ionization of atomic beams of two elements, the ionization potential of one of the beams being known (the reference potential)[91]. Indeed, from (6.5) or (3.20) we obtain for this ratio

$$\lg\left(\frac{i_1}{i_2}\right) = \lg\left(\frac{\nu_1 A_1}{\nu_2 A_2}\right) + \frac{5040}{T}(V_2 - V_1). \tag{11.1}$$

When $\alpha \ll 1$, the quantity $\varphi^*$ does not depend on $V$ (see §6), the ratio $A_1^*/A_2^*$, according to (6.6), being equal to $A_1/A_2$. If the ratio of the fluxes $\nu_1/\nu_2$ remains constant in the measurements and the ratio $A_1/A_2$ does not depend on $T$, then the plot $\ln(i_1/i_2) = f(1/T)$ will be a straight line, its slope being determined by the difference $V_2 - V_1$. The smaller the error in the determination of the difference $V_2 - V_1$, the closer are the magnitudes of the reference ionization potential and the potential to be found. Ordinarily, one knows, for the reference element, not only the value of $V_2$, but also the excitation levels, from which one can readily calculate the temperature dependence of the statistical sum ratio $A_2$ of the ion and the atom. If the excitation levels of the ion and atom of the element in question are unknown, then the quantity $V_2 - V_1$ can be found from (11.1) only to the slope of the relationship $\ln(A_1/A_2) = f(1/T)$. It should be remembered that, if the excited states of the ion and neutral particle lie close to the ground level, then, due to the compensation effect, the ratio $A = Q_+/Q_0$ will depend only weakly on $T$. This method was used, with the mass-spectrometer technique (the experimental layout is similar to that of Fig. 15), to determine the ionization potential of the atoms of uranium, thorium and rare-earth elements[(91–96)] presented in Table 4.

TABLE 4

| | $V$ (eV) | | | $V$ (eV) | | | $V$ (eV) | |
|---|---|---|---|---|---|---|---|---|
| | I | II[96] | | I | II[96] | | I | II |
| La | 5.61 ± 0.03[92] | 5.55 ± 0.05 | Eu | 5.68 ± 0.03[94] | 5.64 ± 0.05 | Tm | 6.14 ± 0.06[92] | 6.03 ± 0.04[96] |
| Ce | 5.60 ± 0.05[93] | 5.54 ± 0.06 | Gd | | 6.16 ± 0.05 | Yb | | 6.04 ± 0.04[96] |
| Pr | 5.42 ± 0.04[92] | 5.40 ± 0.05 | Tb | 5.98 ± 0.02[93] | 5.82 ± 0.03 | Lu | 5.41 ± 0.02[94] | 5.32 ± 0.05[96] |
| Nd | 5.51 ± 0.02[92] | 5.49 ± 0.05 | Dy | 5.80 ± 0.02[94] | 5.82 ± 0.03 | Th | 6.95 ± 0.06[93] | |
| Pm | | ~ 5.55 | Ho | 6.19 ± 0.02[94] | 5.89 ± 0.03 | U | 6.08 ± 0.08[91] | 6.22 ± 0.06[95] |
| Sm | 5.70 ± 0.02[94] | 5.61 ± 0.05 | Ez | | 5.95 ± 0.03 | | | |

Equation (11.1) may also be used for the determination of the ionization potentials of molecules in the case of no chemical reactions occurring on the surface. Thus, from the temperature dependence of the ionic-current ratio of bismuth and aniline $C_6H_5NH_2$, in ionization on an oxidized tungsten surface, can be determined the ionization potential of the aniline molecules[79].

The method of surface ionization in mass spectrometry may further be used to determine the value of $V$ in cases where the flux $\nu(T) = \nu_0 + \nu_+$ of particles of a given chemical composition, which depends on $T$, desorbs thermally as a result of chemical reactions on the surface (see § 9). For this purpose, the temperature dependence of the desorbing fluxes $\nu_0(T)$ and $\nu_+(T)$, or of quantities proportional to these fluxes, must be measured independently. This can be done in a mass spectrometer with a double ion source (Fig. 17) by using an electron beam of constant intensity and energy to ionize the desorbing flux $\nu_0(T)$ of neutral particles. The ionic current $i$ from the surface, which is detected in the mass spectrometer, is proportional to the flux $\nu_+(T)$ and the ionic current $i'$, from the source with electronic impact ionization, is proportional to the quantity $\nu_0(T)/\sqrt{T}$ (see § 10), so

$$\alpha = \frac{\nu_+(T)}{\nu_0(T)} \sim \frac{i(T)}{i'(T)\sqrt{T}}.$$

In ionization on a homogeneous surface, we have

$$\lg\left[\frac{i(T)}{i'(T)\sqrt{T}}\right] = \text{const} + \lg A + \frac{5040}{T}(\varphi + \sqrt{(e\varepsilon)} - V) \qquad (11.2)$$

by (3.14). For ionization on an inhomogeneous surface, if $\alpha \ll 1$ and the accelerating field $\varepsilon$ is sufficient to cancel out the contact field of the patches, (11.2) will retain its meaning, provided $A$ and $\varphi$ are replaced by their effective values $A^*$ and $\varphi^*$. The value of $\varphi$ or $\varphi^*$ can be determined from the temperature dependence of the SI current of atoms with a known $V$. It is assumed here that, throughout the range of temperature variation, the surface coverage is close to zero, which can be independently checked by the constancy of thermoelectronic current from the surface, the molecular beam being chopped successively.

*(ii) Determination of the electron affinity energy eS*

The majority of investigations of negative surface ionization have had the purpose of determining the values of $S$ for some atoms and radicals. In the beginning, it was the only direct method of $S$ determination. Only recently have more accurate techniques been developed, such as the method involving electron photodetachment from negative ions[97] and the method of absorption spectra behind a shock front[98]. In the first experiments on the determination of $S$ use was made of (8.7), which relates the ratio of the thermoelectronic current $i_e$ to the negative-ion current $i_-$ to $T$ (see Fig. 29). For separate measurement of $i_e$ and $i_-$, a magnetic field in a triode with an intermediate grid (the magnetron method) was used and also a method involving the displacement of the current–voltage characteristics by the space-charge field, which is a function of the concentration of thermoelectrons and negative ions. The first method was employed to measure $S$ for the atoms of iodine[4], bromine[99], chlorine[54, 100] and oxygen[101] in ionization of the molecules of these substances on polycrystalline tungsten filaments. Later, the magnetron method was utilized to study surface ionization of a number of organic substances and to determine $S$[102]. The method of displacement of the current–voltage characteristics was used to determine $S$ for the atoms of iodine and bromine[103]. However, because of the poor accuracy, this method has not been used again.

The method of determining $S$ of halogen atoms from the comparison of the currents of negative ions $X^-$ and positive ions $M^+$, which form in the surface ionization of alkali–halide molecules MX, has been described[104–5]. The $X^-$ ionic current was separated from thermoelectrons by a magnetic field. This method appears to be promising in the case of hard-ionization of the atoms making up the molecule ($\alpha_+ \ll 1$, $\alpha_- \ll 1$). Indeed, in ionization on an inhomogeneous surface, it follows from (6.5) and (8.6) that

$$\frac{i_-}{i_+} = \frac{A_-^*}{A_+^*} \exp\left[\frac{e}{kT}(V_M + S_X - \varphi_R - \varphi^*)\right]. \tag{11.3}$$

Here $\varphi_-^* = \varphi_R$ (see § 8). Using (11.3) implies an independent determination of the magnitudes of $\varphi_R$ from the Richardson plots for the thermoelectronic current, and of $\varphi^*$ from the temperature dependence of the positive ion current for hard-ionizing atoms. For ionization on a homogenous surface, $\varphi_R = \varphi^*$, $A_-^* = A_-$ and $A_+^* = A_+$. Equation (11.3) can be used for the determination of the quantity $(V + S)$, provided the particles desorbed from the surface are partially ionized to form both positive and negative ions (for instance, the atoms of Cu, Ag, Sb, Bi)[58]. Equation (11.3) can also be used for particles forming on the surface as a result of chemical reactions, since the reaction yield [the quantity $\gamma$ in (9.1)] is the same for positive and negative ions, and is cancelled in the current ratio $i_-/i_+$.

Since ionization on pure high-melting metals, in the formation of negative ions, relates to the case of hard ionization ($\alpha_- \ll 1$), the principal technique for the determination of $S$ is the mass-spectrometer method which permits the most reliable measurements of the ionic current of a given chemical composition. The mass-spectrometer technique was used[57], for the first time, to study the surface ionization of alkali–halide molecules on polycrystalline tungsten. In these experiments, it was shown that only positive atomic ions $M^+$ of the alkali metal and negative atomic ions $X^-$ of the halogens are desorbed, and that (8.6) is correct for the temperature dependence of $i_-(T)$.

The most precise and reliable values of $S$ are obtained by comparing the currents of two elements ionizing simultaneously on the surface. According to (8.6), the ratio of these currents is expressed by a formula similar to (11.1), viz.

$$\frac{i_1}{i_2} = \frac{v_1 A_{-1}}{v_2 A_{-2}} \exp\left[\frac{e}{kT}(S_1 - S_2)\right]. \tag{11.4}$$

Such a method has been proposed[106] and used to determine the value of $S$ for the atoms of the halogens[106–107], sulfur[108], gold, silver and copper[58]. The measurements were carried out in a mass spectrometer with two evaporators (Fig. 15) producing the beams to be ionized. An attempt was also made to determine $S$ for the CN-radical in the surface ionization of KCN and KCNS molecules[109–10].

Recently, the reference version of the method of measuring the temperature dependence of the negative-to-positive ion current ratio has been used to determine the value of $S$ for the atoms of antimony and bismuth[111]. Thermal desorption of these elements from a tungsten surface occurs in the form of both atoms and molecules, the composition being dependent on the surface temperature $T$. This dependence is not included in (11.2). For the ratio of the currents of the two elements being compared, we obtain, according to (11.2),

$$\left(\frac{i_-}{i_+}\right)_1 \left(\frac{i_+}{i_-}\right)_2 = \left(\frac{A_-^*}{A_+^*}\right)_1 \left(\frac{A_+^*}{A_-^*}\right)_2 \exp\left[\frac{e}{kT}(S_1 - S_2 + V_1 - V_2)\right] \qquad (11.5)$$

from which we can readily determine the unknown value of $S_2$, provided the values of $S_1$, $V_1$ and $V_2$ are known. Silver was chosen as the reference element for these experiments, because the value of $S$ was known precisely from independent measurements of the temperature dependence (8.7) for the current ratio of $Ag^-$ ions and thermoelectrons. Equation (8.7) was used earlier to determine the value of $S$ for hydrogen atoms[112].

As mentioned in § 10, the value of $S$ for W and Re atoms has been evaluated[90] from the temperature dependence of their positive and negative ion currents during evaporation. Table 5 presents a list of

TABLE 5

| | S (eV) | | | S (eV) | |
|---|---|---|---|---|---|
| | I | II | | I | II |
| H | 0.8[112] | | I | 3.12[57] | 3.063 ± 0.003[98] |
| Li | 0.5 ± 1.05[146] | | S | 2.11 ± 0.05[108] | 2.07 ± 0.07[97] |
| F | 3.36 ± 0.07[106, 110] | | Cu | 1.5 ± 0.5[58] | |
| | 3.57[107] | 3.448 ± 0.05[98] | Ag | 2.0 ± 0.2[58] | |
| Cl | 3.60 ± 0.03[110] | | | 1.90 ± 0.15[111] | |
| | 3.77[107] | 3.613 ± 0.003[98] | Au | 2.8 ± 0.1[58] | |
| Br | 3.34 ± 0.02[110] | | Sb | 1.50 ± 0.17[111] | |
| | 3.52[107] | 3.383 ± 0.003[98] | Bi | 1.76 ± 0.16[111] | |

the most reliable values of $S$, which have been determined by the method of surface ionization, using the mass-spectrometer technique (column 1), and by the method of electron photodetachment from negative ions (column 2).

*(iii) Determination of the temperature dependence of vapor pressure and heat of evaporation*

In § 10, we have already considered the measurement of the temperature dependence of the ion currents and neutral particle fluxes in the evaporation of emitter material, from which the heats of evaporation $l_+$, $l_-$ and $l_0$ can be determined. It should be stressed, once more, that these measurements should be carried out with a double ion source mass spectrometer. Generally speaking, this method is applicable to evaporation from the surfaces of both solids and liquids, in the range of relatively low evaporation rates.

The flux density $\nu(T)$ of particles evaporating from a surface is related to the equilibrium vapor pressure $p(T)$ by the formula

$$\nu(T) = \frac{p(T)}{\sqrt{(2\pi MkT)}}, \tag{11.6}$$

$M$ being the mass of the particles. In experiments, one measures not the absolute values of $\nu(T)$, but the values of the ionic currents $i(T) \sim \nu(T)/\sqrt{T}$ recorded by the mass spectrometer, which are proportional to them. Then, according to (11.6), $p(T) = bTi(T)$, $b$ being the proportionality coefficient. Thus, if we measure $i(T)$ by this method, we may also determine from it the relation $p(T)$. For this purpose, one should determine, by means of independent measurements, the calibration coefficient $b$ at some value of $T$. This method simplifies, substantially, the cumbersome work of determining the relationships $p(T)$ in absolute units.

In all cases, both in ionization on homogeneous surfaces (3.18), and in ionization on inhomogeneous surfaces (6.1), the ion current from the surface is proportional to the particle flux incident on it, i.e. $j = e\nu\beta$ (2.2), which is used in the experimental measurements of $\nu(T)$ and also, according to (11.6), of $p(T)$. The molecular beam is directed to the ionizing surface from either an open surface (Langmuir evaporation), or a Knudsen evaporator in the mass spectrometer with an ion source similar to that of Fig. 17. The lower limit of the measured values of $\nu(T)$ is determined by the magnitude of the ionization coefficient $\beta$, and by the detection sensitivity of the ionic current $i$ in the mass spectrometer. The upper measurement limit for $\nu(T)$ is determined by the molecular nature of the particle flux from the evaporator.

Throughout the measurement range of $\nu(T)$, the magnitude of $\beta$ should remain constant. However, it can vary, because of adsorption of residual gases and molecules from the beam. Therefore, the measurements should be carried out in high vacuum and at as high a temperature of the ionizing surface as possible. The most suitable metals for the ionizer are rhenium and iridium, which have the highest values of $\beta$ for pure metals (high values of $\varphi$) and relatively weak chemisorption activity.

From the experimentally measured values of $\nu(T)$, one infers the heats of evaporation $l_0$, and, using calibration at some evaporator temperature $T$, also the absolute values of $p(T)$. This approach has been used[113] to obtain $p(T)$ for a number of alkali halides.

### (iv) *Measurement of heats of desorption*

In cases where one studies the surface ionization of hard-ionizing particles ($\alpha \ll 1$) on a homogeneous surface, one can determine, from the temperature dependence of the ionic current density $j(T)$, the difference $l_0 - l_+$ of the heats of desorption of particles in the neutral and ionic states. Indeed, in accordance with (2.3), when $\alpha \ll 1$, $\beta = \alpha$ and, from (2.2), $j = e\nu\alpha$. The slope of the plots $\ln j = f(1/T)$ is determined, according to (3.4), by the quantity $l_0 - l_+$ or, from (8.1), by the difference $l_0 - l_-$. Knowing the value of one of these heats, as well as the magnitudes of $\varphi$ and $V$ from Schottky formula (3.9), one can determine the value of the other. The magnitude of $l_0$ can be measured by the method of modulated particle fluxes[5] incident on the adsorbent surface.

Sometimes, the quantities $l_+$ can be determined from the temperature dependence of the ionic current in the ionization threshold region. If $\alpha > 1$, then the relation between the first threshold temperature and the ionic current density $j$, at the peak of the ionization curve, is given by (4.10). It follows, from this equation, that the plot of $\ln (j/T_0) = f(1/T_0)$ should be a straight line, with a slope proportional to $l_+$. Such plots have been obtained for the case of ionization of alkali atoms on tungsten and for some other metal–atom systems[20, 114–17].

If $\alpha < 1$, and the adsorption of particles ionizing on the surface reduces the adsorbent work function $\varphi$, then the ionic current curves $\ln i = f(1/T)$ reveal a break in the low temperature region at the temperature $T_0$. As the particle flux $\nu$ increases, the break in the curve shifts to higher values of $T_0$. The break-points in the $\ln i = f(1/T)$ plots, for different values of $\nu$, fit a straight line (Fig. 38), its slope being determined by the magnitude of $l_+$[116]. Such a method is applicable to the determination of $l_+$ in the desorption of alkali-earth and rare-earth elements from the surface of high-melting metals.

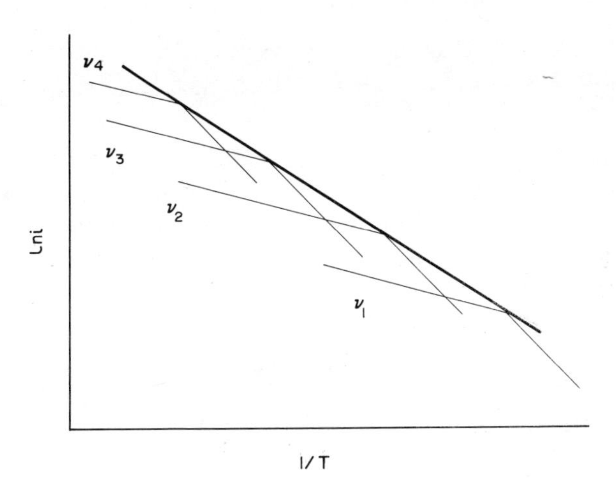


FIG. 38. The graphs of $\ln i = f(1/T)$ at different $\nu$ for the determination of $l_+$ for elements with $\alpha < 1$, whose adsorption is accompanied by a decrease of $\varphi$.

The threshold temperature $T_0$ depends not only on particle flux $\nu$ at $\varepsilon =$ const, but, according to (4.11), is also dependent on $\varepsilon$ at $\nu =$ const[24, 25, 64]. In this case, the quantities $T_0$ and $\sqrt{\varepsilon}$ are related by a linear law, from which one can also determine $l_+$[25].

### (v) *Measurement of heats of molecular dissociation*

Figure 31 shows the plots $i\sqrt{T} = f(T)$ of the ionic currents, which are proportional to the particle fluxes desorbed from the surface of a tungsten ribbon, and represent the products of the catalytic dissociation reaction of alkali-halide diatomic molecules MX. A constant flux $\nu$ of these molecules arrives on the ribbon surface from a special evaporator. The particle fluxes, desorbed from the surface, are ionized by electrons in a combined mass-spectrometer ion source (Fig. 17), or on an auxiliary surface by the SI process. We will consider the reaction of surface dissociation $MX \rightleftarrows M + X$, which enables the heat of dissociation $D_{MX}$ of these molecules to be determined.

If chemical, as well as thermal, equilibrium exists on the surface, then the condition for such equilibrium has the form

$$\frac{N_M N_X}{N_{MX}} = K_0 \exp\left(-\frac{D'_{MX}}{kT}\right). \tag{11.7}$$

where the $N$'s are the surface concentrations of the particles participating in the reaction, and $D'_{MX}$ is the heat of dissociation of the MX molecules on the adsorbent surface. The fluxes desorbing from the surface are related to the surface concentrations of the particles by (3.2), i.e. $\nu_i = K_i N_i \exp(-l_i/kT)$. From the condition for chemical equilibrium we obtain

$$\frac{\nu_M \nu_X}{\nu_{MX}} = \text{const}\,.\exp\left(\frac{l_{MX} - l_M - l_X - D'_{MX}}{kT}\right) = \text{const}\,.\exp\left(-\frac{D_{MX}}{kT}\right). \tag{11.8}$$

The equality $-D_{MX} = l_{MX} - l_M - l_X - D'_{MX}$ can be readily found from a consideration of the cycle. $D_{MX}$ itself comes from the temperature dependence of the flux ratio[29]

$$\frac{\nu_M \nu_X}{\nu_{MX}} \sim \sqrt{T}\frac{i_M i_X}{i_{MX}}.$$

A similar consideration of the reaction $(MX)_2 \rightleftarrows MX + MX$ (the region of low $T$ in Fig. 31) yields the heat of dissociation of dimer molecules[118]. Here the magnitude of $D_{MX}$ can be obtained from a single experimental plot $\nu_{MX}(T)$, curve 2 in Fig. 31, since the fluxes $\nu_M = \nu_X = \nu - \nu_{MX}$.

*(vi) Determination of the diffusion coefficient*

We recall from § 10 that, if the atoms of an impurity desorb from a metal surface predominantly in the ionic state ($\beta \simeq 1$), then the temperature dependence of the thermionic current of the impurity element, according to (10.4), is determined by the temperature dependence of the diffusion coefficient of the impurity atoms in the metal. If the impurity atoms desorb from the surface predominantly in the neutral state, then the temperature dependence of the thermionic impurity current is described by (10.5). Comparing the temperature dependence of the ionic currents in thermionic impurity emission, and in surface ionization of the same atoms impinging on the surface from outside, enables the diffusion coefficient to be determined.

### (*vii*) *Determination of the accommodation coefficient*

If the accommodation coefficient of particles incident on the adsorbent surface is unity, and their lifetime in the adsorbed state is sufficient for the onset of thermal equilibrium, then the distribution of thermally desorbed particles in initial energies should be Maxwellian. The retardation curves, measured in the field of a plane condenser with a homogeneous emitter and collector, will be straight lines, their slopes corresponding to the ion emitter temperature (see § 7).

If, however, the accommodation coefficient of the particles on the adsorbent surface is not unity, which implies that partial elastic and inelastic reflection occurs, or that particles have not acquired the adsorbent temperature during their lifetime on the surface, then the retardation curves will not be straight lines. Only in the case of elastic reflection of all the particles incident on the surface (the zero accommodation coefficient) will the retardation curves be straight lines, but now with a slope corresponding to the temperature of particles impinging on the surface.

If the electrodes of the electrostatic energy analyzer are inhomogeneous, then the conclusion on the coefficient of accommodation can be drawn from a comparison of the retardation curves measured under the same conditions for the particles in question and for particles for which it is known that their accommodation coefficient is unity (for instance, alkali metals).

The retardation curves for particles desorbing in the neutral state can be obtained, in a similar way, after their ionization by electron impact.

*(viii) Investigation of catalyzer properties*

The phenomenon of surface ionization proper represents the simplest catalytic reaction. Some other examples of catalytic reactions on the surface were treated earlier. Here we will consider a few illustrations of the application of surface ionization to the investigation of the properties of catalyzers.

Any chemical process on a surface is associated with electron interactions in the adsorbate layer and between the adsorbate and adsorbent. One of the characteristics of the catalytic activity of the adsorbent is the electronic work function $\varphi$. Surface ionization can readily be used, side by side with other methods, to measure the value of $\varphi$. From the temperature dependence of the ionic current in SI of hard-ionizing elements, (3.14) and (6.5), can be determined the value of $\varphi^*$. In the case of large $\varphi$ and low operating temperatures (for instance, in the case of metal oxides), at which the thermoelectronic emission current is small, this approach has some advantages over the method of Richardson plots and has been used to measure the magnitude of $\varphi$ for tungsten at a monolayer coverage of the surface by fluorine, chromium and bromine[119]. Weiershausen[120], in examining the ionization of Ag atoms, also studied the change of $\varphi$ of tungsten with temperature in the presence of oxygen. If the surface is homogeneous, then the experimental values obtained by the method of surface ionization and the Richardson plots coincided, $\varphi^* = \varphi_R = \varphi$. For inhomogeneous surfaces, these methods enable the practical surface contrast $(\varphi^* - \varphi_R)$, with respect to the work function, to be determined.

Measuring the current–voltage characteristics for the total current in a retarding electric field (retardation curves), just as in the case for the thermoelectronic current, makes it possible to determine the work function of the emitter and collector (the method of contact-potential difference), as well as the changes of $\varphi$ with varying electrode temperature, in the process of adsorption or desorption of foreign particles. The accuracy of the determination of $\varphi$, and its variation, is of the order of a few hundredths of a volt[80].

Appreciable changes in the catalytic properties of the adsorbent, due to adsorption of foreign atoms, are not always associated with a change of the work function. An illustration has been given in § 9 of how low carbon concentrations on Re and Ir, which do not affect the magnitude of $\varphi$, inhibit considerably the dissociation reaction of alkali-halide molecules[72].

In conclusion of this section, it should be said that surface ionization, as a method of studying physicochemical processes and catalyzer properties, can be used to advantage in many areas.

## 12. Molecular Beam Detection and Ion Sources

These are two closely related fields in which the phenomenon of surface ionization has found wide application. In the molecular beam technique, SI is used for beam detection and measurement of the flux density variations in the beams. In the ion source, SI is employed in the production of ionic beams of different intensity and for different applications.

### (i) *Beam detection*

The principal problem here is the sensitivity of detection. It is different for different substances and, in ionization on a given surface, is determined by the magnitude of the ionization coefficient $\beta$, depending, in accordance with (3.17) and (6.1), on the ionization potential $V$ of the particles to be detected. At a given $V$, the greater $T$ and $\varphi$, the greater is the value of $\beta$. Therefore, it is expedient to choose a high-melting material, with a large work function, as the ionizing surface. Rhenium and iridium are metals which meet this requirement best. Figure 39 presents graphs of the relationship $\beta = f(V - \varphi)$ for four values of $T$,

$$A\left(\frac{1 - R_+}{1 - R_0}\right) = 1$$

and $\varepsilon = 0$, which may be used to evaluate the detection efficiency for each given case.

Sometimes, in order to increase $\beta$, one uses an oxidized metal surface, increasing in this way the value of $\varphi$ of the surface. However, an oxidized metal surface can be maintained in a stable condition only at relatively low values of $T$. Indeed, surface oxidation of, say, tungsten raises $\varphi$ by 1–1.5 V, while reducing the operating temperature down to 1500–1600°K. Now what is the net result? An answer to this question may be found from the graphs of Fig. 39. The value of $\beta = 10^{-5}$ can be reached for $V - \varphi = 3.0$ V and $T = 3000$°K, as well as for $V - \varphi = 1.5$ V and $T = 1500$°K. Hence, an increase of $\varphi$ by 1.5 V is equivalent to a decrease

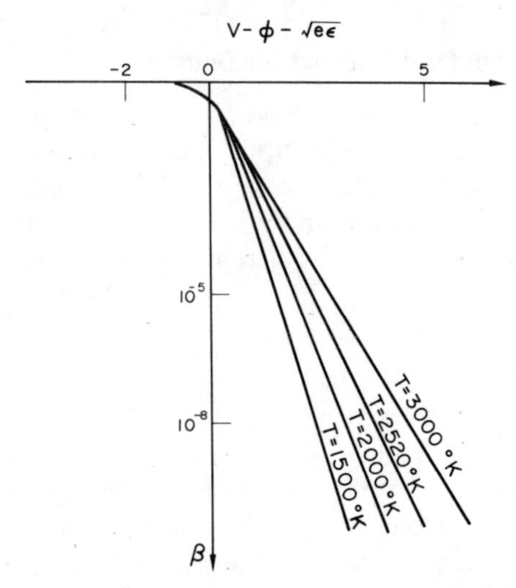


FIG. 39. The graphs of $\beta = f(V - \varphi)$ at $T =$ const.

of $T$ from 3000°K down to 1500°K, so that, for the detection of elements with relatively small $V$ ($V - \varphi < 3$ V), the oxidation of a metal surface is clearly advantageous. To ionize elements with high values of $V$, surface temperatures $T$ should be taken as high as possible. In fact, the same ionization efficiency of $\beta = 10^{-10}$ can be obtained at $T = 1500$°K with $V - \varphi = 3$ V, and at $T = 3000$°K with $V - \varphi = 6$ V, i.e. in the latter case with elements having greater $V$.

Negative surface ionization can turn out to be advantageous in the detection of particles with high $V$, but possessing an electron affinity $S$. In this case, as follows from (8.3), high-melting materials, with low $\varphi$, are preferable for use as an ionizing surface.

To increase the ion-detection efficiency, the accelerating field $\varepsilon$ should be increased, whenever possible. For an ionizing surface, thin filaments, sharpened ribbon edges ("blades"), points sharpened by etching or a set of points are used. According to (3.14), the field $\varepsilon$ has the effect of reducing the magnitude of $V$ by $\sqrt{(e\varepsilon)}$ (see Fig. 6).

Besides the molecules from the beam, the ionizing surface is also bombarded by molecules of the residual gas present in the vacuum chamber. With all the gases contained in air having high $V$, the efficiency of their ionization on the surface is small. Interference may come from the molecules of organic substances evaporating from pump oil and lubricants. To reduce this undesirable contribution to the ionic current, one should improve the vacuum in the detector chamber and freeze out the organic vapors.

If the beam to be detected does not contain easy-ionizing impurities and the background ionic current, produced by the ionization of residual gases, is small compared to the current of the particles in question, then one can collect and measure the total ionic current from the surface. The density of this current ($j = e\nu\beta$) is proportional to the beam intensity $\nu$. However, the isolation from the beam of a component with a given chemical or isotopic composition requires a mass-spectrometer analysis. In this case, the detecting surface serves as the mass-spectrometer ion source. Here, the detection efficiency will decrease by a factor of the mass spectrometer "transmission" coefficient $\gamma$, the detected current being $j = e\nu\beta\gamma$.

In all cases, the detector sensitivity is determined, not only by the limiting value of the ionic current density $j$, but also by the signal-to-noise ratio of the detector. In an SI detector, the principal source of noise, besides the fluctuations in the beam density, is the thermionic emission (mainly of $Na^+$ and $K^+$) of ions present on the ionizing surface of the detector. Recommended measures of control of this noise are the purification of the ionizer material from impurities by its prolonged heating in high vacuum and special methods of preparation of the detecting surfaces[121–2].

The application of mass-spectrometer separation of the ions to be detected favors an improved signal-to-noise (S/N) ratio and, in this way, compensates for the loss in detector sensitivity because of the incomplete collection of the ionic current from the surface. An improved S/N ratio can also be reached by the modulation of beam density $\nu$ (for instance, by chopping the beam periodically), with a subsequent synchronous detection of the ion beam.

### (*ii*) *Ion sources*

Surface ionization has found wide use in the development of ion sources for various applications. Among them can be listed sources for isotopic and chemical mass-spectrometer analysis; sources to produce ion beams for the study of ion-molecular reactions of interest in the investigation of ions with solid surfaces; sources for doping the surface layer with small concentrations of some elements; powerful ion sources for magnetic isotope separators and for the production of exhaust thrust in low-power engines.

All that has been said above about the efficiency of molecular beam detection also applies to the efficiency of ion production in surface ionization sources, where the particles to be analyzed arrive as molecular beams. Distinctive features of SI sources are the localization of the ion-producing spot on the surface and the low spread of initial ion energies, which essentially simplifies the development of efficient ion optics that are used to produce ion beams. In addition, the ion composition is usually simple (mass spectra with few lines), only singly charged ions being formed.

The field where SI sources are most widely used is in the isotopic analysis of elements and chemical analysis by the isotopic dilution method. Up to 1959, fifty-four elements had been analyzed for isotopic composition with these sources(123). Moreover, many elements produce negative ions on a surface with an efficiency sufficient for isotopic analysis to be carried out. Easy-ionizing elements ($\beta \simeq 1$) can be analyzed in samples of down to $10^{-15}$ g. The low limits for the analysis of other elements can be evaluated by means of the graphs in Fig. 39.

The sample to be analyzed can be deposited on the surface, which is ordinarily an oxidized metal, in the form of a solution of any chemical compound containing the element in question. The heating of such a thermoanode results in the thermal decomposition of the sample (it can be a salt), with subsequent evaporation (partially in the form of ions) and diffusion into the surface layer of the anode. The source is capable of prolonged operation, which is sustained by the diffusion of the element to be analyzed to the surface. Sources of this type can readily be made to occupy many positions, so that a large number of samples can undergo isotopic analysis without impairing the vacuum in the mass spectrometer[124]. One of the drawbacks of these so-called single-filament sources is the effect of fractionation in evaporation and diffusion, resulting in some dependence of the isotopic ratio in question on the duration of the source operation, which is particularly essential in the analysis of light elements. In many cases, one can use, for the isotopic analysis, the emission of ions in the evaporation of the emitter proper (see § 10).

Strong ion beams can be produced by using sources with a separate evaporator containing an amount of the substance to be ionized. Such a system provides prolonged (hundreds of hours) stable emission of ions. The sources are designed in such a way as to use the radiation from the thermoanode for partial heating of the evaporator[125]. Stable, high-density ionic current can be maintained for tens of hours by thermionic emission from the specially prepared catalyzers mentioned in § 10.

Consider now methods of producing the highest ion fluxes in the sources designed for prolonged operation (isotope separators, ion engines). The principal requirement to be met by sources of this type is the maximum utilization factor of the working substance. As an illustration, we will consider the ionization of the most easy-ionizing element, viz. cesium. On pure, and particularly on surface-oxidized high-melting metals (tungsten, rhenium), cesium is ionized almost completely ($\beta \simeq 1$). In order that the utilization factor of cesium be likewise close to unity, the consumption of the metal, during the operation of the source, should mainly occur through the evaporation (desorption) of cesium ions from the ionizing surface. Ionization on porous emitters satisfies this requirement best of all. Figure 40 shows a layout of such a source. The porous emitter 1 is maintained at the desired temperature $T_3$ by means of the heater 4. Cesium vapor, from evaporator 3, via duct 2, emerges to the surface of the porous ionizer by way of bulk and pore surface diffusion. The temperature is maintained so that $T_3 > T_2 > T_1$. The rate of cesium atom diffusion to the emitting surface of ionizer 1 depends on the structure of the porous ionizer (pore size) and, at the given temperature $T_3$, can be controlled within a fairly broad range by varying the cesium vapor pressure in the chamber. The porous ionizer is made of compacted metal powder plates with uniform grain size. The plates are reinforced with wire gauze for mechanical strength. The ionizer can also be made of thin wires or plates with diffusion channels perpendicular to the ionizer surface. Such ionizers provide current densities of up to 0.1 A/cm$^2$ with practically an unlimited lifetime.

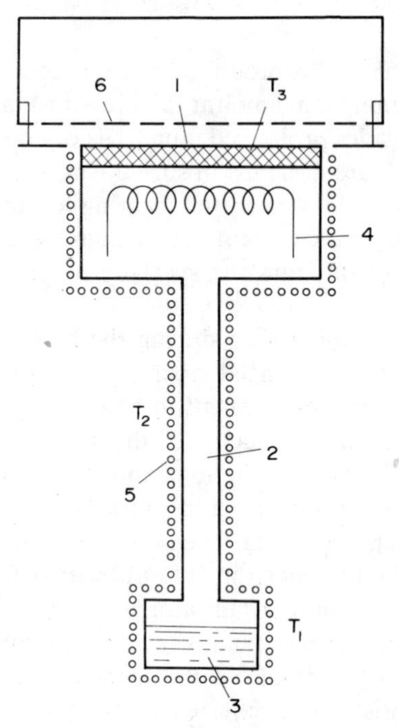


FIG. 40. The layout of an ion source with a porous emitter.

The theoretical analysis of the operation of a porous ion source is very complicated[126–9] and is carried out for the idealized case of straight pores perpendicular to the ionizer surface. As an example, let us briefly consider the ionization of the atomic cesium flux passing through a narrow rectangular slit with a homogeneous wall surface (Fig. 41). The slit is of height $l$, length $b$ and width $a$, $a \ll l \ll b$. The cesium vapor pressure at the entrance to the slit at $x = l$ is $p$, which is the pressure in the chamber (Fig. 40), and at the exit at $x = 0$ it is zero (vacuum). We have in the slit an outgoing Knudsen flux of atoms $\nu_V$, and a surface diffusion flux by the walls $\nu_S$. Since $a \ll l$, we have a strongly directional flux $\nu_V$ at the slit exit; for $l/a = 75$ the halfwidth of the flux distribution is 1.5°, as compared to 120° for a cosine distribution of flux density[130].

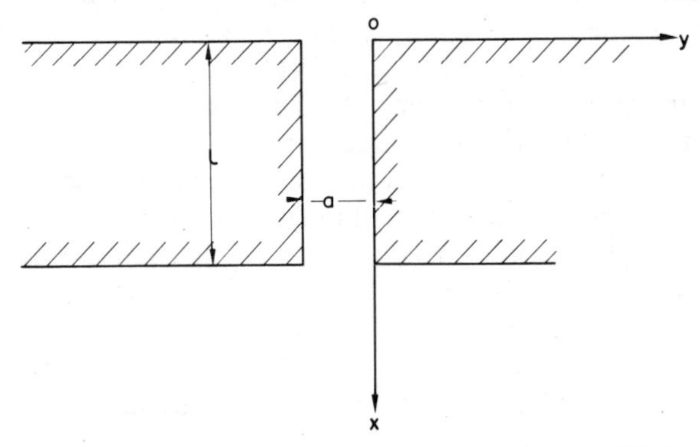


FIG. 41. The diagram of the rectangular slit for the analysis of the diffusion flux of cesium atoms.

The external accelerating field $\varepsilon$ penetrates inside the pore by $x \simeq a$. For $x \gg a$, in the pore, we should have a quasi-equilibrium thermal plasma with a low degree of ionization $\alpha$. The directed flux $\nu_V$, emerging from the slit, is probably weakly ionized. The surface diffusion flux $\nu_S$ desorbs within $a \geqslant x \geqslant 0$ and $0 \leqslant y < \infty$, in both the atomic and ionic states, the degree of ionization of the desorbed particles $\alpha$ being determined by (3.14) for the surface ionization. The ionization coefficient of the flux emerging from the slit and from its surface is

$$\beta_\Sigma = \frac{\nu_S^+}{\nu_S^+ + \nu_S^0 + \nu_V} = \frac{\beta}{1 + \nu_V/\nu_S} \leqslant \beta \qquad (12.1)$$

where $\beta$ is, as earlier, the surface ionization coefficient. From (12.1) it follows that, in order to improve the ionization efficiency in the pore, one should insure conditions for which $\nu_v \ll \nu_s$. This may be achieved by reducing the slit width $a$. Thus, we come to the conclusion that the principal mechanism of ion production may only be the desorption of ions from the slit surface in the region $x < a$ and $y > 0$. The surface area for $x < a$ receives, besides the diffusion flux, a flux of particles from the volume of the pore, whereas, within the region $y > 0$, only those particles are desorbed that arrive by way of surface diffusion.

Consider desorption from the end face of the pore into the region $y > 0$. At a given surface temperature $T$, the rates of desorption of ions $\nu_+$ and atoms $\nu_0$ are determined by (4.2) and depend on the coverage $\theta$ of the surface by adatoms at each point $y$ on the surface. The magnitude of $\theta$ depends on $y$, and is maximum at $y = 0$, decreasing with increasing $y$. Moving over the surface from higher values of $y$, it is seen, from the consideration of the desorption isotherm in Fig. 10, that the rate of desorption $f(\theta) = \nu_0 + \nu_+$ reaches a maximum at $\theta_1(y_1)$ and a minimum at $\theta_2(y_2)$.

To evaluate the region of effective ionization, one should know the function $\theta(y)$. In the steady state, for each element of the pore end face area, $y$ to $(y + dy)$ wide and 1 cm long, the atomic diffusion flux is $N_0 D_s\,(d^2\theta/dy^2)$, $D_s$ being the diffusion coefficient. For steady state, this flux is equal to the desorbed flux $f(\theta)$, i.e.

$$D_s \frac{d^2\theta}{dy^2} - C\theta\left[\exp\left(-\frac{l_+}{kT}\right) + 2\exp\left(-\frac{l_0}{kT}\right)\right] = 0. \qquad (12.2)$$

To obtain a numerical solution of this equation, and determine the function $\theta(y)$, we assumed the following values for the parameters in (12.2), which, for $\theta \leqslant 0.1$, approximately correspond to the cesium–tungsten pair; $D_s = D_0 \exp(-l_M/kT)$, $D_0 = 0.3\ \text{cm}^2/\text{sec}$, $l_M = 0.7$ eV (we took $D_s$ to be dependent on $\theta$): $C = 2.5 \times 10^{13}\ \text{sec}^{-1}$, $l_0 = (2.8 - 1.8\theta)$ eV and $l_+ = (2.08 + 7\theta)$ eV. Figure 42 presents the graphs of $\theta(y)$ for the two values of surface temperature, $T = 1300$°K (curve 1) and $T = 1600$°K (curve 2). Also presented there, for the corresponding values of $\theta(y)$, are the graphs of the degree of surface ionization $\alpha(y)$, calculated from the formula $\alpha = \frac{1}{2}\exp[(l_0 - l_+)/kT]$ for $T = 1300$°K (curve 3) and $T = 1600°K$ (curve 4). It follows from these graphs that, if, at the slit boundary at $y = 0$, the coverage $\theta = 0.1$, then it will decrease to $\theta = 10^{-6}$ at a distance $\sim 5.2 \times 10^{-4}$ cm at $T = 1300°K$ and $\sim 1.8 \times 10^{-4}$ cm at $T = 1600°K$.

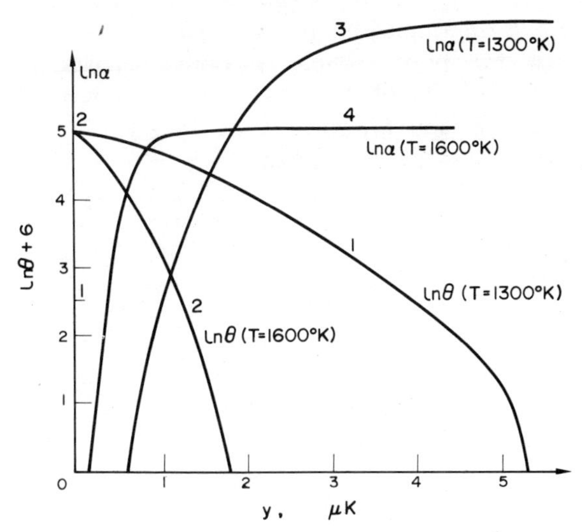


FIG. 42. The functions $\theta(y)$ and $\alpha(y)$ in the surface diffusion of cesium atoms along the edge of the slit.

Figure 43 presents the distribution of the ionic current density $j_+(y)$ in mA/cm$^2$ emitted in the region $y > 0$ for $T = 1300°K$ (curve 1) and $T = 1600°K$ (curve 2) for the calculated distribution $\theta(y)$ (Fig. 42). In the calculations, we assumed that $CN_0 = 8 \times 10^{26}$ cm$^{-2}$ sec$^{-1}$. It follows from these graphs that the region of effective ionization ($\alpha \gg 1$) represents a strip parallel to the slit and a few microns wide. The strip width decreases with increasing temperature $T$. The maximum coverage in this region is 0.01–0.02. The average ionic current density at $T = 1600°K$ may reach a few hundreds of milliamps from 1 cm$^2$. Thus, we come to the conclusion that, if the ion source represents a set of parallel slits, then the optimum separation between the slits depends on $T$, and is approximately equal to the double width of the region of effective ionization. At such separations between the slits, one should obtain the maximum ion-current density in the ion source.

In order that the volume diffusion flux $\nu_V$ be much smaller than the surface diffusion flux $\nu_S$ (12.1), the slit width $a$ should be taken as small as possible. If $\nu_V \ll \nu_S$, then the magnitude of the flux $\nu_S$, at the given temperature $T$, is determined by the equilibrium cesium, coverage $\theta_{eq}$ of the back side of the porous source. The optimum coverage $\theta = 0.01$–$0.02$ on the emitting side of the source can be reached by properly adjusting the cesium vapor pressure in chamber 3 (Fig. 40). The smaller the slit height $l$, the lower is this pressure.

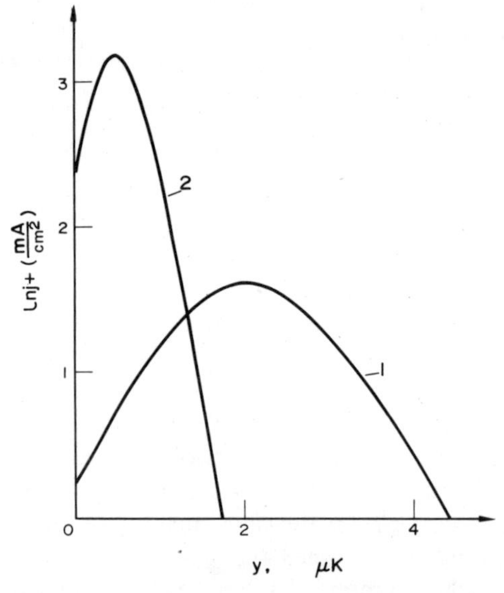


FIG. 43. The functions $j_+(y)$ of the desorbed current density of $Cs^+$ ions.

If we gradually increase the cesium vapor pressure $p$ in the chamber (increase $\theta_{eq}$), at a given temperature of the porous emitter, then the region of optimum ion emission in desorption should shift to the exit side of the slits. The ion emission will first appear at the slit edges (from the depth to which the field $\varepsilon$ penetrates) and, later on, shift to the bridge between the slits. At a further increase of $p$, the coverage of the emitting surface will exceed the optimum value, and the ionic current will start to decrease. If we now decrease $p$, the ion-emission pattern will change to the reverse sequence.

Now if, at a given cesium vapor pressure $p$, we gradually decrease the emitter temperature $T$ from a high value (at which $\theta = 0$), then effective ionization will first appear at the slit edges, shifting later on to the bridges, after which the ionic current will start to decrease, because of increasing $\theta$ (decreasing rate of desorption $f(\theta)$) and the resulting decrease of the degree of ionization $\alpha$. Thus, a temperature ionization threshold should also exist, in the case of porous emitters. Experiments support the above qualitative picture of ion emission[131].

As already mentioned, porous ion sources are usually fabricated by compacting metal (tungsten) powder. In this case, there are no diffusion channels of regular shape. Even if such through channels did exist, they would be twisted and of variable cross-section. The size of the bridges between the ends of such channels on the emitting surface can vary, the conditions of optimum emission for different pores being different. The pattern of emission will be more complicated than in the above case of ideal pores. However, here again, in order to obtain optimum ion current, the optimum size of the grains should depend on $T$ and be of the same order of magnitude as the separation between the adjacent slits. In practice, the optimum regime of a porous ion source is found by properly adjusting the magnitudes of $p$ and $T$ at which the ion current collected will be maximum.

## 13. Surface Ionization and Some Aspects of Electronic Engineering

### (i) *Prebreakdown conduction in interelectrode gaps in a vacuum*

This is a very important and complicated problem in electronic engineering, and has been considered in a number of monographs and reviews, for instance, ref. 132. As is well known, if the potential difference $U$ between two electrodes in a vacuum is increased, an electric breakdown will occur in the gap at some value $U_d$, i.e. a spark or arc discharge will form between the electrodes. The magnitude of $U_d$ depends on the material of the electrodes, their treatment and geometrical shape. At $U \ll U_d$, an electric current will start to flow in the gap (the prebreakdown conduction). its strength increasing with increasing $U$. This conduction produces an undesirable load on voltage generators and limits the operating voltage across the electrodes of high-tension equipment.

At residual gas pressures between the electrodes of below $10^{-5}$ torr, the contribution from the volume processes, involving the formation of charged particles in the volume, is negligible. Hence, the principal role in prebreakdown conduction is played by the emission of charged particles from the electrodes which is caused by the electric field $\varepsilon$ at the electrode surface. In the ideal case of an absolute vacuum and absolutely pure metal electrodes, the only process by which charged particles can be emitted, at low electrode temperatures, is field emission from the cathode. Emission of positive ions from the anode material can take place at fields of two orders of magnitude higher.

In real systems, there are particles adsorbed on the surface of the electrodes, whose material can contain impurities, including easy-ionizing alkali metals. The adsorbate can also contain organic substances, which have been adsorbed from the gas, or come from the electrode polishing with various pastes and cleaning with organic solvents. Desorption of positive ions of many organic molecules (see Fig. 35) and alkali metals (Fig. 22) can occur from the electrodes, at room temperature, for fields $\varepsilon = 10^5$–$10^6$ V/cm. The field emission of positive ions can start at even a lower $U$ than that of electrons. Figure 44 presents a current–voltage characteristic of the prebreakdown current between plane electrodes separated by 4 mm in a vacuum. At low $U$, the interelectrode current has the shape of separate pulses (microdischarges) and, at $U \approx 130$ kV, transforms into a continuous prebreakdown current, which increases steeply with increasing $U$ (curve 1). If we suppress the electron component of the prebreakdown current, by applying a strong magnetic field (curve 2), then the interelectrode current due to ion emission will, at first, be comparable to that of electrons, the current retaining the microdischarge nature. A suggestion has been made[134] that the ionic component of the microdischarge current is caused by the desorption of the ions of easy-ionizing particles from the surface by the electric field. Let us consider this problem in more detail.

Since the electrode surface is inhomogeneous with respect to the quantities $\varphi$ and $l_+$, and is rough on the microscopic scale, the emission of ions with increasing $U$ initially occurs at the points with minimum $l_+$ and maximum field $\varepsilon$ (microprojections). According to (3.1), if $l_+$ decreases with decreasing concentration of adsorbed atoms [electropositive adsorbate, see (4.1)], then the rate of ionic desorption $\nu_+$ will at first increase sharply, the desorption proceeding essentially as an explosion of the adsorbed layer at the given active-surface area.

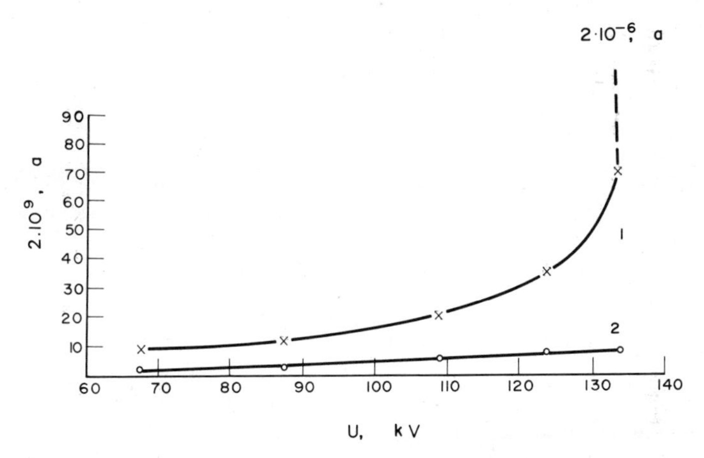


FIG. 44. The $V$–$I$ characteristics of the prebreakdown current $i$ in a vacuum between plane electrodes at $d = 4$ mm[133].

The initial pulse of ion emission will become amplified by the following secondary processes at the electrodes: (a) Positive ions from the anode which strike the cathode surface with energy $eU$ will produce secondary-electron emission and possibly emission of negative ions. Another process to be considered here is the photoeffect from the cathode caused by bremsstrahlung X-rays, which are produced by secondary electrons impinging on the anode. Let us denote the total coefficient of secondary emission of electrons and negative ions on the cathode by $\gamma$.

Then, if the initial pulse contained $n_+$ positive ions, $\gamma n_+$ secondary electrons and negative ions will be formed at the cathode. (b) $\gamma n_+$ electrons and negative ions striking the anode surface will generate tertiary charged particles, through electron–ion and ion–ion emissions and the conversion of negative ions into positive ones. If the total coefficient of "tertiary" emission at the anode is $\sigma$, then the pulse of tertiary particles from the anode to the cathode will be $\sigma\gamma n_+$.

Repeated interaction of the charged particles with the electrodes will result in an amplification of the initial pulse of positive ions into a microdischarge pulse of the interelectrode current. The number of charged particles $N_1$ transferred in such a pulse when $\sigma\gamma < 1$ is

$$N_1 = n_+(1 + \gamma + \gamma\sigma + \gamma^2\sigma + \gamma^2\sigma^2 + \ldots) = n_+\left(\frac{1 + \gamma}{1 - \gamma\sigma}\right). \qquad (13.1)$$

In a similar way, the pulse due to $n_-$ negative ions, desorbed by the field $\varepsilon$ from the cathode, will produce a flux of $N_2$ charged particles between the electrodes:

$$N_2 = n_-\left(\frac{1 + \sigma}{1 - \gamma\sigma}\right). \qquad (13.2)$$

According to published data, for $U$ of a few kilovolts, $\sigma$ is about an order of magnitude smaller than $\gamma$. Hence, the principal contribution to conduction, at the microdischarge stage, is due to pulsed desorption of positive ions from the anode surface.

It should be noted that current pulses can heat the microspots on the electrode surface where they impinge. According to (3.1), this should affect both the rate of desorption $v_+$ and the rate of impurity diffusion to the surface. On the other hand, the adsorption of particles from the gas phase should favor regeneration of desorption-active spots on the surface.

At high densities, the main mechanism of the interelectrode current, namely, the breakdown and the conduction immediately preceding it, is, of course, electron field emission from the active spots (i.e. those with low local values of $\phi$ and locally high $\varepsilon$) on the cathode surface. However, at the microdischarge stage, the ion-emission current can favor, as a result of ionic etching of the surface, the formation of active centers of electron field emission on the cathode.

### (*ii*) *Halogen leak detectors*

The effect underlying the operation of these widely used instruments is the sharp increase of electrical conductivity between a heated platinum anode and the surrounding cylindrical cathode when a tiny amount of a volatile halogen-containing substance is added to atmospheric air. For example, to locate leaks in sealed refrigerator systems, with freon as the working medium, air from near the suspected parts of the system is blown through the gap between the electrodes. The current in the diode circuit will increase sharply if the air contains even a tiny amount of escaping freon. The sensitivity of leak detectors is sufficiently high to permit the location of leaks with rates as low as 0·5 g freon per year.

A study of the thermionic emission of platinum in vacuum[135–9] has shown that the ions emitted are predominantly $Na^+$ and $K^+$. Here, halogen-containing gases are conducive to the emergence of impurity atoms to the platinum surface, with their subsequent desorption in the form of ions[138]. The mechanism for the activation of thermionic emission remains unclear. Probably the work function of platinum is smaller in air than the ionization potentials of potassium and sodium and the ionization coefficient of the atoms is also small. In the presence of halogens, the work function and the coefficient increase, as does the thermionic current. It is also probable that this effect is caused by chemical etching of the platinum surface and the associated opening of the micropores (see § 10).

In the course of operation of a leak detector, platinum is depleted with impurity metals, with a resulting decrease of sensitivity. It has been reported[139] that this "aging" effect can be substantially suppressed, if the ceramic rod, on which the platinum filament is usually wound, is saturated with potassium hydroxide solution. This also indicates the predominant role of the emission of alkali-metal ions in leak-detector operation.

*(iii) Compensation of electron space charge*

Surface ionization is used for compensation of the electron space charge in vacuum thermionic heat-to-electricity converters, as well as in production of the so-called quiescent plasma. With the compensation conditions being similar in the cases, we will consider this problem as related to the converter.

A converter is basically a diode with plane parallel electrodes at a separation $d$. The cathode is heated to the temperature $T$, the anode being maintained at lower temperature. Figure 45 presents, schematically, a volt–ampere characteristic $j(U)$ of the current, not limited by

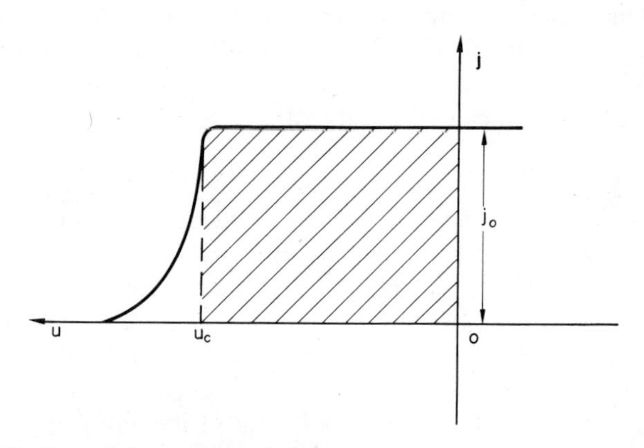


FIG. 45. The $V$–$I$ characteristic of the thermoelectronic current in a vacuum diode at total compensation of the space-charge field.

space charge, under conditions that $\varphi_c > \varphi_a$ and the thermoelectronic emission from the anode (inverse current) is small compared to that from the cathode. The maximum electric power collected into the diode output circuit from 1 cm$^2$ of cathode area is approximately $jU_c$ (which is the shaded area in Fig. 45), where $U_c = (\varphi_c - \varphi_a)$ is the contact-potential difference. The theory of thermoelectronic converters has been treated by several authors(140–4).

In our analysis of the conditions of compensation, we will make the following assumptions:

(a) electrode surfaces are homogeneous;

(b) positive ions form only by surface ionization on the cathode of the atoms directed, for this purpose, into the interelectrode gap;

(c) the mean free path $\lambda$ of electrons and ions is much larger than $d$, so that ionization in the volume may be neglected. It is assumed here that the electric field throughout the gap is zero and that the oscillatory mode of compensation does not become established (i.e. the single-flight compensation regime).

Under these conditions, the electrons and ions travel from the cathode to anode with thermal velocities. The condition for compensation reduces to the equality of the concentrations of the electrons $n_e$ and ions $n_+$ in each element of the gap volume. The ratio of fluxes, $\nu_e$ and $\nu_+$, or currents, $j_e$ and $j_+$, for compensation ($n_e = n_+$) is

$$\frac{j_+}{j_e} = \frac{\nu_+}{\nu_e} = \frac{n_+ v_+}{n_e v_e} = \sqrt{\frac{m}{M}}, \tag{13.3}$$

$M$ and $m$ being the ion and electron mass, respectively. The ion flux $\nu_+$ is produced by ionization, on the cathode, of the atom flux $\nu$ incident on 1 cm$^2$ of cathode area, so that $j_+ = e\nu_+ = e\nu\beta$. The flux $\nu$ is related to the vapor pressure in the volume by the formula $\nu = P/\sqrt{(2\pi MkT_g)}$ ($T_g$ is vapor temperature). Inserting $\nu$ into (13.3) we obtain that, in order to compensate the space charge, the vapor pressure $p_c$ should be

$$p_c = \frac{j_e}{e\beta}\sqrt{(2\pi mkT_g)} = 4.13 \times 10^{-6}\frac{j_e\sqrt{T_g}}{\beta} \tag{13.4}$$

where the numerical factor arises because $p_c$ is expressed in torr and $j_e$ in A/cm$^2$.

To obtain high conversion efficiencies, at the given cathode temperature $T$, the current density $j_e$ should be made as high as possible. For this purpose, the magnitude of $\varphi_c$ should be small. If the condition $e(V - \varphi_c) \gg kT$ is satisfied, $V$ being the ionization potential of the compensating gas atoms, then $\beta \simeq \alpha = A \exp [e/kT(\varphi_c - V)]$. Inserting this expression into the one for the thermoelectronic current density $j_e$, (6.2), the compensation condition (13.4) can be rewritten as:

$$\varphi_c = \tfrac{1}{2}V + \frac{T}{10080} (\lg B + 2 \lg T + \tfrac{1}{2} \lg T_g - \lg p_c - \lg A - 5.38), \quad (13.5)$$

thus relating the quantities $\varphi_c$ and $p_c$ corresponding to compensation at the chosen values of $V$ and $T$. It follows from (13.5) that, at the chosen values of $T$ and $p_c$, the smaller is $V$ the smaller is the magnitude of $\varphi_c$. Therefore, from the viewpoint of compensation of the electron space charge in the diode, one should fill it with cesium vapor, because its atoms have the lowest ionization potential $V$. The cathode temperature $T$ should also be as high as possible, while at the same time satisfying the requirement of a long service life of the converter (i.e. low evaporation of the cathode material). The magnitude of $p_c$ should satisfy, at a given $\varphi_c$ and $T$, the condition for single-flight compensation, for which $\lambda \gg d$. The smaller is $d$, the greater is the magnitude of $p_c$.

The electrodes of a cesium-vapor converter are always coated with a film of adsorbed cesium. The adsorption of cesium practically always results in a decrease of the electrode work function. Since the anode temperature, in an operating converter, is much lower than the cathode temperature, the coverage $\theta$ of the anode by cesium is larger than that of the cathode. Accordingly, with electrodes of the same material, the work function of the cathode during operation will be larger than that of the anode. In this case, the adsorption of cesium is accompanied by a decrease of $\varphi_c$ and an increase of the contact-potential difference $(\varphi_c - \varphi_a)$ between the cathode and anode· which results in an increase of the diode output power.

Of practical interest for the fabrication of the cathodes are high-melting metals (Mo, Ta, W, Re) and some rare-earth metal compounds with boron, nitrogen and carbon. Consider, as an illustration, the conditions for compensation at different $T$ in a diode with a tungsten cathode. The density of the thermoelectronic saturation current from the cathode is determined by (6.2), i.e.

$$\lg j_e = \lg B + 2 \lg T - \frac{5040}{T} \varphi_c(\theta). \quad (13.6)$$

Here, the ionic current density, multiplied by the compensation factor $\sqrt{(M/m)}$, will, according to (3.1), be

$$\lg\left(\sqrt{\frac{M}{m}}j_+\right) = \lg\left(e\sqrt{\frac{M}{m}}\right) + \lg(CN) - \frac{5040}{T}l_+(\theta). \qquad (13.7)$$

For compensation of the space charge the right-hand sides of (13.6) and (13.7) should, according to (13.3), be equal. The cathode coverage by cesium $\theta_c$, which meets the condition for compensation, can be determined graphically by plotting $j_e$ and $\sqrt{(M/m)}j_+$ as functions of $\theta$. To do this, one should know the dependence on $\theta$ of $\varphi(\theta)$, $l_+(\theta)$ and $CN$. They have been measured for the system cesium–tungsten[17], and extrapolated for $\theta < 0.5$, by means of the following formulae[145]:

$$e\varphi = 4.64 - e\Delta\varphi; \, l_+ = \frac{2.8}{1 + 0.714\theta} - 0.75 + e\Delta\varphi;$$

$$e\Delta\varphi = 10.679\theta - 22.982\theta^2 + 42.53\theta^3 - 34.91\theta^4; \qquad (13.8)$$

$$\lg(CN) = 26.7 + 2.13\,(\theta - \tfrac{1}{2}\theta^2) + \frac{0.4343}{1-\theta} + \lg\frac{\theta}{1-\theta}.$$

Figure 46 presents the graphs of (13.6) and (13.7), for three values of the cathode temperature T. Also shown is the plot of $\alpha(\theta) = A\exp\{e[\varphi(\theta) - V]/kT\}$. Calculations were performed with $B = 60$ A/cm² deg², $A = \frac{1}{2}$. The abscissae of the intercepts of the graphs $\lg j_e$ and $\lg(\sqrt{(M/m)}j_+)$, at different $T$, correspond to three values of $\theta_c$, which satisfy the condition for compensation.

Current densities, of practical importance for thermoelectronic converters, lie around $j_e \sim 10$ A/cm², which, according to the graphs of Fig. 46, corresponds to $\theta_c \simeq 0.33$, $T \simeq 1800$°K and $\alpha(\theta_c) = 10^{-4}$–$10^{-5}$. Taking for the cesium-vapor temperature $T_g = 1200$°K, which is about the mean value for the cathode and anode, we obtain from (13.4) the value for $p_c \simeq 15$–150 torr. Knowing $p_c$, one can easily evaluate the separation $d$ between the cathode and anode which corresponds to the single-flight compensation regime. For tungsten electrodes, the coverage of $\theta_c \simeq 0.33$ corresponds, according to (13.8), to $\varphi(\theta_c) \simeq 2.25$ V. Under the same conditions, the anode should apparently have close to a monolayer coating, and $\varphi_a = 1.3$–$1.5$ V. In this case, the contact-potential difference between the cathode and anode will be $U_c \simeq 0.8$–$1.0$ V.

It follows, from the above analysis, that the regime of compensation is determined by the thermoelectronic and adsorption characteristics of the cathode and cesium. Any factor bringing about a change of these characteristics also affects the performance of the converter in the

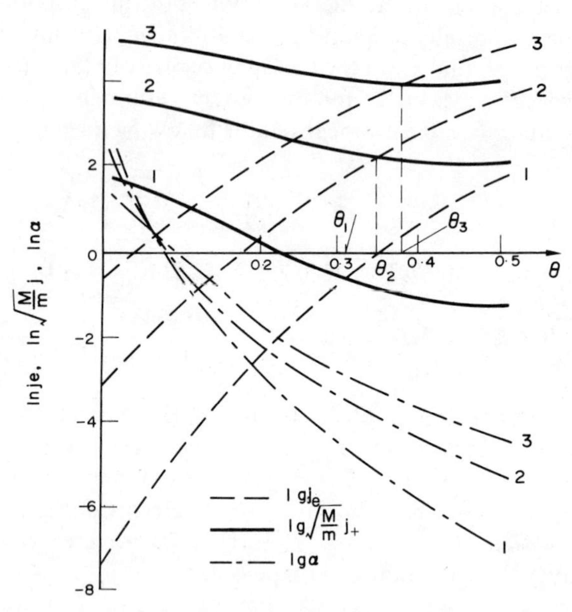


FIG. 46. The graphs of $\lg j_e$ (A/cm$^2$), $\lg\sqrt{(M/m)}j_+$ (A/cm$^2$) and $\lg \alpha$ as functions of $\theta$. 1. $T = 1500$°K; 2. $T = 2000$°K; 3. $T = 2500$°K.

compensation regime. Thus, for instance, the adsorption of electronegative gases (oxygen, fluorine) on the tungsten surface involves the formation of a W–O–Cs type surface layer, which has lower values of $\varphi(\theta)$ and $l_+(\theta)$ than the W–Cs layer. Adding oxygen or fluorine to cesium can result in an improvement of the converter performance.

If (13.3) is not satisfied, then an excess space charge of electrons (if $j_e > \sqrt{(M/m)}\, j_+$) or of ions (if $j_e < \sqrt{(M/m)} j_+$) will exist in the cathode–anode gap. The additional potential barrier of the space-charge field will reduce the excess ion or electron emission from the cathode down to values satisfying the compensation requirement.

In the case considered, the cesium vapor reduces the work functions of the cathode and anode, while at the same time changing the contact-potential difference between the electrodes. It also provides, by way of surface ionization, an ion flux from the cathode, which compensates the space charge. All these functions of the cesium vapor are interrelated. However, one can also affect the work functions of the cathode and anode independently, by bringing to their surface (from the volume or by diffusion) electropositive atoms with greater heats of desorption than that of cesium (for example, atoms of alkali-earth and rare-earth metals). Such an additional independent action may possibly produce an improvement in the performance of the thermoelectronic converters.

### (*iv*) *Generation of low-frequency noise*

Relatively slow variations in the cathode emissivity of electron tubes produce noise in the low-frequency ($< 10^3$ Hz) range (the flicker noise). In tubes with metal cathodes operating in the space-charge limited current regime, one observes low-frequency current "bumps" (the anomalous flicker-noise). In 1925, Johnson[(147)] explained this noise as being due to the heated cathode emitting pulses of positive ions amounting to $10^5$–$10^6$ ions per pulse. When trapped in the potential well, formed by the field of the electron space charge between the cathode and the space-charge grid, these ions oscillate, thus producing compensation of the space charge and a bump of electron current. In the single-flight regime, the degree of compensation is proportional to the number of positive ions in the pulse and to $\sqrt{(M/m)}$ (13.3), increasing proportionally with the number of ion oscillations in the potential well.

The origin of ionic pulsed emission, mainly of sodium and potassium, has already been discussed in § 10. Pulses of the ion emission of sodium and potassium have been observed experimentally, the number of ions per pulse being $10^4$–$10^6$ and the pulse duration below 1 $\mu$sec[(148–50)].

The anomalous flicker effect, in tubes with space charge, can also originate from the emission of ions from the grids, which can arise, for instance, as a result of surface ionization of organic molecules (see § 9)[76].


## Acknowledgements

This article is based, in part, on material prepared by Prof. E. Ya. Zandberg and the present author for a monograph[5]. The availability of this material greatly simplified the task of writing the review, for which the author wishes to express his sincere gratitude to Prof. E. Ya. Zandberg.

I am also indebted to Prof. S. G. Davison for his assistance in publishing the manuscript.


## REFERENCES


1. K. H. Kingdon and I. Langmuir, *Phys. Rev.* **21**, 380 (1923).
2. H. Ives, *Phys. Rev.* **21**, 385 (1923).
3. N. D. Morgulis, *Sov. Phys. JETP* **4**, 684 (1934).
4. P. P. Sutton and J. Mayer, *J. Chem. Phys.* **3**, 20 (1935).
5. E. Ya. Zandberg and N. I. Ionov, *Poverkhnostnaya Ionizatsiya* (*Surface Ionization*), Nauka, Moscow (1969).
6. K. H. Kingdon and I. Langmuir, *Proc. Roy. Soc.* A**107**, 61 (1925).
7. G. M. Pyatigorskii, *Sov. Phys. Tech. Phys.* **34**, 1144 (1964).
8. L. N. Dobretsov, *Elektronnaya i Ionnaya Emissiya* (*Electronic and Ionic Emission*), Gostekhizdat, Moscow, 1952.
9. L. N. Dobretsov, *Sov. Phys. Tech. Phys.* **35**, 535 (1965).
10. E. Ya. Zandberg, *Sov. Phys. Tech. Phys.* **33**, 743 (1963).
11. A. I. Ansel'm, *Sov. Phys. Doklady* **3**, 329 (1934); *Sov. Phys. JETP* **4**. 678 (1934).
12. R. W. Gurney, *Phys. Rev.* **47**, 479 (1935).
13. H. Grover, *Phys. Rev.* **52**, 982 (1937).
14. N. I. Ionov, *Sov. Phys. Tech. Phys.* **39**, 716 (1969).
15. V. I. Vedeneyev, L. V. Gurvich, V. N. Kondrat'yev, V. A. Medvedev and E. L. Frankevich, *Potentsialy Ionizatsii i Srodstvo k Elektronu* (*Ionization Potentials and Electron Affinity*), U.S.S.R. Academy of Sciences, Moscow, 1962.
16. H. Altertum, K. Krebs and R. Rompe, *Z. Phys.* **92**, 1 (1934).
17. J. B. Taylor and I. Langmuir, *Phys. Rev.* **44**, 423 (1933).
18. J. D. Levine and E. P. Gyftopoulos, *Surface Science* **1**, 171 (1964).
19. J. A. Becker, *Adv. Catalysis* **7**, 135 (1955).

20. E. F. CHAIKOVSKII and G. M. PYATIGORSKII, *Sov. Phys. Tech. Phys.* **34**, 1092 (1964).
21. E. F. CHAIKOSKII, G. M. PYATIGORSKII and G. V. PTITSYN, *Sov. Phys. Tech. Phys.* **35**, 1132 (1965).
22. J. D. LEVINE and E. P. GYFTOPOULOS, *Surface Science* **1**, 349 (1964).
23. N. I. IONOV, E. N. LEBEDEVA and M. A. MITTSEV, *Sov. Phys. Tech. Phys.* **39**, 1893 (1969).
24. N. I. IONOV, *Sov. Phys. Tech. Phys.* **36**, 2200 (1956).
25. E. YA. ZANDBERG, *Sov. Phys. Tech. Phys.* **27**, 2583 (1957).
26. N. I. IONOV, *Sov. Phys. Tech. Phys.* **39**, 721 (1969).
27. M. J. DRESSER and D. E. HUDSON, *Phys. Rev.* **137A**, 673 (1965).
28. E. YA. ZANDBERG, N. I. IONOV and A. YA. TONTEGODE, *Sov. Phys. Tech. Phys.* **35**, 1504 (1965).
29. N. I. IONOV and M. A. MITTSEV, *Sov. Phys. Tech. Phys.* **35**, 1863 (1965).
30. G. N. SHUPPE, E. P. SYTAYA and R. M. KADYROV, *Sov. Phys. Izvestiya Ser. Fiz.* **20**, 1142 (1966).
31. E. P. SYTAYA, M. I. SMORODINOVA and N. I. IMANGULOVA, *Sov. Phys. Solid State* **4**, 1016 (1962).
32. E. F. CHAIKOVSKII and G. V. PTITSYN, *Sov. Phys. Tech. Phys.* **35**, 1158 (1965).
33. F. L. REYNOLDS, *J. Chem. Phys.* **39**, 1107 (1963).
34. W. SCHROEN, *Z. Phys.* **176**, 237 (1963).
35. E. YA. ZANDBERG and N. I. IONOV, *Sov. Phys. Usp.* **67**, 581 (1959).
36. A. I. GUBANOV and N. D. POTEKHINA, *Sov. Phys. Tech. Phys.* **34**, 1888 (1964).
37. N. D. POTEKHINA, *Sov. Phys. Tech. Phys.* **35**, 317 (1965).
38. E. YA. ZANDBERG and A. YA. TONTEGODE, *Sov. Phys. Tech. Phys.* **35**, 325 (1965).
39. E. YA. ZANDBERG and N. I. IONOV, *Sov. Phys. Tech. Phys.* **28**, 2444 (1958).
40. E. YA. ZANDBERG, *Sov. Phys. Tech. Phys.* **30**, 1215 (1960).
41. VON HASSO MOESTA, *Z. Naturforsch.* **16**a, 288 (1961).
42. N. I. IONOV, E. N. LEBEDEVA and M. A. MITTSEV, *Sov. Phys. Tech. Phys.* **39**, 1905 (1969).
43. E. YA. ZANDBERG and V. I. PALEEV, *Sov. Phys. Tech. Phys.* **35**, 2092 (1965).
44. V. I. PALEEV, V. I. KARATAEV and E. YA. ZANDBERG, *Sov. Phys. Tech. Phys.* **36**, 1459 (1966).
45. G. J. MUELLER, *Phys. Rev.* **45**, 314 (1934).
46. N. I. IONOV, *Sov. Phys. JETP* **18**, 96 (1948).
47. H. SCHELTON, *Phys. Rev.* **107**, 1553 (1957).
48. E. YA. ZANDBERG, N. I. IONOV, V. I. PALEEV and A. YA. TONTEGODE, *Sov. Phys. Tech. Phys.* **32**, 503 (1962).
49. N. I. IONOV and V. I. KARATAEV, *Sov. Phys. Tech. Phys.* **32**, 626 (1962).
50. C. E. BERRY, *Phys. Rev.* **78**, 597 (1950).
51. P. ZAZULA and H. WILHEMSSON, *Arkiv Fysik.* **24**, 511 (1963).
52. J. B. HUDSON and J. S. SANDEJAS, *Surface Science* **15**, 27 (1969).
53. E. YA. ZANDBERG, A. YA. TONTEGODE and F. K. YUSIFOV, *Sov. Phys. Solid State* **12**, 1740 (1970).
54. K. J. MCCALLUM and J. E. MAYER, *J. Chem. Phys.* **11**, 56 (1943).
55. F. M. PAGE, *Trans. Farad. Soc.* **56**, 1742 (1960).
56. N. I. IONOV, *Sov. Phys. JETP* **17**, 272 (1947).
57. N. I. IONOV, *Sov. Phys. JETP* **18**, 174 (1948).
58. I. N. BAKULINA and N. I. IONOV, *Sov. Phys. Doklady* **155**, 309 (1964).

59. E. YA. ZANDBERG and V. I. PALEEV, *Sov. Phys. Doklady* **190**, 562 (1970).
60. E. F. CHAIKOVSKII, P. T. MEL'NIK and G. I. PYATIGORSKII, *Sov. Phys. Tech. Phys.* **40**, 225 (1970).
61. W. H. RODEBUSH and W. F. HENRY, *Phys. Rev.* **39**, 386 (1932).
62. M. J. COPLEY and T. E. PHIPPS, *J. Chem. Phys.* **3**, 594 (1935).
63. J. E. HENDRICKS, T. E. PHIPPS and M. J. COPLEY, *J. Chem. Phys.* **5**, 868 (1937).
64. E. YA. ZANDBERG, *Sov. Phys. Tech. Phys.* **30**, 206 (1960).
65. E. YA. ZANDBERG, *Sov. Phys. Tech. Phys.* **28**, 2434 (1958).
66. N. D. POTEKHINA, *Sov. Phys. Tech. Phys.* **35**, 1658 (1965).
67. N. D. POTEKHINA, *Sov. Phys. Tech. Phys.* **40**, 680 (1970).
68. E. YA. ZANDBERG and A. YA. TONTEGODE, *Sov. Phys. Tech. Phys.* **35**, 1115 (1965).
69. S. DATZ and E. H. TAYLOR, J. *Chem. Phys.* **25**, 395 (1956).
70. E. YA. ZANDBERG and A. YA. TONTEGODE, *Sov. Phys. Tech. Phys.* **37**, 2101 (1967).
71. E. YA. ZANDBERG and A. YA. TONTEGODE, *Sov. Phys. Tech. Phys.* **38**, 763 (1968).
72. E. YA. ZANDBERG and A. YA. TONTEGODE, *Sov. Phys. Tech. Phys.* **40**, 626 (1970).
73. N. I. IONOV and M. A. MITTSEV, *Sov. Phys. Doklady* **152**, 137 (1963).
74. M. A. MITTSEV, *Sov. Phys. Tech. Phys.* **35**, 2117 (1965).
75. G. H. PALMER, *Advances in Mass Spectrometry*, Pergamon Press, 1959.
76. E. YA. ZANDBERG and N. I. IONOV, *Sov. Phys. Doklady* **141**, 139 (1961).
77. I. N. BAKULINA, E. YA. ZANDBERG and N. I. IONOV, *Sov. Phys. Tech. Phys.* **35**, 562 (1965).
78. E. YA. ZANDBERG, U. KH. RASULEV and B. N. SHUSTROV, *Sov. Phys. Doklady* **172**, 885 (1967).
79. E. YA. ZANDBERG and U. KH. RASULEV, *Sov. Phys. Tech. Phys.* **38**, 1798 (1968).
80. E. YA. ZANDBERG and U. KH. RASULEV, *Sov. Phys. Tech. Phys.* **38**, 1793 (1968).
81. E. YA. ZANDBERG and U. KH. RASULEV, *Sov. Phys. Doklady* **178**, 327 (1968).
82. E. YA. ZANDBERG, U. KH. RASULEV and M. R. SHARAPUDINOV, *Sov. Phys. Doklady* **185**, 381 (1969).
83. E. YA. ZANDBERG and U. KH. RASULEV, *Sov. Phys. Doklady* **187**, 877 (1969).
84. E. YA. ZANDBERG, U. KH. RASULEV and M. R. SHARAPUDINOV, *Teor. i Eksep. Khimiya* **6**, 328 (1970).
85. O. W. RICHARDSON, *Proc. Roy. Soc.* **89**, 507 (1913).
86. J. CORNIDES, *Naturwiss.* **45**, 125 (1958).
87. A. P. KOMAR and V. P. SAVCHENKO, *Sov. Phys. Solid State* **4**, 1346 (1962).
88. H. F. WINTERS, D. R. DENISON, D. G. BILLS and E. E. DONALDSON, *J. Appl. Phys.* **34**, 1810 (1963).
89. M. D. SCHEER and J. FINE, *J. Chem. Phys.* **42**, 3645 (1965).
90. M. D. SCHEER and J. FINE, *J. Chem. Phys.* **46**, 3998 (1967).
91. I. N. BAKULINA and N. I. IONOV, *Sov. Phys. JETP* **36**, 1001 (1959).
92. N. I. IONOV and M. A. MITTSEV, *Sov. Phys. JETP* **38**, 1350 (1960).
93. N. I. IONOV and M. A. MITTSEV, *Sov. Phys. JETP* **40**, 741 (1961).
94. N. I. ALEKSEEV and D. L. KAMINSKII, *Sov. Phys. Tech. Phys.* **34**, 1521 (1964).
95. G. R. HERTEL, *J. Chem. Phys.* **47**, 335 (1967).
96. G. R. HERTEL, *J. Chem. Phys.* **48**, 2053 (1968).
97. L. M. BRANSCOMB and S. I. SMITH, *J. Chem. Phys.* **25**, 587 (1956).
98. R. S. BERRY and G. W. REIMANN, *J. Chem. Phys.* **38**, 1540 (1963).
99. P. M. DOTY and J. E. MAYER, *J. Chem. Phys.* **12**, 28 (1944).
100. J. J. MITCHELL and J. E. MAYER, *J. Chem. Phys.* **8**, 282 (1940).
101. M. METLAY and G. E. KIMBELL, *J. Chem. Phys.* **16**, 774 (1948).

102. A. L. Farragher and F. M. Page, *Trans. Farad. Soc.* **63**, 2369 (1967).
103. G. Clockler and M. Calvin, *J. Chem. Phys.* **3**, 771 (1935); *ibid.* **4**, 492 (1936).
104. N. I. Ionov, *Sov. Phys. Doklady* **28**, 512 (1940).
105. V. M. Dukel'skii and N. I. Ionov, *Sov. Phys. JETP* **10**. 1248 (1940).
106. I. N. Bakulina and N. I. Ionov, *Sov. Phys. Doklady* **105**, 680 (1955).
107. T. L. Bailey, *J. Chem. Phys.* **28**, 792 (1958).
108. I. N. Bakulina and N. I. Ionov, *Sov. Phys. Doklady* **116**, 41 (1957).
109. I. N. Bakulina and N. I. Ionov, *Sov. Phys. Doklady* **99**, 1023 (1954).
110. I. N. Bakulina and N. I. Ionov, *Zh. Fiz. Khim.* **38**, 2063 (1959); *ibid.* **39**, 157 (1965).
111. E. Ya. Zandberg and V. I. Paleev, *Sov. Phys. Doklady* **190**, 562 (1970).
112. V. I. Khvostenko and V. M. Dukel'skii, *Sov. Phys. JETP* **37**, 651 (1958).
113. B. H. Zimm and J. E. Mayer, *J. Chem. Phys.* **12**, 362 (1944).
114. E. Ya. Zandberg, V. I. Paleev and A. Ya. Tontegode, *Sov. Phys. Tech. Phys.* **32**, 208 (1962).
115. E. F. Chaikovskii and G. V. Ptitsyn, *Sov. Phys. Tech. Phys.* **35**, 528 (1965).
116. E. F. Chaikovskii, G. M. Pyatigorskii and G. V. Ptitsyn, *Sov. Phys. Tech. Phys.* **35**, 1493 (1965).
117. N. I. Ionov, E. N. Lebedeva and Ts. S. Marinova, *Sov. Phys. Tech. Phys.* **39**, 1323 (1969).
118. B. Ya. Kolesnikov, A. M. Kolchin and G. M. Panchenkov, *Sov. Phys. Tech. Phys.* **40**, 868 (1970).
119. K. F. Zmbov, *Adv. Mass Spect.* **3**, 765 (1968).
120. W. Weiershausen, *Ann. Physik.* **7**, F. 15, 150 (1965).
121. J. W. Frazer, R. P. Burns and G. W. Barton, *Rev. Sci. Instr.* **30**, 370 (1959).
122. E. F. Greene, *Rev. Sci. Instr.* **32**, 860 (1961).
123. J. H. Beynon, *Mass Spectrometry and Its Applications to Organic Chemistry*, Elsevier, Amsterdam, 1960.
124. N. I. Ionov and V. I. Karataev, *Zaw. Labor.* **5**, 621 (1957).
125. M. M. Bredov, *Sov. Phys. Tech. Phys.* **20**, 476 (1950).
126. A. T. Forrester and R. C. Speiser, *Astronautics* **4**, 34 (1959).
127. D. Zuccaro, R. C. Speiser and J. M. Teem, in *Electrostatic Propulsion*, Academic Press, New York, 1961.
128. G. M. Nazarian and H. Schelton, in *Electrostatic Propulsion*, Academic Press, New York, 1961.
129. T. M. Reynolds and L. W. Kreps, NASA Techn. Note D-871, Aug. 1961.
130. K. F. Smith, *Molecular Beams*, Methuen, London, 1955.
131. G. Cuskevies, *J. Appl. Phys.* **38**, 4076 (1968).
132. I. N. Slivkov, V. I. Mikhailov, N. I. Sidorov and A. I. Nastyukha, *Elektricheskii Proboi i Razryad v Vakuume* (*Electric Breakdown and Discharge in Vacuum*), Atomizdat, Moscow, 1966.
133. V. I. Gordeenko, Thesis, Phys. Techn. Inst. Acad. Sci. Ukraine; SSR, Kharkov, 1965.
134. N. I. Ionov, *Sov. Phys. Tech. Phys.* **30**, 561 (1960).
135. E. I. Agishev and Yu. I. Belyakov, *Sov. Phys. Tech. Phys.* **29**, 1480 (1959).
136. Ya. M. Fogel', L. P. Rekova and V. Ya. Kolot, *Sov. Phys. Tech. Phys.* **32**, 1259 (1962).
137. L. P. Rekova, S. S. Strel'chenko, Ya. M. Fogel, and Hoang-Sin-Shen, *Rad. i. Elektronika* **9**, 144 (1964).

138. L. P. REKOVA, YA. M. FOGEL' and A. P. ALEKSANDROV, *Sov. Phys. Tech. Phys.* **35**, 1642 (1965).
139. V. I. KARPOV, L. E. LEVINA and L. D. MURAV'YOVA, *Sov. Phys. Tech. Phys.* **35**, 1662 (1965).
140. A. I. ANSELM, *Thermoelektronny Vakuumny Termoelement* ("*Thermoelectronic Vacuum Cell*), Academy of Sciences of U.S.S.R., Moscow, 1951.
141. L. N. DOBRETSOV, *Sov. Phys. Tech. Phys.* **30**, 365 (1960).
142. N. I. IONOV, *Sov. Phys. Tech. Phys.* **30**, 1210 (1960).
143. N. D. MORGULIS, *Termoelektronny* (*Plazmenny*) *Preobrazovatel Energii* (*Thermoelectronic Plasma Energy Converter*), Atomizdat, Moscow, 1961.
144. N. RAZOR, *Adv. Energy Conversion* **2**. 545 (1962).
145. A. T. FORRESTER, *J. Chem. Phys.* **42**, 972 (1965).
146. M. D. SCEHEER and J. FINE, *J. Chem. Phys.* **50**, 4343 (1969).
147. J. B. JOHNSON, *Phys. Rev.* **26**, 71 (1925).
148. W. W. LINDEMANN and VAN DER ZIEL, *J. Appl. Phys.* **28**, 448 (1957).
149. R. E. MINTURN, S. DATZ and E. H. TAYLOR, *J. Appl. Phys.* **31**, 876 (1960).
150. S. DATZ, R. E. MINTURN and E. H. TAYLOR, *J. Appl. Phys.* **31**, 880 (1960).

# SHORT-TERM INTERACTIONS BETWEEN CELL SURFACES*

LEONARD WEISS and JAMES P. HARLOS

*Department of Experimental Pathology,*
*Roswell Park Memorial Institute, Buffalo, N.Y. 14203, U.S.A.*

## Contents



## 1. Introduction

Contact interactions between cells may be usefully classified as:

*Short-term interactions*, which include the approach of cells to other cells or non-living surfaces, and contact and/or adhesion between all or part of the proximating surfaces.

* Much of our own work incorporated in this paper was supported by a Helen Bucholtz Memorial Grant for Cancer Research from the American Cancer Society (P-403-C).

*Longer-term interactions*, which include changes in the metabolic and/or locomotory behavior of the contacting cells, changes in their local microenvironment through synthetic, leakage and/or autolytic processes, cell death, and separation of the contacted parts of cells.

Cell contact occurs when parts of two cell surfaces come close enough together to form mutual chemical or physical adhesive bonds. Thus, from the physical viewpoint, the short-term interactions, which may culminate in cell adhesion, involve two different, although related, problems; those to do with contact, and those relating to adhesion *per se*.

The longer-term interactions are, by definition, also dependent on cells having previously made contact with each other. Thus, the key to understanding many cellular interactions lies in addressing the central problem of cell contact. In this review, we have attempted to define some of the biophysical aspects of the problems relating to cell contact. It will be appreciated that many of the difficulties inherent in the physical analysis of cell contact are common to surface interactions in general, but are further complicated by the very nature of the biological systems studied. Detailed descriptions of a number of the biological systems, which constitute the motivation of this review, have been given elsewhere[(1)], and will not be reiterated in any detail here.

The interactions between parts of two approaching cells, leading to their mutual adhesion, may be initially classified in terms of the distances separating the interacting regions of their surfaces, since these distances will determine the theoretical approach. It should perhaps be emphasized that the cellular systems themselves will not "recognize" such arbitrary separations, and that there is a continuity of interaction. The main advantage of the classification is that, for descriptive purposes, we can concentrate on the leading terms.

The interactions will be considered as follows (with the approximate distances of separation indicated in parentheses);

1. Longer-range interactions ($>15$ Å).
2. Short-range interactions ($<15$ Å).
3. Very short-range interactions ($<5$ Å).

## 2. Longer-range Interactions

### (i) *DLVO theory*

As the surfaces of all cells from vertebrates, so far examined, carry a net negative charge, their interactions have been considered in terms of colloidal theory[1, 2] as generally described by Derjaguin and Landau[3] and Verwey and Overbeek[4] (DLVO theory). Essentially DLVO theory considers the total interaction energy between two particles, $V_T$, as the summation of electrostatic energies of repulsion, $V_R$, and energies of attraction, $V_A$, which are largely due to forces of the London–Van der Waals type, and partly due to electrostatic attraction.

DLVO theory has validity in biological systems, when the cellular particles are separated by distances which are large compared with the effective width of the ionic double-layer surrounding them. This is not unexpected, since the focal point in DLVO theory is the electrical double-layer. The surface of the central particles enter the theory only as constants for defining the boundary conditions. Many of the modifications of the Poisson–Boltzmann equation are concerned with corrections for ionic size, charge, hydration, dielectric parameters, etc., for constituents of the double-layer. In order to comment critically on the use and misuse of DLVO theory in analyzing cell-contact interactions, we shall briefly indicate some of the major considerations used in deriving it.

When a uniformly charged particle is immersed in an ion-containing dielectric medium, an electrical double-layer is formed around it. The electric potential field, $\psi$, surrounding such a particle is given by Gauss' law

$$\operatorname{div}(\varepsilon \operatorname{grad} \psi) \equiv \mathbf{\nabla} . (\varepsilon \mathbf{\nabla} \psi) = -4\pi\rho \tag{1}$$

where $\varepsilon$ is the dielectric factor, $\rho$ the space-charge density and

$$\mathbf{\nabla} \equiv \mathbf{i}\partial/\partial x + \mathbf{j}\partial/\partial y + \mathbf{k}\partial/\partial z.$$

Equation (1) may be expanded to

$$\varepsilon \mathbf{\nabla}^2 \psi + (\mathbf{\nabla}\varepsilon) . (\mathbf{\nabla}\psi) = -4\pi\rho. \tag{2}$$

For an ion-containing system, where the $i$th ion has charge $Z_i$ and concentration $N_i$, the space-charge density is defined as

$$\rho = e \sum_i Z_i N_i \tag{3}$$

where $e$ is the unit of electric charge. $Z$ is the algebraic value, i.e. magnitude and sign, of the ion charge. When only the magnitude of the charge is required, it will be given as $q_i = |Z_i|$. If only a single binary salt is present, where $q_1$ and $q_2$ are the cationic and anionic charges, the space-charge density is given as

$$\rho = e(Z_1N_1 + Z_2N_2) = e(q_1N_1 - q_2N_2). \tag{4}$$

Near to the surface of the central charged particle, $\rho$ will be non-zero, since ions of opposite charge to the surface will be attracted to it, and those of like charge repelled. At distances far from the particle surface, $\rho$ will be zero, since overall electroneutrality must be maintained. If the dielectric parameter $\varepsilon$ is assumed to have a constant value $\varepsilon_0$, independent of position, then eqn. (2) can be simplified to the Poisson equation.

$$\nabla^2\psi = -4\pi\rho/\varepsilon_0. \tag{5}$$

The value of $\varepsilon_0$, the dielectric constant, is that of the region of space in which the electric potential exists. Due to the asymmetric charge distributions present, this may not be numerically equal to that of the bulk phase solvent.

It is also convenient to assume that, at distances far from the charged particle, the ions have "bulk" phase concentration, $N_i^0$, and that the ion concentrations at other points are determined by the positional potential energy of the ion. It is further assumed that the relationship is given by the Maxwell–Boltzmann distribution

$$N_i = N_i^0 \exp(-V_i/kT) \tag{6}$$

where $k$ is Boltzmann's constant, $T$ the absolute temperature and $V_i$ the potential energy of the $i$th ion. In general, $V_i$ should reflect all ion interactions, including those of the central particle and the effects of these charged bodies on the medium. In the limit of low ionic concentrations, however, it is assumed that the effect of the central particle predominates, and that $V_i$ may be given as

$$V_i = Z_ie\psi. \tag{7}$$

It must be remembered that $\psi$ is an algebraic quantity having both magnitude and sign, e.g. $-20$ mV. In the case of a binary electrolyte, with the surface-charge of the central particle being negative, the space-charge density may be written as

$$\rho = N^0 eq\,[\exp(-qe\psi/kT) - \exp(qe\psi/kT)]. \tag{8}$$

Inserting this into eqn. (5) gives

$$\nabla^2\psi = \frac{8\pi N^0 eq}{\varepsilon_0} \sinh(qe\psi/kT). \tag{9}$$

Letting $y = qe\psi/kT$, we obtain

$$\nabla^2 y = \frac{8\pi N^0 e^2 q^2}{\varepsilon_0 kT} \sinh y \equiv \kappa^2 \sinh y. \tag{10a}$$

Since $\sinh\ y \simeq y$, for $y \ll 1$, we can also obtain the linearized form

$$\nabla^2 y = \kappa^2 y. \tag{10b}$$

$\kappa^{-1}$ has the units of length, and is commonly known as the Debye–Hückel length. Either eqn. (10a) or eqn. (10b) may be used (if $y \ll 1$), depending upon the boundary conditions; if eqn. (10a) cannot be integrated in closed analytical form, eqn. (10b) is usually used.

It is apparent that the application of DLVO theory to biological systems will be, to some degree, limited by the validity of the assumptions introduced in the formulation of eqn. (10). The two principal assumptions have been the use of the dielectric *constant*, $\varepsilon_0$, and the introduction of a point charge distribution, the Boltzmann distribution. Sparnaay[5] has examined the effects of the inclusion of dielectric variation and ionic size in the case of interacting flat plates; his results indicate that, at biological ionic concentrations ($\sim 10^{-2}$ N), the error in the computed uncorrected electric field potential is 10–20% too high at any given distance. The effects these corrections would have on particle interactions have yet to be determined.

In any event, it is apparent that, at salt concentrations appropriate for biological systems, the neglect of specific ion behavior may limit the validity of DLVO theory in biological interactions. Using a statistical mechanical approach, Buff and Stillinger[6] show that the potential energy of an ion is not only dependent on the electrostatic energy, but should also include corrections for the local activity coefficient for that ion and a correction for image forces. Therefore, eqn. (7) may be expanded to

$$V_i = Z_i e\psi - \frac{Z_i^2 e^2 \exp(-2\kappa r_i)}{4\varepsilon} - \frac{Z_i^2 e^2 \kappa}{2\varepsilon} \tag{11}$$

where $\kappa$ is now a distance-dependent Debye parameter and $r_i$ is the distance from the surface. The solution of such a problem is nontrivial and would involve using an iterative approach. However, the application of these corrections is an uncertain task since short-range forces are still neglected. The addition of these forces would increase the magnitude of the problem many-fold without assuring that their inclusion would critically improve the value of DLVO theory for biological interactions. We are therefore forced to assume that at long distances simple DLVO theory is asymptotically correct. At short distances its validity is more doubtful and, as will be shown, very dependent on the choice of parameters.

The repulsive interaction energy, $V_R$, may be given by the free energy change involved in bringing two double-layers from infinity to within a distance $d$ of each other:

$$V_R = \Delta G = G_d - G_{\text{inf}}. \tag{12}$$

The free energy of a double-layer system, $G$, is given by

$$G = -\tfrac{1}{2}\sigma\psi_0 \tag{13}$$

where $\psi_0$ is the surface potential and $\sigma$ is the surface charge. The surface charge is related to the electric potential by

$$\sigma = -(4\pi)^{-1}\varepsilon(\nabla\psi)_{x=0}. \tag{14}$$

One particular solution of eqn. (10), which will be needed, is the electric field about a spherical particle, obtained from eqn. (10b),

$$\psi = \psi_0(a/r)\exp[-\kappa(r-a)] \tag{15}$$

where $a$ is the radius of the particle and $r$ the distance from the center of the particle. Another solution is that for the electric field of two parallel plates separated by a distance $d$ obtained by Hogg, Healy and Furstenau[7], namely,

$$\psi = \psi_{01} \cosh(\kappa x) + \frac{\psi_{02} - \psi_{01}\cosh(\kappa d)}{\sinh(\kappa d)} \sinh(\kappa x). \tag{16}$$

This equation was used to obtain, by the method of Derjaguin, the interaction energy between two spherical particles:

$$V_R = \frac{\varepsilon}{4}\frac{a_1 a_2}{a_1 + a_2}\{(\psi_{01} + \psi_{02})^2 \ln[1 + \exp(-\kappa H)] + (\psi_{01} - \psi_{02})^2 \times \ln[1 - \exp(-\kappa H)]\} \tag{17}$$

where $\psi_{01}$ and $\psi_{02}$ are the surface potentials of particles 1 and 2, respectively, and $a_1$ and $a_2$ the corresponding radii of curvature. $H$ is the distance of closest approach of the particles and $\varepsilon$ the dielectric constant.

Since the constituent atoms of the surfaces are capable of interacting with each other by London–Van der Waals interactions, an attractive term is included in the total interaction energy. This problem has been examined by Hamaker[8] who showed that for the case where $\kappa a \gg 1$, the attraction energy between two isolated particles is

$$V_A = -\frac{A}{6}\frac{a_1 a_2}{a_1 + a_2}\frac{1}{H} \tag{18}$$

where $A$ is the Hamaker constant. The total interaction energy is then given by

$$V_T = V_R + V_A. \tag{19}$$

The question is often raised about the distance dependence in the attractive energy evaluation of DLVO theory calculations. These interactions are postulated to be of the Van der Waals–London type, which have an $r^{-6}$ dependence, while the formulae typically used, eqn. (18), has inverse dependence on distance. The question is, what happens to the $r^{-6}$-dependence? Equation (18) is an approximation, which holds only as long as the radius of curvature of the layer particle is very much greater than the distance of closest approach; thus the functional dependence of $V_A$ on $a_1$, $a_2$ and $H$ is not as simple as it appears in eqn. (18).

The London–Van der Waals interactions between two atoms having the same polarizability, and separated by a distance $r_{ij}$, are given by

$$V_A = -\lambda/r_{ij}^6 \tag{20a}$$

where $\lambda$ is the polarizability factor. In the case of a number of atoms, summations for all pairs of atoms are made giving

$$V_A = -\sum_i \sum_j \lambda/r_{ij}^6. \tag{20b}$$

If the density of atoms is large, and can be considered to be in two groups, the summation may be replaced by integration over volumes, i.e.

$$V_A = -\int_{V_1}\int_{V_2} \frac{\lambda \rho \, dV_1 dV_2}{r_{12}^6} \tag{20c}$$

where $r_{12}$ now represents the distance between the center of the infinitesimal volumes $dV_1$ and $dV_2$ and $\rho$ is the number of atoms/cm$^3$. This is the integral form of eqn. (18) and it is apparent here that the $r^{-6}$ dependence still exists. In order to perform the integration over volume, it is convenient to make the volumes parametrically dependent on the distance $r_{12}$. In this procedure, the radii of curvature and the distance between the centers of the spheres are introduced as parameters and, in the course of the integration, the $r_{12}$ parameter is absorbed in the volume evaluation. In effect, one is left with some volume function, which we shall leave unspecified, involving the $r_{12}^6$ divided by $r_{12}^6$, leaving the remainder a function of radii of curvature and distance of closest approach.

The above method of obtaining Van der Waals interaction energies and, therefore, values of Hamaker's constant, involves difficulties of both a practical and fundamental nature. The foremost practical difficulty is due to the large number of atomic interactions to be considered, and the parameters required for their evaluation. Of greater importance are those problems inherent in the theory. In brief, these are the neglect of more than two-body interactions[9] and the *ad hoc* method of determining the effect of the environmental media. Thus, the method of summing pair-wise interactions, while providing easy conceptual insight into attractive interactions, does not provide a fundamentally sound method of obtaining these energies.

One method for correcting the above assumptions, as well as including retardation effects in a rigorous fashion, has been proposed by Lifshitz[10, 11]. This approach treats the components of the system as continua. Since Van der Waals forces arise from the spontaneous oscillation of electrons around atom centres, they can be presumed to result from variations in the electromagnetic field generated by these oscillations. In the case where the region of strong interaction is greater than the distance between atom centers, Lifshitz, following Casimir and Polder[12], proposed that this time-varying electromagnetic field, extending over the entire system, could be frequency analyzed in terms of the theory. Recently, however, Ninham and Parsegian[13–15] have such complete data has imposed a limitation on the practical application of the theory. Recently, however, Ninham and Parsegian[13–15] have shown that the Lifshitz approach can be applied with reasonable success, even when complete spectral data are not available. In particular, they have treated the case of single thin films of water between hydrocarbons[14] as well as triple-layer films[15]. Their results are important for at least two reasons. The first is that they indicate a method of obtaining Van der Waals forces, using macroscopic parameters, which can be determined by either experiment or calculation. Since Van der Waals energies can be related to the Hamaker constant, a theoretical method of obtaining that parameter is not available, independently of the empirical or experimental approach. Furthermore, Ninham and Parsegian's calculations allow the relative importance of contributions of the various frequency components to be determined.

Although some aspects of longer-range interactions can be summed up by eqns. (17) and (18), it must be emphasized that a major problem lies in obtaining numerical values to use in these equations. We will next discuss some of these difficulties, and indicate their effects on computed interactions.

### *(ii) Surface potential*

This parameter may not be measured directly, and most of the currently used estimates are based on measurements of cell electrophoretic mobility. When a cell moves through an electrolyte solution, under the influence of an electric field, part of the diffuse electrical double-layer of ions moves with the cell, while the outer region remains with the bulk-phase of the environment. The interface between these two ionic regions is known as the hydrodynamic slip-plane, and the potential at this plane, with respect to an electrode an infinite distance away in the environment, is the zeta-potential, $\zeta$. Cell electrophoretic mobility, $V$, is related to $\zeta$ by the Helmholtz–Smoluchowski equation:

$$V = \frac{\zeta \varepsilon}{4\pi\eta} \tag{21}$$

where $\varepsilon$ and $\eta$ are the respective dielectric constant and viscosity in the region of the hydrodynamic slip-plane.

Although Haydon[16] has suggested that for oil/water systems, with surface potentials of less than 50 mV, surface potential is approximately equal to $\zeta$-potential, in the case of cells this assumption is doubtful. Equation (21) is strictly applicable to impenetrable particles of "easy" shape. The more penetrable a particle is by counter ions, the more inaccurate will be estimates of $\zeta$ obtained from measurements of electrophoretic mobility[17], leading, in the extreme case of penetrability, to a two-fold overestimate of $\zeta$. The effects of surface shape on the $\zeta$/mobility relationship currently presents intractable problems.

Another problem lies in determining the distance of the hydrodynamic slip-plane from the plane of surface potential. It can be appreciated that the location of the hydrodynamic slip-plane will depend partially on the viscous behaviour of the medium; it would appear unlikely that the distance is more than 10 Å. On the other hand, over distances of less than 10 Å from charged surfaces, in conventional considerations of viscosity, involving bulk-phase, the average parameters are unlikely to be realistic due to water structures and related problems. Hence, the exact location of the slip-plane is uncertain. As will be shown, this uncertainty may be a cause of very considerable error in computing values of surface potential from experimentally derived estimates of zeta-potential.

If we assume a linear Poisson–Boltzmann distribution of ions, and that eqn. (15) adequately represents the electric potential around a spherical particle, with $r = a + d$, where $d$ is the distance between the particle surface and the slip-plane and $a$ is the particle radius, then

$$\psi_0 = \zeta[1 + (d/a)] \exp(\kappa d). \tag{22}$$

Thus, for estimates obtained from measurements of electrophoretic mobility of $\zeta$-potential, the computed surface potential will vary according to the distance of the hydrodynamic slip-plane from the plane of surface potential. In practical terms, surface potentials may be considerably higher than calculated. In Fig. 1 we have computed the relationship between the actual surface potential and the distance of the slip-plane from the plane of surface potential ($d$), for different estimates of $\zeta$-potential. For example, a cell having a $\zeta$-potential of $-25$ mV would have an actual surface potential of approximately $-60$ mV, if $d = 9$ Å, but $\psi_0$ has a value of only approximately $-30$ mV, if $d = 3$ Å. Thus, use of $\zeta$-potential will lead to underestimates of potential energies of repulsion ($V_R$), as indicated in eqn. (18).

Another serious problem, stemming from the use of electrophoretic mobility data in computing interaction energies between cells, is that these give, at best, rather crude estimates of average surface charge density: they do not give any detailed microscopic information about the distribution of ionogenic groups at the cell surface. It has been suggested, on a variety of indirect experimental evidence[18], that charged groups may not be smeared out uniformly over the cell surface, but that they may well be arranged in zones of higher than average, or lower than average, density. A wide variety of observations on living cells by light microscopy, and on fixed specimens under the electron microscope, show that only small regions of opposing cell surfaces are involved in initial contact reactions. When living cells are examined, filopodial extensions from one cell, which are at or beyond the limits of optical resolution (radius $<0.5$ μ), are often seen exploring the more flattened region of another. Electron microscopic examination suggests that the diameters of the tips of these cellular projections may only be of the order of several hundreds of angstrom units. Thus, if heterogeneous distribution of cell surface charge occurs, with regions of different charge densities having comparable areas to those of the tips of the exploring filopods, the electrostatic repulsion encountered by the filopods will vary with the region of the other cell which it explores. If the differences in charge density between various regions of a cell surface are great, then the differences in repulsion energy will be great. In addition, the progressively greater these postulated regional differences in surface-charge density are, then the progressively less

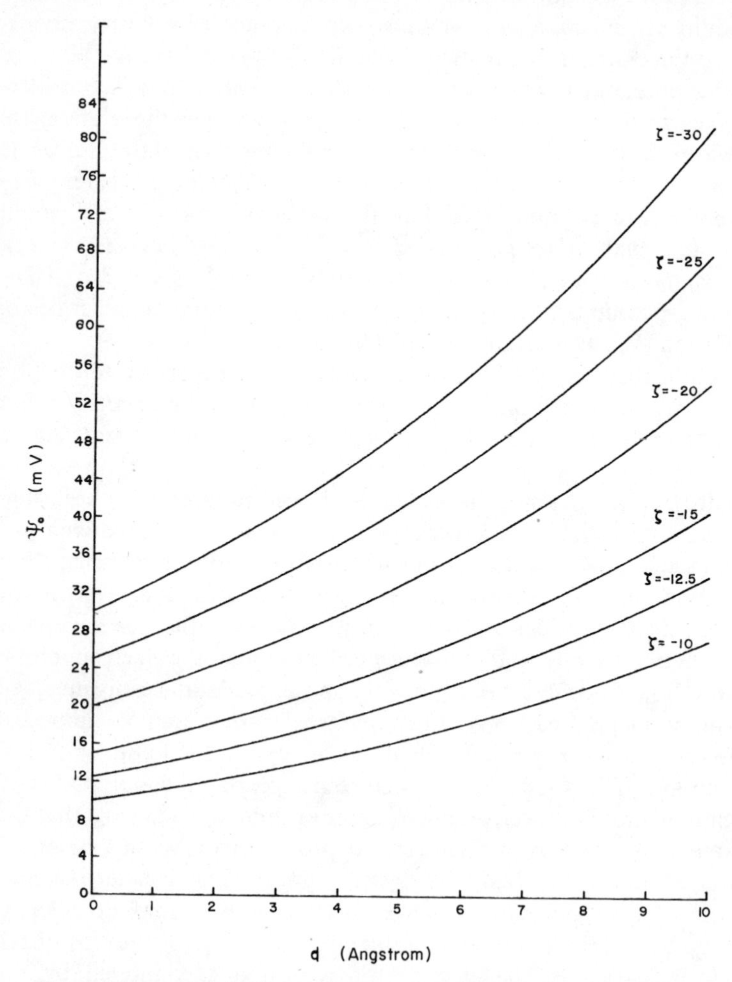


FIG. 1. The computed relationship eqn. (22) between surface potential, $\psi_0$, and zeta-potential, $\zeta$, for different values of $\zeta$, when the distance, $d$, between the hydrodynamic slip-plane and plane of surface potential varies over 0 to 10 Å.

accurate will be computations of interaction energies based on average macroscopic measurements of the electrical nature of the cell periphery.

As local variations in cell surface-charge density are possibly of great importance, their effects can be gauged by first rewriting eqn. (17) in terms of surface-charge density. Since $Q$, the total surface charge, can be related to the surface potential by

$$Q = a\varepsilon(1 + \kappa a)\psi_0 \tag{23a}$$

and the charge per cm$^2$ of a spherical particle is

$$q = Q/4\pi a^2, \tag{23b}$$

then

$$\psi_0 = 4\pi a q/\varepsilon(1 + \kappa a), \tag{24a}$$

giving, with eqn. (17),

$$V_R = \frac{4\pi^2}{\varepsilon}\left(\frac{a_1 a_2}{a_1 + a_2}\right)\left\{\left(\frac{a_1 q_1}{1 + \kappa a_1} + \frac{a_2 q_2}{1 + \kappa a_2}\right)^2 \ln[1 + \exp(-\kappa H)] + \left(\frac{a_1 q_1}{1 + \kappa a_1} - \frac{a_2 q_2}{1 + \kappa a_2}\right)^2 \ln[1 - \exp(-\kappa H)]\right\}. \tag{24b}$$

A simulated approach of a filamentous process of 500 Å radius, from one cell with a surface-charge density $1.2 \times 10^{13}$ e.s.u. per cm$^2$ ($\psi_0 = -25$ mV) toward a flattened part (radius 1 μ) of another cell is shown in Fig. 2. The surface-charge density of the regions approached by the filopod are varied over the range of 0 to $2.2 \times 10^{13}$ e.s.u. per cm$^2$ ($\equiv \psi_0 = 0$ to $-50$ mV) and, as expected, the resulting potential energy barriers to contact vary enormously.

Detailed information about the charge density of different regions of cells with respect to contact is obviously required before further progress can be made. It is possible that cells stained with electron dense, positively charged colloids may give this information when examined under the electron microscope[19–21].

Gingell[22] has examined a model which, in some features, approximates membrane interactions. He considered the effects of keeping surface charge, rather than surface potential, constant and allows the

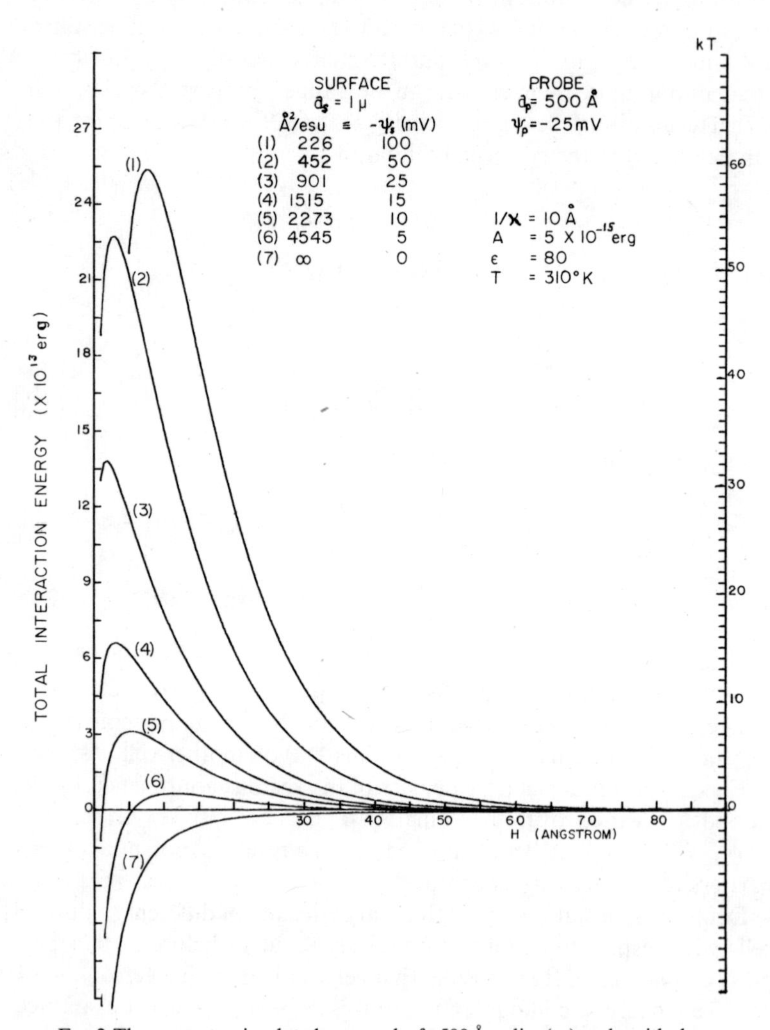


FIG. 2. The computer simulated approach of a 500 Å radius ($a_p$) probe with charge density of 1 e.s.u. per 900 Å$^2$ of surface toward comparatively flat ($a_s = 1$ μ) regions of another cell of the indicated varying surface-charge density. $V_T$ was computed from eqns. (18) and (24).

charge bearing surfaces to be permeable to counter ions. These conditions make surface potential a function of the distance between the charge bearing surfaces of the membranes, and introduce a parametric dependence on the permeability of the surface and the depth of the permeable layer. While at short distances, i.e. $<20$ Å, in physiological solutions, the computed increase in surface potential is impressive, other considerations, which will shortly be discussed, make the use of DLVO theory of questionable validity in this situation. At longer distances, the approach from infinity to 20 Å separation would cause, at most, a rise of $-5$ mV in surface potential. Thus, in the case of a surface potential "at infinity" of $-25$ mV, this rise would only amount to 20% and, in view of the other uncertainties about cell surface potential, is trivial. Hence, in dealing with cell surface potential in relation to contact phenomena, it would appear that average values for the whole cell, even if calculable, are of progressively less value as two cells get closer together.

(*iii*) *Radius of curvature*

Very many observations with the light microscope have revealed that contact between cells is made through the agency of fine peripheral extensions, in which a flattened region of one cell appears to be explored by a low radius of curvature "probe" extended from another. Recent detailed studies on mouse peritoneal macrophages by light microscopy, cinemicroscopy and scanning electron microscopy reveal that these cells settle and attach to glass surfaces initially by the protrusion of "very fine finger-like processes, followed by veils"[23].

It can be seen from eqn. (17), as pointed out originally by Bangham and Pethica[24], that contact between two cells is energetically facilitated by reduction of the radii of curvature of the relevant regions of the cell surface. This effect is shown in Fig. 3, where the computed interaction energies at 15 Å separation are plotted for a region of a cell having a radius of curvature of $10^4$ Å, which is being approached by probes of varying radii of curvature from another cell. It can be seen that, in the case where all of the approaching surfaces have potentials of $-25$ mV, only probes with radii of less than approximately 20 Å can make contact under the influence of Brownian motion (i.e. 1.5 kT).

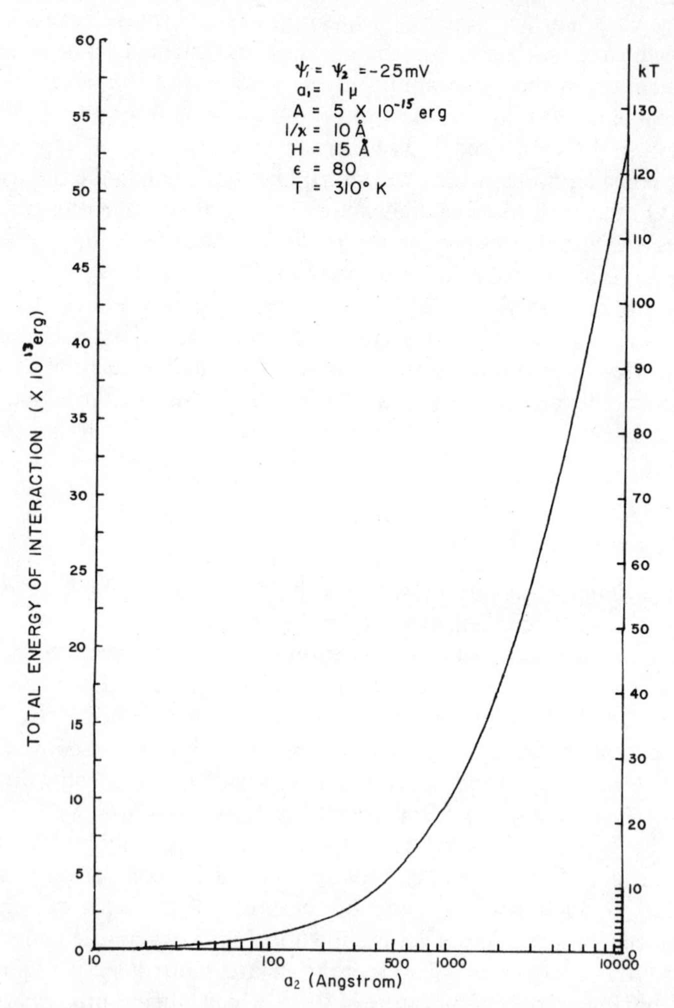


FIG. 3. The computed interaction energies eqns. (17) and (18) between a flat $a_1 = 1\ \mu$ region of one cell and probes of varying radius ($a_2$) from another. Both cell surfaces carry a potential of $-25$ mV.

Such projections are too fine to be bounded by cell membrane, which is probably approximately 100 Å thick and, hence, could be classified as "hairs".

The realization that cells may actively push out extensions from themselves into their environment introduces yet another complication, in that we have virtually no experimental data on the protrusion or locomotor pressures exerted by cells. In the case of the amoeba *Chaos chaos*, an attempt was made to relate its locomotor pressure of 1 to $2 \times 10^3$ dynes/$cm^2$ to computed interaction energies between the contacting surfaces of two amoebae[25]. The results suggested that considerations of locomotor pressures might well be relevant to the problem.

The possible role of an actomyosin-like contractile protein in the cell periphery in adhesive processes has been discussed by Jones[26], who suggests that it may be involved in "intermolecular bonding between adjacent cells"[27]. The inhibition of cell aggregation by antibodies directed against actomyosin has been taken as evidence in support of this hypothesis[28]. However, antibody-blocking leads to notoriously equivocal interpretations in this context, as the antibody molecules may act by masking other sites located close to the antigen, against which they are directed. Alternatively, inhibition of aggregation in these systems may be irrelevant to normal cell aggregation mechanisms.

Thus, cellular probes can presumably fall into two broad classes: those which are essentially tubular, and which may be extended toward another cell by virtue of internal pressure, and those finer processes which are hairlike, and which are subject to Brownian agitation. As discussed earlier, the relevance of either type of propulsive force to contact is very dependent on probe radius of curvature. Current estimates of this parameter are dependent on electron micrographs made from thin sections, or by replication or scanning techniques. Some idea of the complexity of cell surface shape may be obtained from the electron micrographs in Figs. 4 to 8. It must be remembered that the cell periphery is rich in mucopolysaccharides having high specific hydrodynamic volumes, and that in many cells a varying but generally high proportion of their surface charge is carried by carbohydrates—the sialic acid moieties, which determine the location of the plane of surface potential. Fixation and dehydration, preparatory to electron microscopy, is expected to cause a collapse of this ionogenic material, and thus lead to underestimates of the diameters of cell extensions. In the case of the HeLa cells[29], shown in Fig. 4, a minimal value of 500 Å for the radius of curvature appears to be compatible with data for other cells, and possibly indicates a useful general value in computing interactions between cells separated by not less than 15 Å. In the cases covered by Fig. 3, such a 500 Å probe would require to be protruded with a pressure exceeding $15 \times 10^4$ dynes/cm$^2$.

### (*iv*) *Dielectric constant*

At present, the theoretical treatment of the dielectric properties of cellular materials in a physiological environment is rudimentary. In addition, owing to technical difficulties, very few reliable and direct measurements have been made on these materials.

In cellular interactions of the type discussed here, it is highly probable that the dielectric constants of the regions within a few angstrom of the cell surface are different from that of the bulk-phase. Within these regions, the concentration of adsorbed macromolecules will be higher than in the environment. In addition, structured-water is expected near to the involved surfaces[30] and, as structure acts as a restraint on orientation, its dielectric constant is expected to have a value below 30, compared with the value of 80 for bulk-phase water. The role of water will be discussed again later. From the purely physical viewpoint of the interaction of two static, impenetrable surfaces, it could be argued that the adsorbed material and structured-water are part of the cell periphery, and that measurements of distances between cells, and their surface potentials, should be taken from the outermost plane of the absorbents. In this case, bulk-phase values would be in order. However, it is currently impossible to define this outer plane, and its dynamic state introduces still further complications. Since the changes in the dielectric properties of the adsorbents, to those of the relevant parts of the bulk-phase proper, will not be abrupt, as parts of two cell surfaces approach each other, three main dielectric regions must be considered: those of the transitional regions extending out from the cells, and that of the bulk-phase proper. The range will cover the low values for structured-water, to the higher values of 180 for serum[31]. Without resorting to rather dubious calculations, it can be seen that the closer two cell surfaces become, the greater will be the ratio of the widths of the transitional regions to that of the bulk-phase value for dielectric constant.

### (*v*) *The Hamaker constant*

This depends approximately on the ionization potentials, polarizabilities and density of the atoms in the interacting surfaces, and in the medium separating them. A reasonable value of A for the cell periphery would appear to be between $5 \times 10^{-16}$ and $5 \times 10^{-14}$ ergs[32]. Experimental work has suggested values of $10^{-15}$ to $3 \times 10^{-14}$ ergs for interacting leucocytes[33], and $0.5 \times 10^{-14}$ ergs for cells interacting with glass surfaces[18]. The computed effects of the value of the Hamaker constant, over the range of $5 \times 10^{-16}$ to $5 \times 10^{-14}$ ergs, are shown in Fig. 9. We have simulated the approach of two probes having radii of curvature of 500 Å and 1 μ, respectively, toward a region of another cell having a radius of 1 μ, where all surfaces have a potential of −25 mV.

From the data given in Fig. 9, it can be seen that regions of cell surface having radii of curvature of 1 μ, or more, approaching each other, are insensitive to the effects of variation in $A$. At 15 Å separation, the highest value of $A$ ($5 \times 10^{14}$ ergs) still leaves the cells with a repulsion barrier of $3 \times 10^{-12}$ ergs (approximately 70 kT) to overcome before contact can be made. This corresponds to a pressure of over $8 \times 10^5$ dynes/cm$^2$, which is some three orders of magnitude higher than the locomotor pressure of the amoeba *Chaos chaos*. In the simulated approach of a probe having a radius of curvature of 500 Å, to the same surface as before, at 15 Å separation there is a repulsion energy corresponding to 6 kT, where $A = 5 \times 10^{-14}$ ergs, compared with a value of 12.5 kT, where $A = 5 \times 10^{-16}$ ergs. This indicates that, in the case of 500 Å radii probes approaching a comparatively flat surface, there is little sensitivity to the value of $A$. For the lowest value of $A$ ($5 \times 10^{-16}$ ergs), a cellular probe needs to be extruded with a pressure of approximately $3.06 \times 10^4$ dynes/cm$^2$, to overcome a repulsion barrier of $5.5 \times 10^{-13}$ ergs. Pressures of this order appear feasible.

Studies of electron micrographs made with conventional replication and scanning techniques lead to the conclusion that the smallest cellular filopodial protrusions seen have a radius of not less than 500 Å. If the

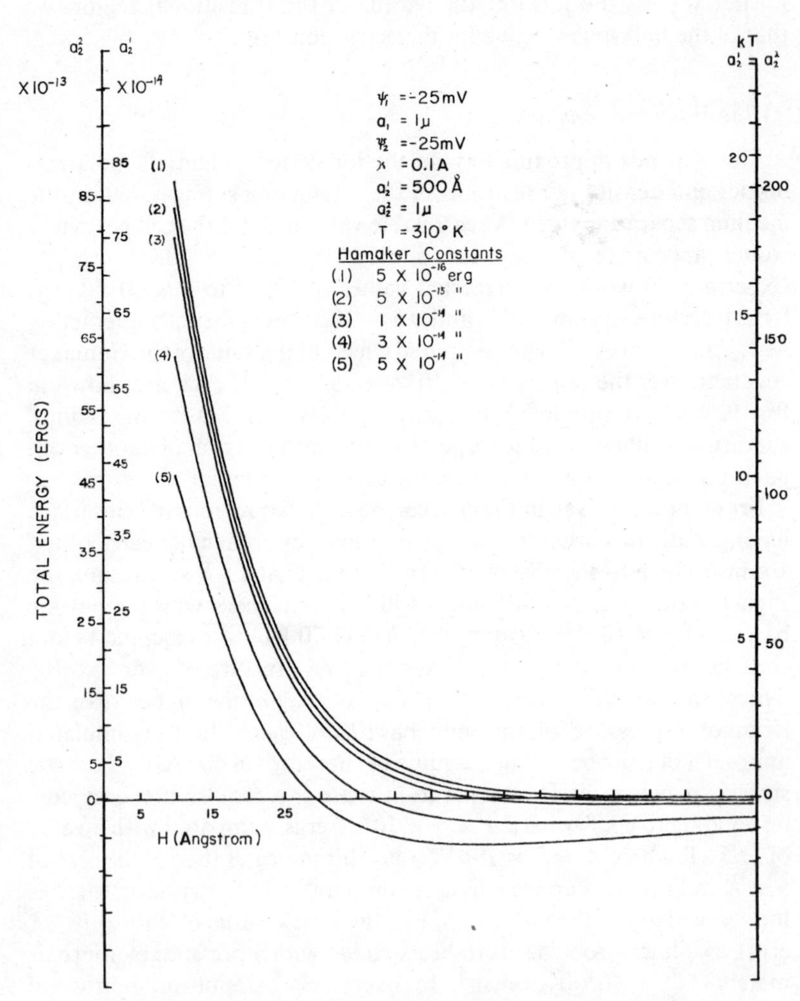


FIG. 9. A computed simulation [eqns. (17) and (18)] of the approach to a flattened region ($a_1 = 1$ μ) of cell surface by probes of radius 500 Å ($a_2'$) or 1 μ ($a_2^2$), for the different indicated values of the Hamaker constant ($A$).

FIG. 4. Extensions from a cultured HeLa cell ( × 20,400) visualized by electron microscopy by use of a shadowing technique. Reproduced from Fisher and Cooper[29] by courtesy of the authors and editors of the *Journal of Cell Biology*.

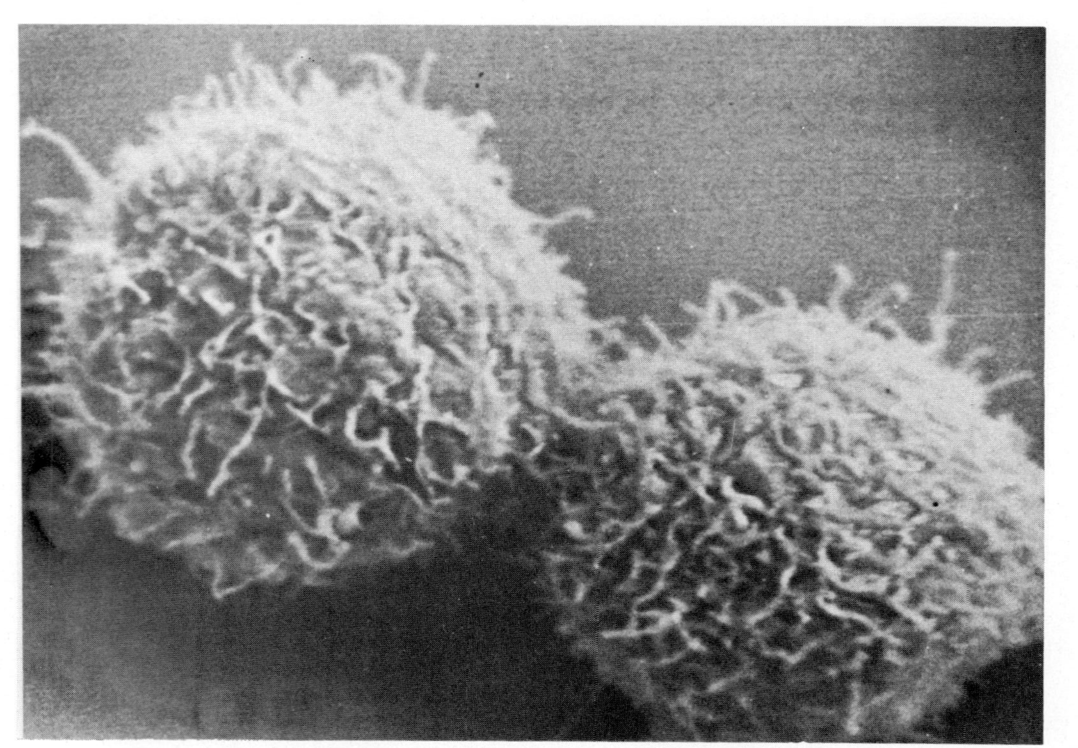

FIG. 5. Surfaces of Landschutz ascites tumour cells visualized by the scanning electron microscope ( × 5775). Reproduced by kind permission of the authors and publishers from Pugh-Humphreys and Sinclair, *J. Cell Sci.* **6**, 477 (1970).

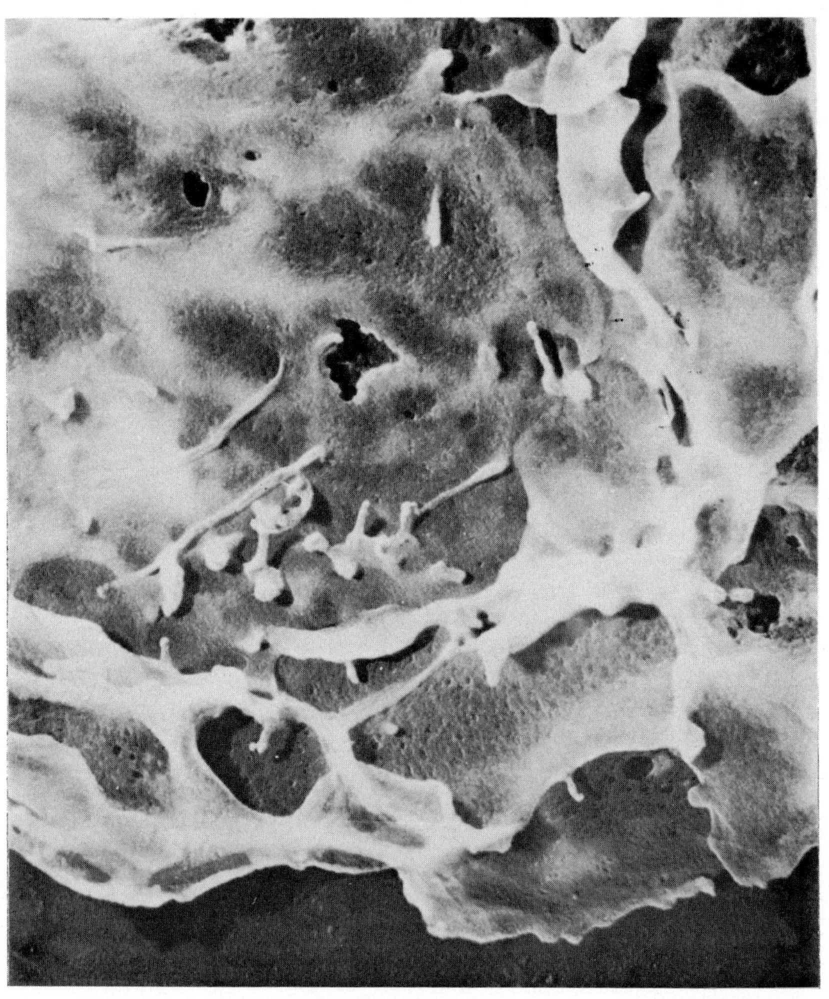

FIG. 6. The surface of a "fibroblast" taken from a 9-day-old chick embryo visualized by surface replication electron microscopy ( × 13,600). Reproduced by kind permission of the authors and publishers from Pugh-Humphreys and Sinclair, *J. Cell Sci.* **6**, 427 (1970).

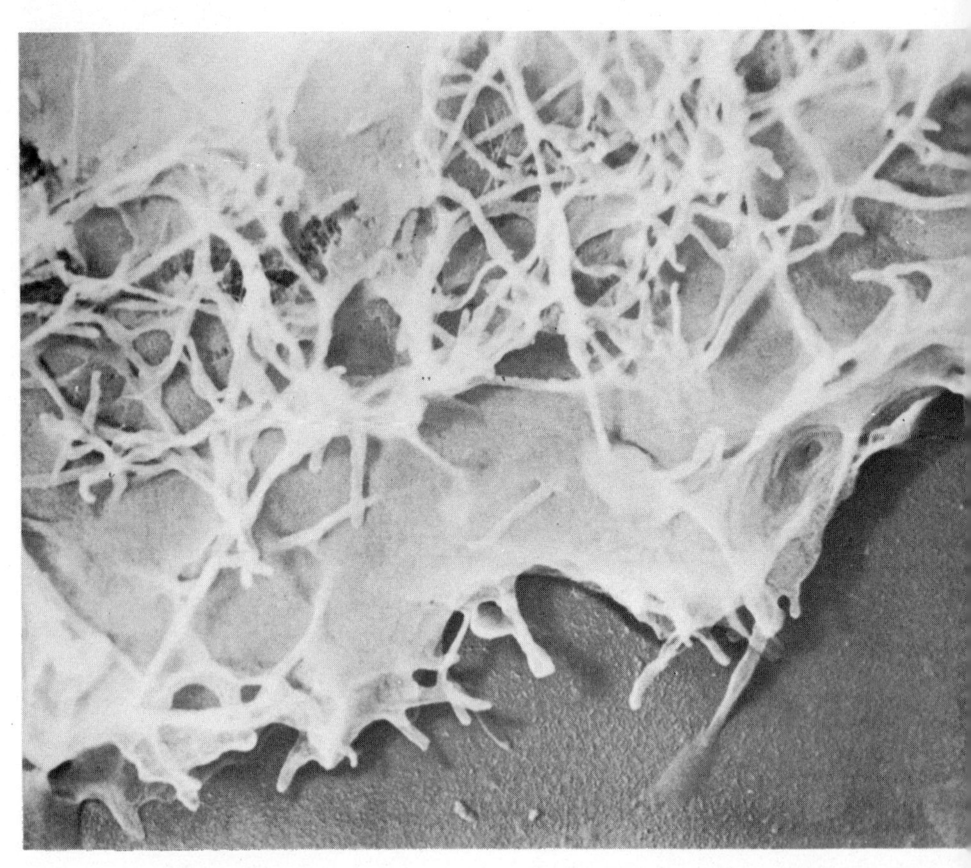

FIG. 7. The surface of a cultured HeLa cell visualized by the surface replication electron microscope ( × 13,125). Reproduced by kind permission of the authors and publishers from Pugh-Humphreys and Sinclair, *J. Cell Sci.* **6**, 477 (1970).

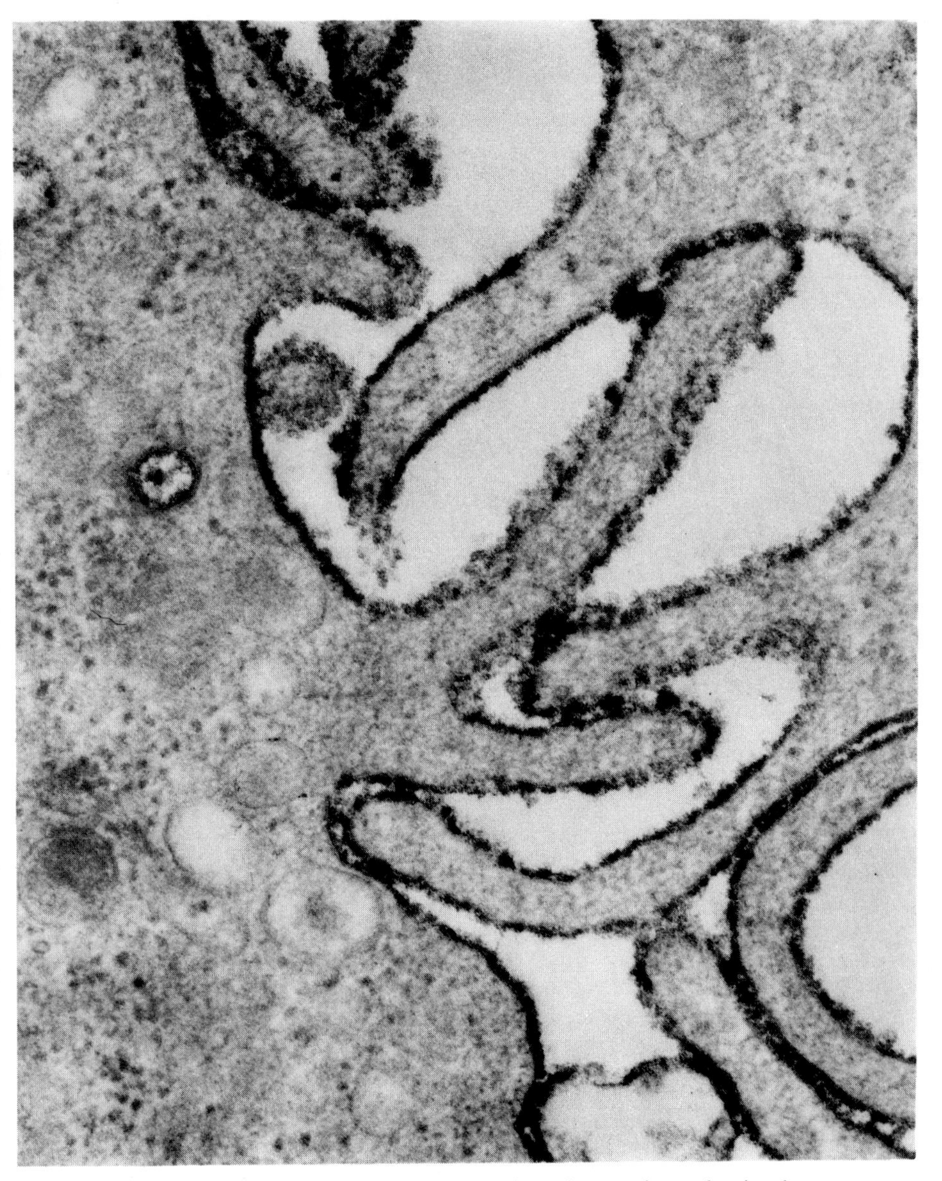

FIG. 8. The adjacent edges of two mouse peritoneal macrophages showing the complicated shapes of their interacting surfaces. Taken from a thin section by conventional electron microscopy ( × 81,000). Reproduced by kind permission of the authors and publishers from Carr, Everson, Rankin and Rutherford, *Z. Zellforsch.* **105**, 339 (1970).

cell membrane complex were simply folded on itself, so that its thickness were its radius, then a generally accepted estimate would be approximately 100 Å. However, as such a tight folding has *never* been observed to the best of our knowledge, we will take a figure of 500 Å as the lowest macroscopic radius. The protrusion extended from one cell could presumably make contact with a similar extension from another cell, although this appears to be an uncommon event. The more usually observed process is for a filopodial projection from one cell to make contact with an apparently relatively flat surface of another. If we simulate the situation of probes of varying radii from one cell approaching a flat surface of another and, by the use of eqn. (17), compute the total interaction energy between two regions of cell surface having a potential of $\psi_0 = -20$ mV, in a physiological environment, our computations which are underestimates (see above) enable us to predict that, from a statistical viewpoint (fraction making contact $= \exp(-V_T^{max}/kT)$, contact is energetically unlikely. If we select other values of the macroscopic radius of curvature of the approaching probe, as shown in Fig. 10, which indicates the maximum total interaction energy, $V_T^{max}$, between extensions of radius 100 to 1000 Å and the "flat" surface of another cell, both with surface potentials of $-20$ mV, it is seen that contact would be expected to occur only once in approximately $10^{10}$ attempts, between probes of 500 Å radius and the surface. Thus, if these estimates are really valid for cell surfaces, from an electrostatic viewpoint, contact and cell adhesion should be a comparatively uncommon occurrence. We may therefore ask whether contact is simply and directly related to macroscopic membrane conformations of the type seen under the electron microscope. If the charges are distributed homogeneously over cell surfaces, and the charge density is high enough and the contacting regions flat enough, a cell will not be able to make contact, under physiological conditions, more frequently than approximately once in $10^2$ attempts with a flat surface, even if it could bend its own membranes into probes of 100 Å radius. Consequently, the presence or absence of microvilli of approximately 500 Å radius may well be irrelevant to cells making contact with a "flat" homogeneously charged surface. This may well account for Cooper and Fisher's[34] observation that cells with many microvilli may be less adhesive to glass than cells without microvilli. Compared with cells, glass probably carries homogeneously

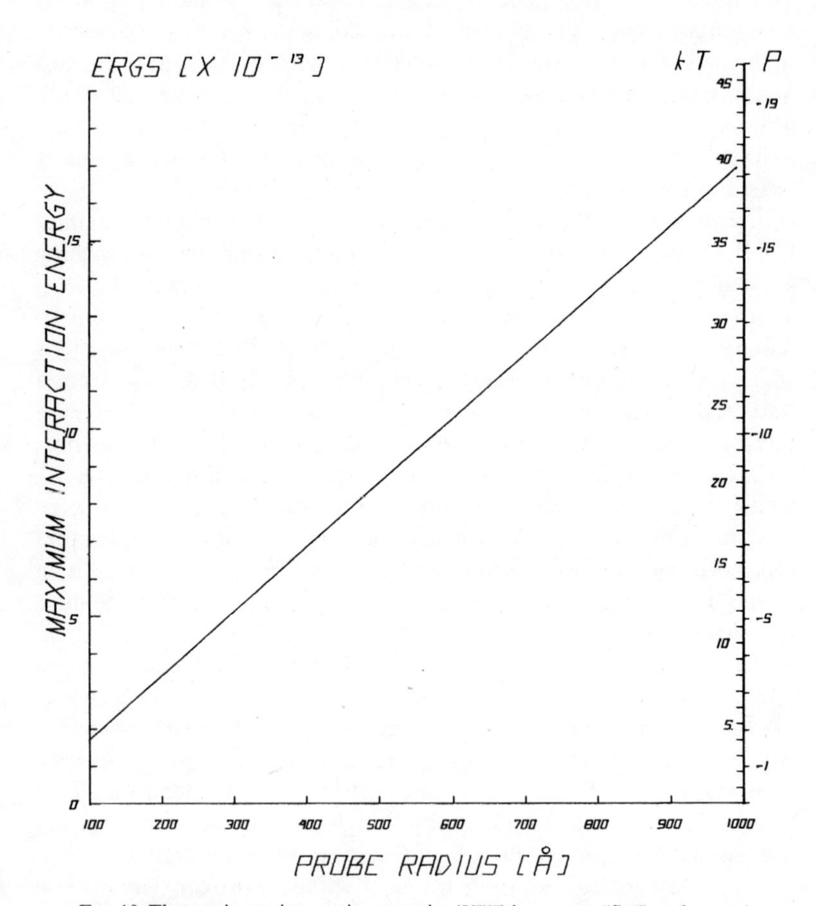


FIG. 10. The maximum interaction energies ($V_T^{max}$) between a "flat" surface and an exploring cellular probe of variable radius. Both surfaces have potentials ($\psi_0$) of $-20$ mV; $1/\kappa = 10$ Å; $\varepsilon = 80$ and $A = 5 \times 10^{-15}$ ergs. The probability of contact occurring by means of non-active cell movement is given by exp ($V_T^{max}$/kT), and is shown on the right-hand ordinate in logarithms to the base 10; i.e. $-10$ implies that contact will occur once in $10^{10}$ attempts.

distributed surface charges. It can therefore be appreciated that failure of contact between cells and glass is readily explicable in terms of electrostatic repulsion; contact between cells and glass is less readily accounted for. If a heterogeneous distribution of charges exists at a cell surface, then the low radius of curvature of an exploring projection from another cell will enable it to oppose to regions of low enough charge density, if the projection is presenting a smaller area than the low charge density region.

### (*vi*) *Positively charged groups*

In our treatment, so far, we have emphasized the interactions between net negatively charged cell surfaces, and between these surfaces and their environment. It will be appreciated that the knowledge that mammalian cells carry a net negative charge at their surfaces does not give any indication of the presence or absence of positively charged groups in this region. This is of some importance, since the electrostatic interactions between negatively charged regions of one cell with positively charged regions of another will be attractive. Such cationic sites would produce perturbations in the electric field surrounding a predominantly negatively charged surface and, finally, polar environmental molecules will orient differently in the field between positively and negatively charged surfaces than between two carrying negative charges. This latter consideration will affect the dielectric properties of the medium separating two cell surfaces.

As the literature relating to cell surface cationic groups is confusing, mainly due to technical problems, the topic is considered important enough to warrant a short review here. Of the possible positively charged groups present in the cell periphery, Gasic *et al.*[19] listed the side-chain amino-groups of lysine and hydroxylysine, terminal protein α-amino-groups, imadazolic groups of histidine residues, guanidino-groups of arginine, phospholipid and glycolipid amines. Up until recently, it was considered that the presence of surface amino-groups had been demonstrated in measurements of cell electrophoretic mobilities in solutions of varying pH. At pH values higher than 9, the *p*K of amino-groups is approached and the observation that, under these conditions, the mean anodic mobilities of a number of different types of cell increased was interpreted as favoring the reaction $—NH_3^+ \rightarrow —NH_2$ [34]. However, the cinemicrographic studies of Pulvertaft and Weiss[35] revealed that, at high pH, cells tend to disrupt, with extrusion of cellular contents, and subsequent difficulties in relating the surface structures of these cells with their normal counterparts result. Cells treated with the lower aldehydes, which react specifically with amino-groups, have shown little or no change in mean electrophoretic mobility[36, 37], but once again the aldehydes produce lethal changes in cells, which are difficult or impossible to interpret[38]. Another approach has been to react cells with aromatic reagents, which exhibit various degrees of specificity for amino-groups, and to measure their electrophoretic mobilities before and after treatment. In the case of human erythrocytes, neither reaction with p-toluene sulphonyl chloride[39], nor 2,4-dinitrofluorobenzene[36], produced evident loss of surface positivity. Later experiments, with both erythrocytes and a variety of nucleated cells, also revealed no change in their electrophoretic mobilities after reaction with 2,4,6-trinitrobenzene-sulphonic acid, 2-chloro-3,5-dinitropyridine or 3-chloro-3,5-dinitro-benzoic acid[37]. These latter negative findings were considered to be of some significance, since the reagents were used at non-lethal concentrations and produced no obvious cell trauma. In contrast to this work, Mehrishi[40] has demonstrated increases in the net surface negativity of a number of cells treated with either citroconic anhydride or 2,3-dimethylmaleic anhydride, under "mild" conditions, and has interpreted his findings in terms of cell surface amino-groups. Although these observations have been repeated[41], it should be noted that the treatment described by Mehrishi makes cells leak, and an alternative explanation for the increased negativity is that the cells adsorb the leakage products to their surfaces, and that the electrophoretic mobilities merely reflect the charge characteristics of the adsorbed material.

Thus, the position at the moment is that the presence of amino-groups (or other likely cationic groups) has not been unequivocally demonstrated at the cell/environment interface by electrokinetic methods. This does not mean that they are not present. Rather, the negative findings can only be interpreted in terms of the sensitivity of the method and the reagents used. In the case of the measurements made in this laboratory on cells reacted with 2-chloro-3,5-dinitrobenzoic acid, which not only removes surface amino-groups, but would also substitute an ionized carboxyl group at the reaction site, the observations of no change in cell electrophoretic mobility may indicate that less than 2–3 % of the ionogenic groups at the cellular electrokinetic surface carry a positive charge. However, it is hardly necessary to emphasize that the biological importance of positively charged groups, if they are present, could well be disproportionate to their average relative density at the cell surface.

Taken at face value, the failure to detect positively charged groups by electrophoretic techniques simply means that they are not reflected at the hydrodynamic slip-plane created when a cell moves through its suspending fluid in the electrophoresis apparatus. In suspending fluids of physiological nature (0.145 M NaCl), this implies that ionogenic groups more than approximately 10 Å deep to the hydrodynamic slip-plane will not be detectable. Hence, although it is expected that cationic groups are present within the *peripheral regions* of cells, they are not present in detectable amounts at the cell *surface*, and it is at this cell/environment interface that they would directly influence contact interactions of the type considered here.

### (*vii*) *Water*

In any colloidal system in an aqueous phase, the interactions between the particles and the water must be considered, since variations in water structure are expected to affect colloid stability. A helpful general discussion on this topic has been given by Clifford and Pethica[42], who point out that the comparatively recent advances in the nuclear magnetic resonance (NMR) techniques have allowed direct examination of the bonding and molecular motions of water protons, and that much progress has been made in applying the technique to colloidal systems. In accounting for the modifications of the properties of water in colloidal systems, not only must the surface interactions be considered, but also the size of the water domains, which are determined by the type of system.

The influence of the different colloid systems themselves on water may be illustrated by a few specific examples. NMR studies on silica/water systems[43] show that adsorbed water, present in sub-monolayer quantities, does not freeze and that its molecular mobility depends mainly on its hydrogen-bond interactions with surface hydroxyl groups. At low water concentrations in micellar systems of carboxylic acid soaps, water is essentially immobile, which may be attributed to its existence in a solid hydrate, which is hydrogen-bonded to the head-groups. In contrast, when water is present in these systems in excess of the amounts forming hemihydrates, it behaves as a liquid, although it is still different from bulk-phase water[(42)]. Water associated with proteins also shows distinct behaviour. Thus, in amounts up to 7%, water associated with keratin or collagen exists as hydrogen-bonded, tightly bound single molecules. In excess of 7%, the water exists in small clusters in the protein matrix, where its behavior is different from both the tightly bound and bulk-phase water[(44, 45)]. The effects of surface interactions and water domain size on water structure are also seen in concentrated polyvinyl acetate sols. Above a limiting particle concentration, there is increase in the relaxation rate of water protons, which suggests that close approach of the polymer particles has a cooperative effect on the water between them. It is not clear whether this cooperative effect is due to changes in water structure in small spaces, or is due to the effect of dispersion forces on hydrogen-bond energy levels. Whatever the detailed explanation, the effect is related to a reduction in the flocculation rate of the polymer particles.

### (*viii*) *DLVO theory and experimental data*

So far, we have been concerned with a theoretical treatment of colloid stability, and with an appraisal of the problems in applying this approach to cellular systems. It is well known that there are inherent limitations in the DLVO theory[(46)]; nonetheless, the theory has undoubted predictive value in many non-living systems[(47)]. A pertinent question is whether or not *experiment* has shown that DLVO theory has a predictive value in cell contact interactions.

Theories of colloid stability were invoked by Puck[(48)], Puck and Tolmach[(49)] and Dirkx, Beumer and Beumer-Jochmans[(50)] to explain various aspects of the analogous situation of bacteriophage adsorption to bacteria. Some experimental findings, particularly those showing the dependence of adsorption on environmental ionic strength and valency, appeared to be compatible with colloid theory. This led to the view that initial contact between the virus particles and bacterial surfaces was regulated by mutual electrostatic repulsion, and that subsequent adhesion involved a coulombic interaction between ionized phosphates, and positively charged amino-groups at their surfaces. A similar approach was used in studying the adsorption of viruses to mammalian cell surfaces by Valentine and Allison[51, 52]. These workers quantitated virus adsorption, when the surface potentials at the interacting surfaces were presumably altered by varying the environmental pH, ionic strength and valency. It should be noted that in none of these experiments were estimates made of the surface potentials based on actual measurements of electrophoretic mobilities. Valentine and Allison interpreted their own data along much the same lines as those who had worked with bacteriophage. More recently, Weiss and Horoszewicz[53] reinvestigated some of the physical aspects of virus adsorption to mammalian cells, with respect to the electrokinetic nature of the involved surfaces. The electrophoretic mobilities of three types of cultured mammalian cells were studied, to determine whether these properties could be correlated with the adsorption of EB virus, as observed by electron microscopy and immunofluorescent techniques. As both viruses and cells carried a net negative surface charge, computations were made of the interactions between them, using experimentally determined electrokinetic data and DLVO theory. The results showed quite clearly that, in these defined situations, knowledge of the various *macroscopic* surface parameters did not enable predictions to be made on whether the viruses adsorb to cell surfaces or not. Modification of the net surface charge of the cells, and/or the viruses, with neuraminidase, ribonuclease or with reagents reacting with amino-groups did not demonstrably affect adsorption, which was also apparently insensitive to calcium concentration.

Further examples of situations in which the contact interactions of cells with surfaces did *not* correlate with the computed results involved cells and glass surfaces[18, 54] and gingival cells and tooth surfaces[55]. In other cellular systems, Wilkins, Ottewill and Bangham[33] obtained a partial agreement between their electrophoretic studies and flocculation data on sheep leucocytes; however, a major problem in assessing this work lies in gauging the amount of artifactual damage inflicted on the cells.

Attempts have been made to affect contact phenomena by lowering the net surface negativity of cells by treating them with neuraminidase. Thus, previous incubation of macrophages with neuraminidase enhanced phagocytosis of plastic particles[56] and bacteria[57]. It would be tempting to attribute enhancement to reduction in potential energy barriers to contact and, indeed, following treatment with neuraminidase there was an approximate 50% reduction in the net surface negativity of the monocytes[56]. However, it cannot be assumed that the only relevant cellular property changed by incubation with neuraminidase is surface charge, since previous work on other cells has shown that such treatment can make the cell periphery significantly more deformable[58], which could also presumably influence contact. Other workers have tried to determine the role of surface potential in cell contact by studying the effect of neuraminidase- and/or ribonuclease-induced reduction in surface negativity on the reaggregation of enzymatically disassociated embryonic cells. No change was observed following neuraminidase-[59, 60] or ribonuclease-treatment[61]. Contrary to the views expressed by Kemp[59], these negative findings do not constitute a test of the relevance of colloid theory to cell aggregation. As pointed out by Weiss[18], experiments such as these really only test, at most, the relevance of cell electrophoretic mobility data to cell contact. As discussed at length, this parameter is too macroscopic to account for microscopic phenomena, except by coincidence. A second major obstacle preventing the unequivocal interpretation of these types of data is that, under optimal cultural conditions, the cells are often able to replace enzymatically removed surface groups within hours.

(*ix*) *Molecular considerations*

Another possibility to be considered is that contact does not occur between the visualized macroscopic "micro"-villi at all, but occurs at a microscopic level between compatible molecular projections from microvilli, or other peripheral regions of a cell. Such molecules could presumably arise as a result of synthesis and/or exudation of macromolecules of the type envisaged by Weiss[62] in his "ground mat" or "microexudate" hypothesis. It is therefore pertinent to consider how impending contact between cells can trigger synthesis and/or leakage of macromolecular "glues" between them.

## 3. Short-range Interactions

Interactions of this type cover the range of $2/\kappa$ (*ca.* 15 Å) down to where the distance between the surfaces of the two cells is approaching the distance between the molecules in the individual surfaces.

In our considerations of longer-range interactions, it was stressed that a major difficulty in hindering their enumeration was that, at best, only average bulk values could be given for many of the relevant parameters. When two approximating cell surfaces are separated by less than 15 Å, significant overlap of their diffuse electrical double-layers will begin. In anthropomorphic terms this implies that, instead of one cell surface "seeing" a diffuse homogeneous electrical field surrounding another cell, discrete charges on it now begin to "see" the discrete electronic charges on the other cell, which give rise to the field. These charges may be translated into ionic arrangements, structured-water or other rearrangements of the medium.

As with any detailed discussion of interactions at a molecular level, it is necessary to know the chemical identities of the interacting molecules and ions at the cell surfaces, as well as their spatial arrangements. This detailed knowledge is presently lacking and thus, once again, most experimental data are macroscopic, when really microscopic details are required. In this section we will treat the cellular surfaces as interacting molecular species. This treatment will, of necessity, be general, because the required details of the molecules themselves are largely unknown. However, even if we cannot provide specific answers, we can perhaps define the relevant questions, and indicate an approach to solve them.

### (*i*) *A general quantum mechanical approach*

The treatment of cellular interactions, as a problem of interacting molecular species, may be approached in various ways. The most general would be a quantum mechanical treatment, in which the problem is reduced to an energy calculation of a system comprised of a large number of nuclei and electrons. The lowest energy, i.e. the most stable configuration for the interacting molecules, could then be obtained by making the calculation self-consistent in both electronic and nuclear distribution.

Although a formal quantum mechanical approach to the problem requires all of the molecules in the interacting surface regions to be defined in terms of the distribution of all of their electrons in the fields of their nuclei, a great deal has been accomplished in related studies by use of approximations[63–66]. One approximation is to consider the electrons in two groups. In the first are those electrons which, together with the atomic nuclei, constitute fixed, unpolarizable "cores", about which are distributed electrons of the second group. An even more limited approximation, which has yielded useful results, is to include σ-electrons in the "cores", and to consider in detail only the π-electrons[67].

The consideration of only π-electrons can be further approximated for descriptive purposes. In the Hückel-type theories, each electron is assumed to be distributed in a molecular orbital, to which is assigned an energy dependent only on its own constituent atomic orbitals; the detailed distribution of electrons in other molecular orbitals is not considered. An alternative to the Hückel theories is the antisymmetrized molecular orbital (ASMO) formulation, which includes all energetic interactions, and which involves more complex computations. In both the Hückel and ASMO approaches, values must be ascribed to the energies appearing in the various calculations. In the so-called "zero differential overlap" approximation, the energies arising from the overlap of electronic orbitals on neighboring atoms are assumed to be negligible, even though such overlap contributes significantly to the formation of chemical bonds. In a very useful review of the field, Rein and Harris[68] list in the following order of reliability the various approximations in the applications of quantum mechanics to biochemical problems:

(i) All valence electrons, ASMO, retention of near neighbor overlap.
(ii) All valence electrons, ASMO, neglect of overlap.
(iii) All valence electrons, Hückel.
(iv) π-electrons, ASMO.
(v) π-electrons, Hückel.

There are presently two main objections to use of this quantum mechanical approach, one being that we cannot give a complete description of the cell surface at even the molecular level, let alone in subatomic terms. However, let us assume for the moment that the cell periphery is composed of isolatable repeating units, and that this unit structure has been resolved by X-ray diffraction (or other techniques), so that a precise description of relevant peripheral regions could be given at atomic and subatomic levels. Given this information, could we now compute the interaction between two cell surfaces? An initial consideration, based on existing computers, is not encouraging. If we take the very much simpler case of a complete computational analysis of the interactions between two benzene molecules, we would require of the order of $10^7$ units of storage for the various molecular integrals. A very large existing computer might have of the order of $6 \times 10^5$ of fast core storage; peripheral units, such as tapes or discs, might be used, but only with great loss of time. Thus, we would have a lack of storage, which is compounded by the amount of time required to compute, store and recall these integrals for use. This problem would, of course, be magnified many-fold in the more complex case of interacting cellular surfaces. Of these two objections to a quantum mechanical analysis of cell interactions, the most serious is lack of knowledge about the detailed chemistry of the cell periphery; limitations in present-day computers are expected to be overcome first.

*(ii) A perturbation approximation*

In outlining a general quantum mechanical approach, essentially adjacent parts of two interacting cells are considered as one "supermolecule", and their separate identities are ignored. Another approach is to treat parts of two interacting cells as two "supermolecules", in which each cell retains a macroscopic identity. This implies that the forces holding the individual cells together are stronger than the forces imposed on the cells by their environment. In the case of individual cells this is self-evident. Thus, the effects of surrounding media and/or cells may be considered as perturbations to the energy of an isolated cell. This may be regarded as an application of a theorem developed by Hellmann[69] and Feynman [70], which in effect states that, once the electron densities of the interacting units are known, the forces between them may be obtained by classical electrostatic considerations.

One way of treating the perturbation energies is to separate them into components with different distance dependence, namely, (I) electrostatic, (II) polarization and (III) dispersion interactions. Therefore, the total energy of interaction may be given as the sum of the electrostatic, polarization and dispersion energies, i.e.

$$E\,(\text{total}) = E\,(\text{electrostatic}) + E\,(\text{polarization}) + E\,(\text{dispersion}).$$

While the formulae for these interactions could be given, it is preferable to give instead their parametric dependence. The simplest and strongest of these perturbation interactions is the electrostatic one. The electrostatic energy

$$E\,(\text{electrostatic}) = E(r^{-1}, q_a, q_b)$$

depends on the net atom charges, $q_a$ and $q_b$, on each cell surface, and is inversely proportional to the distance separating these charges, which are not equivalent to the valence charges, but reflect the instantaneous distribution of electrons. The polarization energy, due to the net charges on each surface polarizing the bonds on the other surface, is given by

$$E\,(\text{polarization}) = E(r^{-4}, q_a, q_b, \alpha_a^L, \alpha_a^T, \alpha_b^L, \alpha_b^T)$$

where $\alpha^L$ and $\alpha^T$ represent longitudinal and transverse polarizabilities of bonds in the surfaces $a$ and $b$, and $r$ is the distance between a net charge in one surface and the midpoint of a bond in the other. The dispersion, or London, interaction energy is parametrically given by

$$E\,(\text{dispersion}) = E(r^{-6}, I_a, I_b, \alpha_a^L, \alpha_a^T, \alpha_b^L, \alpha_b^T)$$

where $I_a$ and $I_b$ represent the ionization energies of bonds in surfaces $a$ and $b$. $r$ here is the distance between the midpoint of a bond in surface $a$ and the midpoint of a bond in surface $b$. None of the actual expressions for these energies is as simple as implied by the parametric equations, since they incorporate summations over all atoms and/or bonds in each surface, as well as various angular dependencies. The parametric equations indicate that the energies are dependent on parameters reflecting the instantaneous electron and nuclear distribution in each surface.

It is apparent that the *perturbation* approach requires information, such as electron densities and bond ionization energies, which are not available since a quantum mechanical calculation is not feasible for the reasons given above. Nevertheless, some description of cellular interactions may be obtained, if one considers the nature and magnitude of these perturbation forces. The cell is not a rigid structure. This means that the form and shape of a cellular surface is maintained by local chemical bonds and extensive intermolecular interactions ranging in strength from hydrogen bonding to the weaker perturbation forces, similar in nature to those invoked for the intercellular interaction. It is

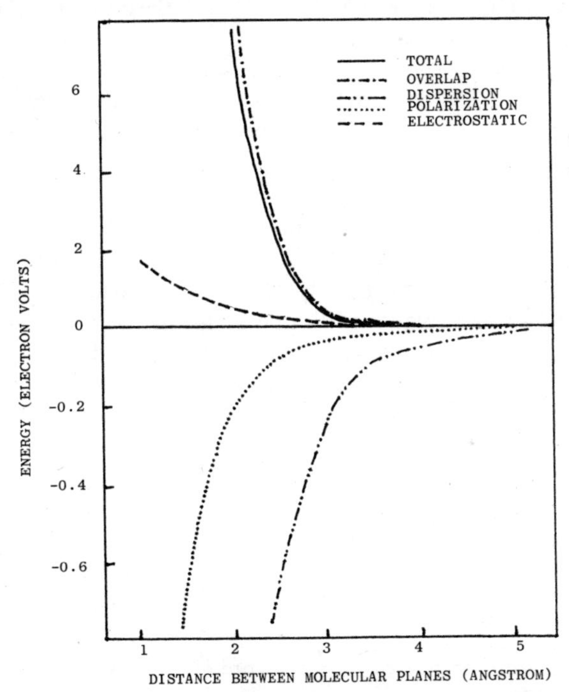


FIG. 11. The component interaction energies between two electrically neutral thymine molecules in the *cis* (head-to-head) configuration as a function of the distance between molecular planes[71].

well known that the perturbation approach is most valid when the perturbation energy is small compared with the total energy. In other words, when the perturbation becomes large, one should consider rearranging the nuclear positions, and therefore the electron densities, to minimize the amount of perturbation, i.e. a recalculation of electron densities with altered nuclear positions. While this would be possible, in principle, in practice it is not. However, it does suggest a molecular description of cellular interaction. As the perturbation forces of one cell surface, interacting with another and with any solvent molecules, approach that of the intracellular forces, the cellular molecules rearrange themselves to balance these opposing forces and minimize the total energy of the system. This will not, in general, destroy the cellular surface, since the local bonds would remain intact. It would, however, modify the structure imposed by perturbation forces within the cell. On a macroscopic level, we could observe this balancing of forces as the distortion, movement and formation of protrusions on the cell periphery. The energy dissipated probably appears as heat, i.e. increased atomic motions.

A computed example, showing the magnitudes of the various perturbation interactions, is given in Fig. 11 for a *cis* (head-to-head) configuration between two electrically neutral thymine molecules[71]. The distances between the planes of the interacting molecules are significant up to approximately 5 Å. It is seen that the total ground state interaction is repulsive, because the stable energy of the system is very low and the repulsion energy, due to overlap of molecular orbitals, is correspondingly high. We should perhaps emphasize that "zero" separation in a colloid approach to analogous interactions corresponds to approximately 3 Å in Fig. 12, where overlap energy becomes significant.

## 4. Very Short-range Interactions

These interactions cover distances from about 5 Å to those of intramolecular interatomic separations. We have already remarked that the distance criteria used in this discussion are somewhat arbitrary, as there is a continuity of interaction. In the case of the very short-range interactions, this arbitrariness becomes increasingly obvious, because, although terms remain which may be identified as electrostatic, dispersion and polarization interactions of the Van der Waals type, the validity of the classical perturbation formulation becomes doubtful over the short distances involved. This is due to the addition of new factors, which may be called the overlap terms, and which include those effects due to the overlap of the electron densities of the two interacting species. These effects could on the one hand lead to repulsion, but on the other hand could also lead to strong attractive forces, due to delocalization and exchange of electrons leading to the formation of new chemical bonds. Thus, many of the interactions operating over a very short range have already been discussed. Those which have not yet been considered may be conveniently grouped as hydrogen-bond interactions and chemical bond formation.

### (*i*) *Hydrogen-bond interactions*

The hydrogen bond is a result of proton sharing between two relatively negative atoms, such as F, N or O. The interaction is, therefore, predominantly ionic in nature. Since cell peripheries contain anionic and cationic groups, parts of two opposing cell surfaces could be linked by means of water bridges utilizing hydrogen bonds. Figure 12 indicates

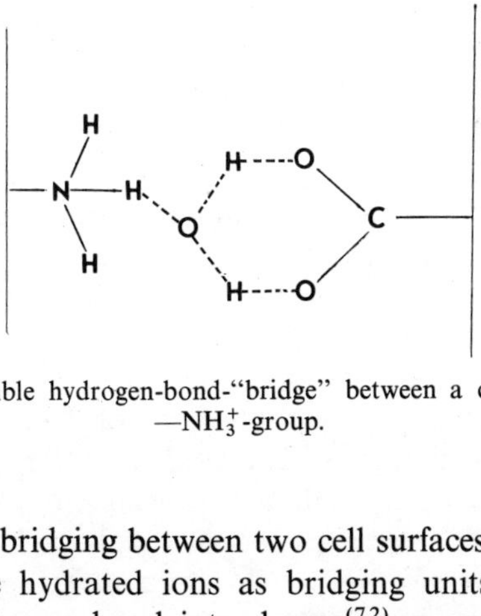

FIG. 12. A possible hydrogen-bond-"bridge" between a carboxyl and an —$NH_3^+$-group.

a hydronium ion bridging between two cell surfaces; other structures may also involve hydrated ions as bridging units. In an aqueous environment, hydrogen-bond interchange[72] may permit separation of two hydrogen-bonded groups by as little as 1.5 kcal per mole, in contrast to the usual estimate of approximately 5 kcal per mole in non-aqueous media. In a dynamic situation, this type of bond is advantageous, since it provides a means of forming temporary adhesions between cell surfaces. In addition, because of their relatively low energies, these structures may be disrupted without damage to the stronger chemical bonds stabilizing individual cell peripheries.

### (ii) *Chemical bond formation*

Since the formation of chemical bonds depends upon interacting atoms sharing electrons, very close approach of the interacting atoms is required. In general terms, a chemical bond is formed when such sharing results in the completion of electronic shells around both atoms. This implies that some electronic orbitals extend over both nuclear centres; these common orbitals constitute the "bond". The net effect of electron sharing is to lower the energy of the shared electrons, making for increased stability. This will be manifest throughout the whole molecular system as a result of consequent nuclear and electronic reorganization. Chemical bonds are not formed when electron sharing would produce no change in internal energy ($\Delta E = 0$). If the electronic orbitals of two opposed atoms are filled, then the repulsion between the orbitals (Born repulsion) will prevent their close approach. Thus, although chemical bonds cannot be formed unless two atoms approach within a center to center distance of approximately 3 Å, close approach can occur without bond formation.

In order to predict whether or not chemical bonds can be formed between two opposed cell surfaces, detailed information is required of the chemical nature of the surface regions. Our lack of knowledge in this respect has already been discussed. However, in the case of very close-range interactions leading to chemical bond formation, the spatial location of the various atoms becomes of increasing importance. Moreover, the possibilities of chemical bond formation in terms of randomness are probably high, partly because of the enormous numbers of atoms present, and partly because of the high number of degrees of freedom in their movement. Therefore, if the opposing surfaces of two cells fail to form mutual chemical bonds, this could as well be due to their inability to approach closely enough together as to their lack of mutual chemical reactivity.

## 5. Surface Free Energy, Surface Tension and Adhesion

The theoretical development stressed so far has been to replace macroscopic parameters with microscopic or molecular parameters, as the distances between interacting cell surfaces decrease. In this section, we wish to examine the relationship between the molecular picture presented previously and the use of macroscopic parameters, such as cell surface tension, surface free energy and energies of adhesion.

The relationship between free-energy changes and the potential energy variations considered above is clear, since free energy is the thermodynamic analogue of the potential energy used in mechanics. The evaluation of the free-energy changes would involve some description of the energy changes involved in the movements of the molecular structure of the cell, and in time-averaging them. These changes may be interpreted as the energies of adhesion.

Surface tension is the macroscopic parameter expressing the balancing of forces experienced by a molecule at a surface. The forces are those of interaction between neighboring atoms and/or molecules in the surface and the environment, and are related to the forces discussed in previous sections. Given these forces, one could, in principle, by various summation and averaging techniques arrive at a value for the surface tension. Such calculations have been made for chemically "pure" solid surfaces[(73)] and have yielded results which are in excellent agreement with experimentally determined values from contact angle measurements[(74)]. However, calculations of the surface tension of a cell are obviously out of the question, if the detailed chemistry is unknown. Experimentally derived estimates of this parameter will be discussed shortly.

In an interesting approach, Steinberg[(75)] has discussed, in terms of their surface free energies, the ability of different cells, cultured within the same aggregate, to sort out, so that like cells lie in continuity. The relevance of work of adhesion to this type of biological organization is also suggested by Carter's[(76)] observation that motile cells will move over surfaces in the direction where adhesion between the cell and substratum surfaces are expected to be greatest. The problem here is in measuring cell adhesion and in establishing causal relationships between adhesion, cell surface free energy and cell behaviour. Many of the so-called measurements of cell adhesion reported in the literature are, in fact, measurements of rates of adhesion, stability ratios or of cell separation, from which the required data cannot be disentangled.

It should perhaps be emphasized that, when the interactions of cells with any surfaces are being considered in terms of surface tension, not only are data required about the cell/substratum interface, but also the cell/medium and substratum/medium interfaces, since the final result will be a balance of all three interactions. Thus, simple measurements of the surface tension at a substratum/medium interface will not enable a prediction to be made about a cell/substratum interaction, except by coincidence. This view is possibly substantiated by the failure of Taylor[77] and Weiss and Blumenson[78] to relate the adhesion of cells to various surfaces to the initial-water-wettability of these surfaces, when the cell/surface interactions occur in the presence of serum, which is amphophilic, and which may well mask underlying surfaces to which it is adsorbed. In contrast to these experiments is the work of Lyman *et al.*[79] on the interaction of platelets with uncharged hydrophobic polymer surfaces, in which a direct relationship has been demonstrated between platelet adhesion and the critical surface tension of the polymers. The interactions were studied with the platelets suspended in blood, constituents of which could adsorb to and contaminate the polymer surface. However, this apparently did not occur. The authors note that in studies of this kind blood/air interfaces should be avoided, as they have "damaging" effects resulting in "unusually heavy" platelet adhesion. It is difficult to extrapolate these experiments to more usual biological situations, where all of the surfaces with which tissue cells come into contact are thought to be charged. It is also currently impossible to obtain measurements of the critical surface tensions of the small areas of cell surfaces, which are actually involved in contact processes.

It may be asked at this stage whether any reliable data exist on surface tension at a cell/medium interface. Although various figures are quoted in the literature, based on deformability experiments[80], these refer to tension at the cell surface, as distinct from true cell surface tension, and the former exceeds the latter by an unspecified amount. Another problem is that in contact phenomena, in general, initially small areas of a cell make contact with small areas of another surface. Therefore, as in the case of cell surface potential, macroscopic measurements are only valid for homogeneous surfaces. Even if methods were available for determining the surface tension of specified areas of cells in a relevant manner, the measurements would only be pertinent to situations in which the whole of the specified area at the surface of one cell reacted with the whole of a specified area at the surface of another.

## 6. Additional Factors

At present, there is no general agreement on the structure of the cell periphery. It is known that mucosubstances are present at the cell surface, and that the act of examining the cell under the conventional electron microscope causes an undetermined amount of structural distortion of this region. Thus, no unequivocal visual observations have been made of cell contact processes at the level of several angstrom separation of the interacting surfaces. This uncertainty in observing contact at micro-ultrastructural level leads in turn to uncertainties in deciding *when* contact has occurred in any particular situation. It is therefore often impossible to decide whether some of the observed responses of interacting cells should be classified as "short term" or "longer term".

Although so far we have considered contact interactions between parts of two cell surfaces, it must not be overlooked that many such interactions involve, or take place in the presence of, whole populations of cells. Thus, short-term interactions between one pair of cells may well be affected by the longer-term interactions resulting from previous contact between other cells, and other parts of the same cells. Although a detailed discussion of longer-term interactions is beyond the scope of the present review, we will briefly discuss some of them which may influence short-term cell interactions.

In two very interesting papers, Gingell[22, 81] has pointed out that if, in the case of two approaching negatively charged cell surfaces, the charges are fixed, there will be an increase in negative surface potential. As discussed earlier, although we consider the general concept correct and of great interest, the doubtful validity of applying DLVO theory to surfaces separated by distances of less than 15 Å, combined with uncertainty of the degree of penetrability of the cell surface by ions, makes Gingell's numerical estimates problematical. In addition, the assumption of homogeneity of surface-charge distribution made by this author is probably unwarrantable. However, if we accept the premise that close approach of two cell surfaces results in an increase of their potential by an unknown amount, it is of interest to consider some of the implications of this change, particularly in the light of Wolpert and Gingell's[82] suggestion that this may act as a transducing mechanism for translating environmental change into a cytoplasmic response.

Direct measurements have revealed that, in many reported cases, transmembrane potentials exist between intracellular electrodes and electrodes placed in the surrounding medium. This electrical potential difference is due to differences in concentration of ions, proteins and other polyelectrolytes between the inside and outside of the cell. All findings show that the inner membrane surface is negative with respect to the outer, and that transmembrane potentials vary in magnitude with the suspending medium and temperature. In the case of human erythrocytes in Ringer's solution at room temperature, Jay and Burton[83] reported transmembrane potentials of 8 millivolts, and Borle and Loveday[84] have reported values of 15 mV for cultured HeLa cells in protein-containing growth medium at 37°C. As we have seen, the surfaces of all mammalian cells, so far examined, are at a negative potential with respect to electrodes placed in their medium, due to the presence of fixed ionogenic groups. Thus, as remarked by Gingell[22], if the surface potential, $\psi_0$, becomes more negative, then the transmembrane potential difference will decrease. Although there may well be other explanations, it is of interest in the present context that, when HeLa cells were transferred from calcium-containing media to a calcium-free medium (which increases their net surface negativity), the potential difference across their membranes fell from $-17{\cdot}1$ to $-7.14$ mV[84].

It will be appreciated that the fixed charges in the cell periphery will only have a direct effect on the electrochemical transmembrane potential if they are asymmetrically distributed across the membrane. Such asymmetry of distribution is indeed suggested by the work of Benedetti and Emmelot[85], which indicated at electron microscopic level that, in at least some mammalian cells, the major part of the sialic acid moieties associated with their membranes are located at the outside surface. Indirect ways in which ionized groups, symmetrically distributed across the cell periphery, may affect permeability and transmembrane potentials will be discussed shortly.

At this stage it may be asked whether there is any evidence at all that variations in surface potential can effect membrane permeability. The evidence comes from studies on artificial membranes and living cells. The former possibly illustrate the direct effects of surface potential on membrane stability and "passive" diffusion processes, the latter are concerned with a more complex situation, in which "active" or "facilitated" diffusion is also involved.

Experiments made on artificial phospholipid vesicles in electrolyte solutions by Bangham *et al.*[86] suggested that the permeability of their membranes to $K^+$ was correlated with the density of fixed negative charges at the outwardly directed hydrophilic heads of the phospholipid molecules. However, more recent studies, reviewed by Papahadjopoulos[87], indicate that permeability to small ions is not directly related to membrane surface-charge densities of *up to one negative charge per phospholipid molecule,* i.e. 60–70 $Å^2$ of membrane surface. Recently, Ohki and Aono[88] have calculated that, in the presence of 0.1 N NaCl, an increase from 1 to 1.25 charges per molecule would favor the transformation of a phospholipid bilayer into cylindrical micelles, which can act as aqueous pores. Although it is not by any means proven that lipid bilayers form a good model for cell membranes[89], it is well accepted that the peripheral regions of all cells, so far examined, contain phospholipids.

In the last decade, the realization that many different cells carry bound sialic acids at their surfaces in an ionized state, together with the ready availability of highly purified sialidase, has led to many experiments to determine the effects of sialidase (neuraminidase) on various cell functions. The interpretation of experiments of this type is complicated by the observation of Nordling and Mayhew[90] that, when cells are incubated with neuraminidase, the enzyme enters them and attacks intracellular substrates, such as on the nuclear membrane, for example, as well as susceptible groups on the outside of the cell. Many reagents enter cells, and react with their internal contents, constantly raising the question of whether an observed effect is due to changes produced on the outside and/or inside of cells. Another often overlooked practical point is that cells often show a time-dependent recovery, after treatment with non-lethal reagents. After neuraminidase treatment, for example, a number of different types of cultured cells show a reduction in their net surface negativity. On being returned to culture, they recover the lost net surface negativity within a few hours, under optimal growth conditions, by replacement of the lost sialic acids. Thus, experiments aimed at determining the effects of many reagents must, of necessity, be short-term. Lefevre[(91)], Widdas[(92)], Rosenberg and Wilbrandt[(93)], and many others, have suggested that, before ions can move across cell membranes, they must first bind to a carrier bearing a charge of opposite sign. Thus, it might be expected that, in addition to other effects on transmembrane movements of ions and molecules, cell surface ionogenic groups might act as carriers.

An interesting effect of neuraminidase-treatment of L1210 leukemia cells on potassium transport was described by Glick and Githens[(94)], who concluded that sialic acids at the cell periphery mediate both the inward and outward diffusion of $K^+$ in these particular cells. However, these experiments were somewhat limited technically. A more exhaustive study of the effects of cell surface potential on transmembrane ion-movements was made by Weiss and Levinson[(95)], who reduced the net surface negativity of Ehrlich ascites cells by ribonuclease- or neuraminidase-treatment, and then measured changes in cationic flux. Neither enzyme treatment affected the $Na^+$ or $K^+$ content of the cells, when their electrophoretic mobilities were reduced by 30–45% following neuraminidase-treatment, or by approximately 30% following incubation with ribonuclease. Experiments with $K^{42}$ revealed a 10–15% reduction in unidirectional $K^+$-fluxes following incubation with neuraminidase, but no change occurred after ribonuclease-treatment. Thus, the effect of treatment in reducing the unidirectional $K^+$-flux appears to be due to loss of sialic acid moieties, as distinct from loss of surface negativity *per se*. However, for technical reasons, a 10–15% reduction in flux does not permit too many unequivocal conclusions to be drawn. The data suggested that surface anionic sites associated with both RNA and sialic acids are not of major quantitative importance in regulating either intracellular $Na^+$ or $K^+$ concentrations, or unidirectionation transmembrane $K^+$-flux. However, our experiments clearly did not enable us to determine whether ion-binding to anionic sites at the cell surface is *not* essential to transmembrane movement, or whether such binding is essential but occurs though the 40% of cell surface net negativity which is not susceptible to ribonuclease or neuraminidase. In view of Mayhew's[96] observation that the density of sialic acid moieties at the cell surface increases during the mitotic cycle, it is of interest that Jung and Rothstein[97] have shown that variations in $Na^+$ and $K^+$ fluxes and content in mouse tumour cells are also related to the mitotic cycle.

Most cells transport $Na^+$ outwards and $K^+$ inwards, against concentration gradients, by means of an energy-requiring process. In common with so many other examples of cellular energy expenditure, ATP and adenosine triphosphatase (Na–K–ATPase) are involved in the active transmembrane movements of these cations[98]. In his chemiosmotic hypothesis, Mitchell[99] has postulated that the free energy, which is necessary for the synthesis of ATP, is derived from a transmembrane electrochemical potential difference by means of a reversible proton translocating ATPase. The recent work of Junge, Rumberg and Schroder[100, 101] on photophosphorylation processes involving the thylakoid membrane indicates that, before one molecule of ATP can be synthesized from ADP, $3H^+$ have to be translocated across a membrane carrying an electrochemical potential difference. An indirect effect of sialic acids on the active transport of ions has been suggested by Emmelot and Bos[102], on the basis of their experiments on neuraminidase-treated isolated surface membranes from rat liver. It appears that removal of the terminal O-glycosidic-linked sialic acid from membrane glycoprotein inhibits the $Mg^{2+}$–ATPase, and inhibits or activates the $Na^+$–$K^+$–$Mg^{2+}$–ATPase, which are both involved in active transport of ions. Emmelot and Bos suggest that it is possible that the removal of sialic acid from a membrane might change the relative orientation of the constituent proteins and phospholipids, thereby affecting the activity of the ATPases which are, in some way, phospholipid dependent. Sialic acids have also been implicated in amino-acid transport by the observation that, following neuraminidase treatment, the accumulation of $^{14}$C-L-alpha-aminoisobutyric acid by HeLa cells is reduced[103]. Thus, in line with Mitchell's hypothesis, it is possible that changes in cell surface potential can produce changes in transmembrane potential, which in turn can affect the formation of ATP, which in turn provides the energy for the movement of cations across the cell membrane. So far, a reasonable hypothetical case can be made out for changes in cell surface potential altering the permeability properties of membranes. Direct experiments have also linked the mechanical properties of the cell periphery[(58)] and phagocytosis[(56)] with the presence of sialic acids. The question remains of whether significant increases in surface potential occur during the contact process, and whether these changes trigger off others.

It was postulated many years ago by Weiss[(62, 104)] that an important part of cell contact was the release of informational macromolecular microexudates from cells. Such exudates were possibly demonstrated by Rosenberg[(105)], who used ellipsometric techniques after cells had made contact with polished metal plates. By the use of $^{51}Cr$ labels the release of proteinaceous materials has been demonstrated from cells making contact with glass[(18)] or with other cells[(106)]. Initial attempts at characterization of proteins released from fibroblasts in culture have been made by Halpern and Rubin[(107)], although the type of release is not specified. It will obviously be of importance to determine whether such release substance can affect the contact reactions of adjacent cells, particularly as it is known that proteinaceous materials released from cells can adsorb to the surfaces of the same or other members of a cellular population, producing alterations in their zeta-potentials[(38)].

Various surface morphological specializations have been observed under the electron microscope, associated with contact between regions of cells making contact. These junctional regions have been described in detail by Farquhar and Palade[(108)] and Kelly[(109)]. It would appear that, in the case of one type of junctional specialization, the desmosome, close approach of two cell surfaces triggers off the formation of plaques between them, followed by the deposition of intracellular fibrils normal to the plaques[(110)]. It is not known how close two cell surfaces must come together before such junctions are formed—whether they are formed before or after contact, as the term is defined by us, or whether the plaques represent initially a sophisticated type of leakage.

The activities of many regulatory enzymes are affected by their environmental cation concentration. In the case of the phosphotransferase systems discussed earlier, the stimulation of their activity by $Mg^{2+}$ and/or $K^{+}$ is often counteracted by $Ca^{2+}$ and sometimes by

$Na^+$[111]. The general impression is gained that enzyme activity within a compartmentalized cell could be markedly influenced by local cation concentrations, and changes in the relative concentrations of these ions. In the case of *E. coli* B and *B. subtilis*, it is known that the concentration of intracellular $K^+$ is a rate-determining factor in protein synthesis[112]. Lubin[113] has shown that, in the presence of amphotericin B, which attaches to their membranes causing selective leakages from sarcoma 180 cells, a depression in protein and DNA synthesis results, which parallels the loss of cell potassium. The depression could be reversed in media containing high levels of $K^+$. Metabolic control has also been linked to ionic environment by Bygrave[114]. Other experiments linking membrane permeability with cell division were described by Mayhew and Levinson[115] on Ehrlich ascites tumour cells. These authors suggested that the cardiac glycoside ouabain reversibly inhibited cell division by a mechanism related to its inhibition of membrane transport system(s), which include not only $Na^+/K^+$, but also amino-acids.

An exciting recent development has been the discovery, by Loewenstein[116] and his colleagues, that electrical coupling occurs between many types of adherent cells through junctions of low electrical resistance. Movements of ions and small molecules take place across these junctions. Although low junctional resistances had previously been demonstrated in nerve tissues, and were concerned with signal transmission, the importance of the experiments of Loewenstein and others lies in the demonstration that these junctions also occur between cells not normally considered to be engaged in signal transmission. Wolpert and Gingell[82] have made the interesting suggestion that coupling may reflect localized increases in membrane permeability, consequent upon increased cell surface potentials arising from the close approach of two charged membranes.

On the one hand, we have limited evidence that cationic fluxes can influence energy requiring cell functions and, although there is at the moment no experimental evidence, cell movement could probably be included among these functions. On the other hand, the work of Loewenstein and others shows that ionic communication between cells occurs through low-resistance junctions. It is therefore tempting to combine both lines of evidence and speculate, in a similar manner to

Loewenstein[116], Lubin[113], and Wolpert and Gingell[82], that the act of cell contact promotes the formation of low-resistance, intercellular junctions, which lead to the exchange of information between cells, which may result in alteration of their social behaviour. Such alterations could possibly include contact inhibition of cell movements[117] and division[118]. It should be noted here that the inhibition in growth exhibited by "normal" contacting cells *in vitro* may not be related to contact, as defined here, at all. Jainchill and Todaro[119] have emphasized that cell growth in culture involves the interaction of at least three factors: the inherent genetic constitution of the cells, the concentration of serum and other nutritional factors in the environment and "proximity inhibition". The latter would appear to be due to the diffusion of growth inhibitory substances produced by the cells themselves[120–122], although physical contact may also normally play an inhibitory role.

Specific changes in the physicochemical nature of the cell periphery are associated with primary embryonic induction, cytodifferentiation, virus infection, carcinogenesis, etc. These changes have been reviewed at lcngth elsewhere[1, 123] and the present remarks will be of a general nature. Eisenberg *et al.*[124] studied the correlation between the electrophoretic mobilities of rat liver cells with their growth rates. Cells isolated from neonatal and regenerating livers had significantly higher mobilities than liver cells from normal adults. It was therefore concluded that cell proliferation is associated with increased surface negativity. Work on parasynchronous cultures has shown that, in the G2 phase of the mitotic cycle, an increase in cellular electrophoretic mobility may occur which is most readily explicable in terms of an increased density of negatively charged surface groups, which in some cases are the carboxylic groups of sialic acids[96]. It has also been demonstrated in some cells that increased surface negativity is associated with increased growth rate and that, in contrast to the changes associated with the mitotic cycle, these changes are due to an increased density of anionic groups associated with RNA at the cellular electrokinetic surface[125].

Another factor to be considered in connection with the dynamic nature of the cell periphery is sublethal autolysis. It was shown that treatment of cells with either high doses of vitamin A[126] or sublethal amounts of antiserum[127] agents, which are known to activate the lysosomes, cause detectable changes in the cell peripheries associated with release of lysosomal hydrolases. Other work demonstrated that non-lethal release of lysosomal enzymes in cultured embryonic mouse limb rudiments could be provoked by exposure to antiserum and resulted in the degradation of intercellular cartilaginous matrix[128]. It was also shown that, in mice, metastasis of spontaneous mammary cancer, a process initially dependent on cell separation, could be promoted by overdosage of vitamin A. These and other results were interpreted to indicate that lysosomal enzymes, released from the tumor cells, attacked their own peripheral regions, facilitating their separation from the parent solid carcinomas[129]. Lysosomal enzymes may be activated and released from many different types of cells by a very wide variety of stimuli acting initially at the cell periphery and often producing changes in permeability[130]. These enzymes may change the peripheral regions of not only the cells from which they are released, but also those of other cells. It would not be surprising, therefore, if such sublethal lytic processes could be triggered off by the permeability changes thought to be associated with the close approach of cells.

Although at present many of the details are obscure, it is not exceeding the bounds of reasonable speculation to consider that contact phenomena with interacting populations of cells cannot be dealt with adequately on the assumption that the cells are inert particles with steady-state, homogeneous, impenetrable surfaces. A major problem in this subject is not only to define the physicochemical properties of the cell periphery, but also to recognize its inherent dynamic state which may be continuously modified by its interactions with other cells, and which in turn may continuously modify those interactions.

In this review we have deliberately omitted any discussion of the fascinating topic of specificity. Many examples may be catalogued of how one cell "recognizes" another cell, and it would appear that at least some of the explanations for these phenomena may be found in the cell periphery[1]. However, we feel that, until some of the problems raised here about cell contact in general are considered, it would be premature for us to deal with this subject.

## References


1. L. Weiss, *The Cell Periphery, Metastasis and Other Contact Phenomena,* North-Holland Press, Amsterdam, 1967.
2. A. S. G. Curtis, *The Cell Surface, its Molecular Role in Morphogenesis,* Logos Press, London, 1967.
3. B. V. Derjaguin and L. Landau, *Acta Phys.-chim. U.R.S.S.* **14**, 633 (1941).
4. E. J. W. Verwey and J. Th. G. Overbeek, *Theory of the Stability of Lyophobic Colloids,* Elsevier Publishing Co., Amsterdam, 1948.
5. M. J. Sparnaay, *Recueil* **77**, 541 (1958).
6. F. D. Buff and F. H. Stillinger, Jr., *J. Chem. Phys.* **39**, 1911 (1963).
7. R. Hogg, T. W. Healy and D. W. Furstenau, *Trans. Faraday Soc.* **62**, 1938 (1966).
8. H. C. Hamaker, *Physica* **4**, 1058 (1937).
9. L. Weiss, ref. 1, p. 170.
10. E. M. Lifshitz, *Zh. Eksp. i. Teor. Fiz.* **29**, 95 (1955); *Sov. Phys. J.E.T.P.* **2**, 73 (1961).
11. I. E. Dzyaloshinski, E. M. Lifshitz and L. P. Pitayesku, *Ad. Phys.* **10**, 165 (1961).
12. H. G. B. Casimir and D. Polder, *Phys. Rev.* **73**, 360 (1948).
13. V. A. Parsegian and B. W. Ninham, *Nature* **224**, 1197 (1969).
14. B. W. Ninham and V. A. Parsegian, *Biophys. J.* **10**, 646 (1970)
15. B. W. Ninham and V. A. Parsegian, *J. Chem. Phys.* **52**, 4578 (1970).
16. D. A. Haydon, *Proc. Roy. Soc.* **A258**, 319 (1960).
17. D. A. Haydon, *Recent Progress in Surface Science,* Academic Press, New York and London, 1964, Vol. 1, p. 94.
18. L. Weiss, *Exptl. Cell Res.* **51**, 609 (1968).
19. G. J. Gasic, L. Berwick and M. Sorrentino, *Lab. Invest.* **18**, 63 (1968).
20. Y. Marikovsky and D. Danon, *J. Cell. Biol.* **43**, 1 (1969).
21. L. Weiss and R. Zeigel, *J. Cell Physiol.* In press (1970).
22. D. Gingell, *J. Theoret. Biol.* **17**, 451 (1957).
23. K. Carr and I. Carr, *Z. Zellforsch.* **105**, 234 (1970).
24. A. D. Bangham and B. A. Pethica, *Proc. Roy. Phys. Soc.* (*Edin.*) **28**, 43 (1960).
25. L. Weiss, *J. Theoret. Biol.* **6**, 275 (1964).
26. B. M. Jones, *Nature* **212**, 362 (1966).
27. B. M. Jones and G. A. Morrison, *J. Cell Sci.* **4**, 799 (1969).
28. B. M. Jones, R. B. Kemp and U. Gröschel-Stewart, *Nature* **226**, 261 (1970).
29. H. W. Fisher and T. W. Cooper, *J. Cell Biol.* **34**, 569 (1967).
30. L. P. Kayshin (Ed.), *Water in Biological Systems,* 1967. Translation from Russian, Consultants Bureau, New York, 1969.
31. W. Pollack, H. J. Hager, R. Reckel, D. A. Toren and H. O. Singher, *Transfusion* **5**, 158 (1965).
32. L. Weiss and R. F. Woodbridge, *Fed. Proc.* **26**, 88 (1967).
33. D. J. Wilkins, R. H. Ottewill and A. D. Bangham, *J. Theoret. Biol.* **2**, 176 (1962).
34. T. W. Cooper and H. W. Fisher, *J. Nat. Cancer Inst.* **41**, 789 (1968).
35. R. J. V. Pulvertaft and L. Weiss, *J. Path. Bact.* **85**, 473 (1963).
36. G. V. F. Seaman and G. M. W. Cook, *Cell Electrophoresis* (edited by E. J. Ambrose), Little Brown, Boston, 1965, p. 48.
37. L. Weiss, J. Bello and T. L. Cudney, *Internat. J. Cancer* **3**, 795 (1968).
38. L. Weiss and T. L. Cudney, *Internat. J. Cancer* **4**, 776 (1969).
39. D. H. Heard and G. V. F. Seaman, *J. Gen. Physiol.* 43, 635 (1960).

40. J. N. MEHRISHI, *Europ. J. Cancer* **6**, 127 (1970).
41. L. WEISS and T. L. CUDNEY, to be published.
42. J. CLIFFORD and B. A. PETHICA, *Symposium Hydrogen Bonding*, in press.
43. J. CLIFFORD and S. M. A. LECCHINI, S.C.I. Monograph No. 25, *Wetting*, 1967, p. 174.
44. H. J. C. BRENDSEN and C. MICHELSEN, *Annals N.Y. Acad. Sci.* **125**, 305 (1965).
45. J. CLIFFORD and B. SHEARD, *Biopolymers* **4**, 1057 (1966).
46. G. A. JOHNSON, S. M. A. LECCHINI, E. G. SMITH, J. CLIFFORD and B. A. PETHICA, *Discuss. Faraday Soc.* **42**, 120 (1966).
47. H. R. KRUYT, *Colloid Science*, Vol. 1, Elsevier, Amsterdam, 1952.
48. T. T. PUCK, *Cold Spring Harbor Symp. Quant. Biol.* **18**, 49 (1953).
49. T. T. PUCK and L. J. TOLMACH, *Arch. Biochem. Biophys.* **51**, 229 (1954).
50. J. DIRKX, J. BEUMER and M. P. BEUMER-JOCHMANS, *Ann. Inst. Pasteur* **93**, 340 (1957).
51. R. C. VALENTINE and A. C. ALLISON, *Biochim. Biophys. Acta* **34**, 10 (1959).
52. A. C. ALLISON and R. C. VALENTINE, *Biochim. Biophys. Acta* **40**, 400 (1960).
53. L. WEISS and J. S. HOROSZEWICZ, to be published.
54. L. WEISS, *Exptl. Cell. Res.* **53**, 603 (1968).
55. L. WEISS and M. E. NEIDERS, to be published.
56. L. WEISS, E. MAYHEW and K. ULRICH, *Lab. Invest.* **15**, 1304 (1966).
57. J. M. ALLEN and G. M. W. COOK, *Exptl. Cell Res.* **59**, 105 (1970).
58. L. WEISS, *J. Cell Biol.* **26**, 735 (1965).
59. R. B. KEMP, *J. Cell Sci.* **6**, 751 (1970).
60. D. E. MASLOW, unpublished data (1969).
61. D. E. MASLOW and L. WEISS, to be published.
62. P. WEISS, *J. Exp. Zool.* **100**, 353 (1945).
63. R. REIN, P. CLAVERIE and M. POLLAK, *Internat. J. Quantum Chem.* **2**, 129 (1968).
64. R. REIN, M. S. RENDELL and J. P. HARLOS, *Molecular Orbital Studies in Chemical Pharmacology* (edited by L. B. KIER), Springer-Verlag, New York, 1970, p. 191.
65. A. PULLMAN, *Biochim. Biophys. Acta*, **87**, 365 (1964).
66. R. REIN, G. A. CLARKE and F. E. HARRIS, *Quantum Aspects of Heterocyclic Compounds in Chemistry and Biochemistry*, The Jerusalem Symposia on Quantum Chemistry and Biochemistry II, Israel Acad. of Sci., Jerusalem, 1970.
67. A. PULLMAN and M. ROSAI, *Biochim. Biophys. Acta* **88**, 211 (1969).
68. R. REIN and F. E. HARRIS, Quantum mechanics in biochemistry, in *Encyclopedia Dictionary of Physics*, Supp. Vol. 4 (edited by J. THEWLIS), Pergamon Press, Oxford (1971), p. 351.
69. H. HELLMANN, *Quantenchemie Deniche*, Leipzig, 1937, p. 285.
70. R. P. FEÝNMAN, *Phys. Rev.* **56**, 340 (1939).
71. R. M. SAYRE, J. P. HARLOS and R. REIN, p. 207 in ref. 64.
72. J. A. SCHELLMAN, *C.R. Lab. Carlsberg. (Ser. chim.)* **29**, 230 (1955).
73. F. M. FOWKES, *Ind. Eng. Chem.* **56**, 40 (1964).
74. W. A. ZISMAN, *Adv. Chemistry* **43**, 1 (1964), Amer. Chem. Soc., Washington.
75. M. S. STEINBERG, *J. Exp. Zool.* **173**, 395 (1970).
76. S. B. CARTER, *Nature* **213**, 256 (1967).
77. A. C. TAYLOR, *Exp. Cell Res. Suppl.* **8**, 154 (1961).
78. L. WEISS and L. E. BLUMENSON, *J. Cell Physiol.* **70**, 23 (1967).
79. D. J. LYMAN, K. G. KLEIN, J. L. BRASH and B. K. FRITZINGER, *Thrombos Diath. Haemorh.* **23**, 120 (1970).
80. J. F. DANIELLI, *Cytology and Cell Physiology*, 1st ed. (edited by G. BOURNE), Oxford Univ. Press, 1942, p. 78.

81. D. GINGELL, *J. Theoret. Biol.* **19**, 340 (1968).
82. L. WOLPERT and D. GINGELL, *Sympos. Soc. Exp. Biol.* **22**, 169 (1968).
83. A. W. L. JAY and A. C. BURTON, *Biophys. J.* **9**, 115 (1969).
84. A. B. BORLE and J. LOVEDAY, *Cancer Res.* **28**, 2401 (1968).
85. E. L. BENEDETTI and P. EMMELOT, *J. Cell Sci.* **2**, 499 (1967).
86. A. D. BANGHAM, M. M. STANDISH and J. C. WATKINS. *J. Mol. Biol.* **13**. 138 (1965).
87. D. PAPAHADJOPOULOS, *Horizons in Surface Science: Biological Applications* (edited by D. F. SEARS and L. PRINCE), Appleton–Century–Crofts, in press.
88. S. OHKI and O. AONO, *J. Coll. Interface Sci.* **32**, 270 (1970).
89. L. WEISS, *Internat. Revs. Cytol.* **26**, 63 (1969).
90. S. NORDLING and E. MAYHEW, *Exptl. Cell Res.* **44**, 552 (1966).
91. P. G. LEFEVRE, *J. Gen. Physiol.* **31**, 505 (1948).
92. W. F. WIDDAS, *J. Physiol.* **118**, 23 (1952).
93. T. ROSENBERG and W. WILBRANDT, *J. Gen. Physiol.* **41**, 289 (1957).
94. J. L. GLICK and S. GITHENS, *Nature* **208**, 88 (1965).
95. L. WEISS and C. LEVINSON, *J. Cell Physiol.* **73**, 82 (1969).
96. E. MAYHEW, *J. Gen. Physiol.* **49**, 717 (1966).
97. C. JUNG and A. ROTHSTEIN, *J. Gen. Physiol.* **50**, 917 (1967).
98. J. C. SKOU, *Physiol. Rev.* **45**, 596 (1965).
99. P. MITCHELL, *Biol. Rev.* **41**, 445 (1966).
100. W. JUNGE, *Europ. J. Biochem.* **14**, 582 (1970).
101. W. JUNGE, B. RUMBERG and H. SCHRODER, *Europ. J. Biochem.* **14**, 575 (1970).
102. P. EMMELOT and C. J. BOS, *Biochem. Biophys. Acta* **99**, 578 (1965).
103. D. M. BROWN and A. F. MICHAEL, *Proc. Soc. Exp. Biol. Med.* **131**, 568 (1969).
104. P. WEISS, *Amer. Naturalist* **67**, 321 (1938).
105. M. D. ROSENBERG, *Biophys. J.* **1**, 137 (1960).
106. L. WEISS and D. MASLOW, to be published.
107. M. HALPERN and H. RUBIN, *Exptl. Cell Res.* **60**, 86 (1970).
108. M. G. FARQUHAR and G. E. PALADE, *J. Cell Biol.* **17**, 375 (1963).
109. D. E. KELLY, *J. Cell Biol.* **28**. 51 (1966).
110. J. OVERTON, *Devel. Biol.* **4**, 532 (1962).
111. A. L. LEHNINGER, *Physiol. Rev.* **30**, 393 (1950).
112. M. LUBIN and H. L. ENNIS, *Biochim. Biophys. Acta* **80**, 614 (1964).
113. M. LUBIN, *Nature* **213**, 451 (1967).
114. F. L. BYGRAVE, *Nature* **214**, 667 (1967).
115. E. MAYHEW and C. LEVINSON, *J. Cell Physiol.* **72**, 73 (1968).
116. W. R. LOEWENSTEIN, *Devel. Biol.* **15**, 503 (1967).
117. M. ABERCROMBIE, *Cold Spring Harbor Sympos. Quant. Biol.* **27**, 427 (1962).
118. H. EAGLE, *Proc. 7th Canad. Cancer Conf.* Pergamon Press, Toronto, 1967, p. 175.
119. J. L. JAINCHILL and G. J. TODARO, *Exptl. Cell Res.* **59**, 137 (1970).
120. H. RUBIN, *Exptl. Cell Res.* **41**, 149 (1966).
121. R. R. BURK, *Growth-regulating Substances for Animal Cells in Culture* (edited by V. DEFENDI and M. STOKER), Wistar Institute Symposium No. 7, Wistar Inst., Philadelphia, 1967, p. 39.
122. J. YEH and H. W. FISHER, *J. Cell Biol.* **40**, 382 (1969).
123. L. WEISS, *The Chemistry of Biosurfaces* (edited by M. HAIR), M. Dekker: New York, in press (1971).
124. S. EISENBERG, S. BEN-OR and F. DOLJANSKI, *Exptl. Cell Res.* **26**. 451 (1962).
125. E. MAYHEW and L. WEISS, *Exptl. Cell Res.* **50**, 441 (1968).

126. L. Weiss, *Biochem. Soc. Symp.* **22**, 32 (1962).
127. L. Weiss, *Exptl. Cell Res.* **37**, 540 (1965).
128. H. B. Fell and L. Weiss, *J. Exp. Med.* **121**, 151 (1965).
129. L. Weiss and E. D. Holyoke, *J. Nat. Cancer Inst.* **43**, 1045 (1969).
130. J. T. Dingle and H. B. Fell, *Lysosomes in Biology and Pathology*, Vols. 1 and 2, North-Holland Publ. Co., Amsterdam, 1969.

# AUTHOR INDEX

# SUBJECT INDEX